DYNAMICS

DYNAMICS

S. NEIL RASBAND
Brigham Young University

A Wiley-Interscience Publication
JOHN WILEY & SONS
New York Chichester Brisbane Toronto Singapore

Library of Congress Cataloging in Publication Data:

Rasband, S. Neil.
Dynamics.

"A Wiley-Interscience publication."
Includes index.
1. Dynamics. I. Title.

QA845.R35 1983 531'.11 83-10593
ISBN 0-471-87398-5

Printed in the United States of America

10 9 8 7 6 5 4 3 2 1

To my family...

PREFACE

That which you have inherited from your fathers, must be earned to be possessed.

GOETHE

The subject of dynamics has long been at the foundation of an education in physics. One important reason is that the problems the theory addresses are often intimately connected with everyday human experience. No complicated meters or apparatus are necessary to observe a wide variety of mechanical phenomena. Consequently, the study of mechanics reinforces and refines our intuition. When a problem is studied in other branches of physics, which may be removed from the realm of direct experience, familiarity with the methods and concepts of classical dynamics helps one feel at home in a strange land. Nowhere else does one encounter with more force the bedrock nature of invariants such as energy, momentum, and angular momentum. The dynamics of classical systems very often provides the language, concepts, and framework in which the physics of nonclassical systems is phrased.

In the last few decades classical dynamics has once again become an area of active research. The contributions have come from both mathematicians and physicists. However, nothing has played a more central role in this development than the advent of high-speed computers. Computers have revealed a richness of behavior that was not even suggested by the solvable analytic problems, which were previously the center of focus. Mathematicians have attacked many of the problems of stability in classical systems and further penetrated in areas first probed by Poincaré. Physicists have been led to consider similar problems by their interest in stability, as applied to accelerators, statistical systems, and fusion plasmas. Advances in our understanding of the dynamics of classical systems have a profound effect in many areas of research just because so many physical models depend on classical concepts.

And lastly, the subject of classical dynamics can be likened to a beautiful symphony, with new depths of feeling and emotion revealed at each additional hearing. A fascinating array of subtleties are encountered at every turn. And just as the beauty of a symphony is intimately interwoven with its orchestration, so it is with mechanics. The language or garb in which the subject is presented can affect our personal attraction to the subject, as well as reveal its universal applicability. The presentation in this work is intended to introduce the reader to a geometrical language, which is useful in many branches of physics.

This book is designed to serve as a text for a one-semester course at the level of a senior or first-year graduate student. The recommended preparation for the reader is an intermediate-level physics course in mechanics and a course in vector spaces and linear algebra.

The exposition of the subject matter employs geometrical methods throughout. Differential geometry is the "natural" language of mechanics and the concepts of differential geometry are developed as a calculational tool, with an emphasis on applications and not rigor. I have always felt that the best physics texts teach applied mathematics, along with the physics. This book is a modest attempt to reemphasize this tradition in the teaching of classical dynamics.

I thank Clair Nielson and Los Alamos National Laboratory for making available to Brigham Young University the software for the TEDI program editor. The TEDI formatting capabilities played a major role in the preparation of the manuscript. I also thank Mark Nelson for encouragement and helpful comments on early versions of the text. Deeply felt gratitude is expressed to Grant Mason, who read the complete manuscript and served as a patient and persistent teacher in many matters of pedagogy. And lastly, I thank my wife Judi, and my children, who have endured the preparation of this work.

S. NEIL RASBAND

Provo, Utah
July 1983

CONTENTS

DYNAMICS

CHAPTER ONE

KINEMATICS

... geometry should be ranked not with arithmetic, which is purely aprioristic, but with mechanics.

GAUSS

This chapter focuses on the fundamental concepts of mechanics: distance, velocity, and acceleration. We establish a mathematical representation of these quantities and consider them in various coordinate systems. In this chapter we focus principally on kinematic issues as preparation for subsequent applications in dynamics.

In discussing relations among objects, it is necessary to have some means of labeling their relative positions. We do this by assigning a coordinate system to space, at least to that part of space where the objects of interest are located. In spaces with curvature one coordinate system is not enough. Generally, however, our discussions are sufficiently local that a single coordinate system suffices.

Naturally, two observers with the task of assigning coordinates to objects will not necessarily make the same assignments. They may select different origins, and they may not agree on which objects are at rest. Thus it seems desirable to seek a description as independent as possible of any particular observer. That is, we generally prefer a description of phenomena that does not depend on the coordinate selection. That does not mean that we can get by without choosing a coordinate system, which we must have if we want concrete numbers. But we prefer a description of phenomena independent of the coordinate system chosen. Only when actual numbers are desired do we introduce coordinates.

It is thus important to be able to relate quantities in one coordinate system to the description of those same quantities in a different coordinate system. For example, how does the velocity of a particle described in one coordinate system relate to the description in a second set of coordinates?

One final remark concerning different observers and their descriptions: Einstein pointed out that different observers use their own clocks and thereby laid the foundations of relativity. However, for classical mechanics, time is measured by an assumed "universal clock," which is the same for all spatial reference systems. The relative speed between observers must be a significant fraction of the speed of light before different clocks can make a difference. One must not attempt to describe the behavior of light and electromagnetic phenomena without first removing this assumption.

1.1. FOUNDATIONS AND COORDINATE-FREE DESCRIPTION

In seeking a mathematical model for our physical space we choose a Euclidean vector space of three dimensions denoted as $\mathbb{R}^3$. We make this choice because it corresponds with experience. We think of positions in space as being given by a triple of real numbers (x^1, x^2, x^3) where $x^1, x^2, x^3 \in \mathbb{R}$.* Often we denote the triple as a vector $\mathbf{x} = (x^1, x^2, x^3)$, or simply as x^i with the understanding that i ranges over the integers 1, 2, 3 that label the different coordinate directions. Moreover, we assume that between any two points in space, labeled by the vectors $\mathbf{x}$ and $\mathbf{y}$, a measurement of distance can be made (using meter sticks or something equivalent). It is usual to write this distance as

$$d(\mathbf{x}, \mathbf{y}) = \left[(x^1 - y^1)^2 + (x^2 - y^2)^2 + (x^3 - y^3)^2\right]^{1/2}.$$

However, because of the local nature of measurement, the fundamental distance relation is more appropriately formulated in differential form. A differential change in the coordinates is denoted dx^i. The measured distance between two points whose coordinates differ by dx^i we denote by ds. This distance measurement, at a point x^i to a nearby point with coordinates $x^i + dx^i$, leads to a determination of the *metric tensor* according to

$$ds^2 = g_{ij}\, dx^i\, dx^j. \tag{1.1}$$

*$r \in \mathbb{R}$ means that r belongs to the set of real numbers $\mathbb{R}$.

(Note that in the preceding expression we are making use of the customary summation convention. Repeated indices are summed as long as one is a superscript and one is a subscript.)

Associated with the choice of coordinates x^i and the differentials dx^i is a set of basis vectors in $\mathbb{R}^3$ that we denote as $\mathbf{e}_1, \mathbf{e}_2, \mathbf{e}_3$, or for brevity simply $\mathbf{e}_i$. These vectors $\mathbf{e}_i$ are defined at each point of our space and *may change direction and length from one point to the next*. The direction of each of these basis vectors at a point $\mathbf{x}$ is determined by the direction of increasing x^i. That is to say, $\mathbf{e}_1$ points in the direction of increasing x^1 with x^2 and x^3 held constant. Corresponding statements hold for $\mathbf{e}_2$ and $\mathbf{e}_3$ also. See Fig. 1.1.

The scale of these vectors is set by the infinitesimal distance relation (1.1). We consider two neighboring points with coordinates differing by dx^i. The distance vector between these points is defined to be

$$d\mathbf{s} = dx^i \mathbf{e}_i, \tag{1.2}$$

where

$$ds^2 = g_{ij}\, dx^i\, dx^j \equiv d\mathbf{s} \cdot d\mathbf{s} = dx^i \mathbf{e}_i \cdot dx^j \mathbf{e}_j = (\mathbf{e}_i \cdot \mathbf{e}_j)\, dx^i\, dx^j.$$

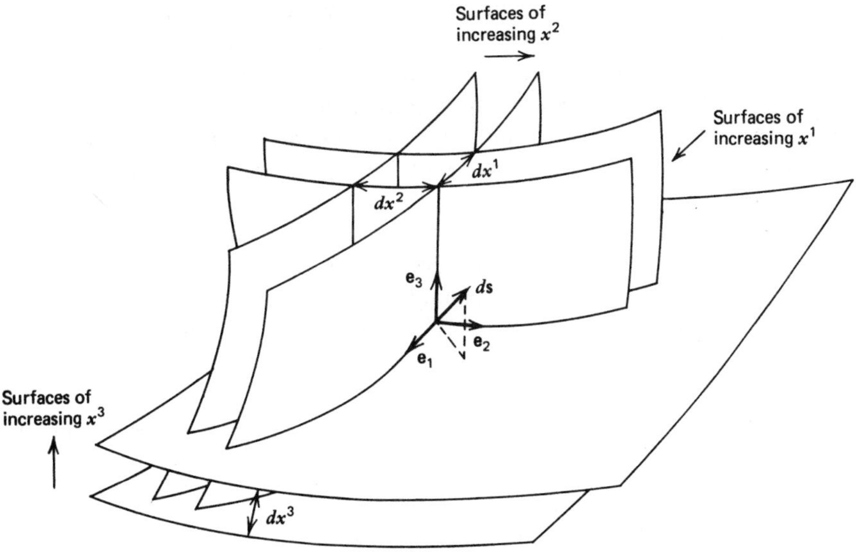

FIGURE 1.1. An intersection of coordinate surfaces at a point. The relationship between the coordinate differentials and basis vectors is indicated.

The preceding relation not only uses the metric to determine the scale of the basis vectors, $|\mathbf{e}_i| = (g_{ii})^{1/2}$, but also defines an inner (scalar) product. The inner product between two basis vectors, defined by

$$\mathbf{e}_i \cdot \mathbf{e}_j = g_{ij}, \tag{1.3}$$

determines the scalar product for an arbitrary pair of vectors $\mathbf{A}, \mathbf{B}$.

$$\mathbf{A} \cdot \mathbf{B} = A^i\mathbf{e}_i \cdot B^j\mathbf{e}_j = (\mathbf{e}_i \cdot \mathbf{e}_j)A^iB^j = g_{ij}A^iB^j. \tag{1.4}$$

It is also frequently convenient to employ the inverse of the metric tensor. This tensor is denoted by g^{ij} (with superscripts) and is defined by the relation $g_{ij}g^{jk} = \delta_i^k$ (the familiar Kronecker delta). From (1.1) we see that g_{ij} is symmetric under index interchange and hence so is g^{ij}.

Combinations such as $g_{ij}B^j$ occurring in (1.4) arise frequently, and for convenience we denote $B_i = g_{ij}B^j$. The metric tensor is used here to "lower" the index on a vector component. Because of the inverse relationship between g_{ij} and g^{ij}, the latter quantity may be used to "raise" indices; that is, $B^i = g^{ij}B_j$. These formal operations of raising and lowering indices, using the metric tensor and its inverse, will be used frequently throughout the book. To distinguish between the components of a vector $\mathbf{B}$ with subscripts and with superscripts, that is, between B_i and B^i, the components are referred to as *covariant* and *contravariant*, respectively. These names are historical and for the present denote little more than whether the component index is up or down. There is, however, some information contained in the position of the index rather than its being merely a matter of convenience. We develop these matters more fully later and for the present simply choose to write coordinates and their differentials with superscripts and vector components either way, going from one to another with the metric g_{ij} or its inverse.

From among the possible coordinate systems we pick one, arbitrary but fixed. Then the position of a particle is given by the vector $\mathbf{x}(t)$, which specifies the position of the particle under study and in general changes with time. The coordinates are $x^i(t)(i = 1, 2, 3)$. The *velocity* at time t_0 is defined to be

$$\mathbf{v}(t_0) \equiv \lim_{\Delta t \to 0} \frac{1}{\Delta t}\left[x^i(t_0 + \Delta t) - x^i(t_0)\right]\mathbf{e}_i \equiv \frac{dx^i}{dt}\mathbf{e}_i = v^i\mathbf{e}_i. \tag{1.5}$$

The *acceleration* vector is defined as

$$\mathbf{a}(t_0) \equiv \lim_{\Delta t \to 0} \frac{1}{\Delta t}[\mathbf{v}(t_0 + \Delta t) - \mathbf{v}(t_0)] = \frac{d}{dt}\mathbf{v}(t)\bigg|_{t=t_0}, \tag{1.6}$$

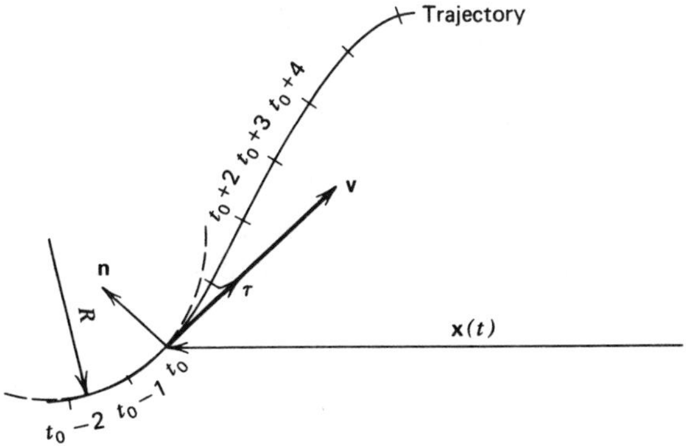

FIGURE 1.2. An illustration of the trajectory of a particle with important vectors of the motion indicated for the specific time t_0. Other time points are noted.

but we note an added complication in computing this derivative that is not present in (1.5). The basis vectors included in (1.5) to define $\mathbf{v}$ may change along the trajectory as well and this change must be considered in (1.6). We deal with this explicitly in the next section.

It is often convenient to think of $\mathbf{x}$, $\mathbf{v}$, and $\mathbf{a}$, which are functions of t, as mappings. For this purpose we let I denote an arbitrary time interval that is a subset of $\mathbb{R}$. Then $\mathbf{x}$, $\mathbf{v}$, and $\mathbf{a}$ are mappings of I into $\mathbb{R}^3$. Consider the vector $\mathbf{x}$:

$$\mathbf{x}: I \to \mathbb{R}^3, \qquad \text{where } \mathbf{x}(t) \in \mathbb{R}^3 \text{ for all } t \in I.^* \tag{1.7}$$

The image of I under the mapping (1.7) is called the *trajectory* of the particle and is illustrated in Fig. 1.2.

Suppose we now measure the distance along a particle trajectory, such as the one in Fig. 1.2, and call it s. Then s is a monotonically increasing parameter along the trajectory, as is t. We think of $\mathbf{x}$ as a function of t through the parameter s. Thus we have

$$\mathbf{v}(t) = \frac{dx^i}{dt}\mathbf{e}_i = \frac{dx^i}{ds}\frac{ds}{dt}\mathbf{e}_i \equiv \frac{ds}{dt}\boldsymbol{\tau}. \tag{1.8}$$

Note that the point of evaluation for derivatives is indicated only when needed for clarity. We call $\boldsymbol{\tau} = (dx^i/ds)\mathbf{e}_i$ the *tangent vector to the trajectory*. It follows from (1.1) that $\boldsymbol{\tau} \cdot \boldsymbol{\tau} = 1$. The quantity $v = (ds/dt)$ is just the speed along the trajectory. Thus $\mathbf{v}(t) = v(t)\boldsymbol{\tau}(t)$.

*One reads $\mathbf{x}: I \to \mathbb{R}^3$ as "$\mathbf{x}$ maps the interval I into the vector space $\mathbb{R}^3$."

We treat the acceleration vector in analogous fashion:

$$\mathbf{a}(t) = \frac{d\mathbf{v}}{dt} = \frac{d^2s}{dt^2}\boldsymbol{\tau} + v\frac{d\boldsymbol{\tau}}{dt}. \tag{1.9}$$

Since $\boldsymbol{\tau} \cdot \boldsymbol{\tau} = 1$ and $d(\boldsymbol{\tau} \cdot \boldsymbol{\tau})/dt = 2\boldsymbol{\tau} \cdot (d\boldsymbol{\tau}/dt) = 0$ we know that $(d\boldsymbol{\tau}/dt)$ is orthogonal to $\boldsymbol{\tau}$, which means that it is normal to the curve. Specifically,

$$\frac{d\boldsymbol{\tau}}{dt} = \frac{d\boldsymbol{\tau}}{ds}\frac{ds}{dt} = \frac{v}{R}\mathbf{n},$$

where $\mathbf{n}$ is the *principal normal* to the curve that lies in the plane formed by $\boldsymbol{\tau}(t + \Delta t)$ and $\boldsymbol{\tau}(t)$ as $\Delta t \to 0$, and R is the *radius of curvature*. These quantities are sketched in Fig. 1.2. Hence

$$\mathbf{a}(t) = \ddot{s}\boldsymbol{\tau} + \frac{v^2}{R}\mathbf{n}. \tag{1.10}$$

We frequently use the notation of a dot over a symbol to denote a differentiation with respect to time. Hence $\ddot{s} = d^2s/dt^2$. Also we will use $\dot{\mathbf{x}}$ interchangeably with $\mathbf{v}$ but note that $\dot{\mathbf{x}} = \dot{x}^i\mathbf{e}_i$ not $d(x^i\mathbf{e}_i)/dt$.

Whereas the velocity is tangent to the trajectory, we note that the acceleration has components both along the tangent and perpendicular to it. These components are given by the linear acceleration $\ddot{s}$ and the centripetal acceleration (v^2/R), respectively.

1.2. STANDARD COORDINATE EXPRESSIONS

The foregoing expressions for $\mathbf{v}(t)$ and $\mathbf{a}(t)$ are in what is sometimes called the "intrinsic" system; that is, they refer to properties of the trajectory itself rather than to a particular coordinate system. In this section we obtain expressions for velocity and acceleration in the familiar cylindrical and spherical coordinate systems. In so doing we will differentiate between "physical" and "coordinate" components and the corresponding basis vectors. We will also introduce the notion of a geometric object and covariant derivatives.

There are several ways in which one may obtain the standard expressions for $\mathbf{v}$ and $\mathbf{a}$. Many of these methods are clever but do not reveal the subtleties associated with coordinate systems. Our methods are such that the reader can readily generalize to other, less familiar coordinate systems.

Consider first the straightforward but tedious method of performing a tensor transformation from cartesian coordinates $x^{\hat{i}}$ to curvilinear coordinates x^i. (We adopt the convention of distinguishing between different coordinate systems by distinguishing the indices—for instance, with primes or carets.) As an example, the transformation law for the contravariant components of the velocity vector $\mathbf{v}$ from the $\{x^{\hat{i}}\}$ coordinates to the components in the $\{x^i\}$ system is given by

$$v^i = \frac{\partial x^i}{\partial x^{\hat{i}}} v^{\hat{i}}. \tag{1.11}$$

The transformation matrix $J^i{}_{\hat{i}} \equiv (\partial x^i / \partial x^{\hat{i}})$ is referred to as the *Jacobian matrix* and must be computed from the functional relationships giving the $\{x^i\}$ coordinates in terms of the $\{x^{\hat{i}}\}$ coordinates. In denoting the components of a matrix such as $J^i{}_{\hat{i}}$, the horizontal position of the index is important since the first slot denotes the row index and the second slot the column index, regardless of its vertical position, that is, whether or not the index is covariant or contravariant. The covariant components of $\mathbf{v}$ (or any arbitrary tensor) transform using the inverse of $J^i{}_{\hat{i}}$, that is, using $(\partial x^{\hat{i}} / \partial x^i)$. This method of tensor transformation as applied in going from cartesian to curvilinear components is a systematic way of obtaining the components of a vector or tensor in a curvilinear coordinate system, but we choose to emphasize here a more geometric view.

Velocity

Cylindrical Coordinates

These coordinates are denoted by (r, θ, z) where $0 \leqslant r < \infty$, $0 \leqslant \theta < 2\pi$, $-\infty < z < +\infty$. Assuming that the infinitesimal distance function is known in these coordinates, we can proceed as follows. From $ds^2 = dr^2 + r^2\, d\theta^2 + dz^2 = g_{ij}\, dx^i\, dx^j = d\mathbf{s} \cdot d\mathbf{s}$ we observe that $\mathbf{v} \cdot \mathbf{v} = g_{ij}(dx^i/dt)(dx^j/dt)$ is given by $\dot{r}^2 + r^2\dot{\theta}^2 + \dot{z}^2$. We then conclude that, since $\mathbf{v} \cdot \mathbf{v} = v_r^2 + v_\theta^2 + v_z^2$ for orthonormal basis vectors $(\hat{\mathbf{r}}, \hat{\boldsymbol{\theta}}, \hat{\mathbf{z}})$,

$$\mathbf{v} = \dot{r}\hat{\mathbf{r}} + r\dot{\theta}\hat{\boldsymbol{\theta}} + \dot{z}\hat{\mathbf{z}}. \tag{1.12}$$

There is of course another way in which we can proceed. We know from (1.5) that the components of $\mathbf{v}$ are given by $v^i = \dot{x}^i = (dx^i/dt)$. Then if we identify the coordinate index labels $(i = 1, 2, 3)$ with (r, θ, z), respectively,

$$\mathbf{v} = \dot{x}^i \mathbf{e}_i = \dot{r}\mathbf{e}_1 + \dot{\theta}\mathbf{e}_2 + \dot{z}\mathbf{e}_3. \tag{1.13}$$

We notice a difference in the middle term between (1.12) and (1.13). This difference stems from the fact that the *coordinate components* and *coordinate basis vectors*, v^1, v^2, v^3 and $\mathbf{e}_1, \mathbf{e}_2, \mathbf{e}_3$ are not the same as those we call the *physical components* and *physical basis vectors*, v_r, v_θ, v_z and $\hat{\mathbf{r}}, \hat{\boldsymbol{\theta}}, \hat{\mathbf{z}}$. Let us contrast these two sets.

Physical	Coordinate
1. Basis vectors are always of unit length and are orthogonal; for example, $\hat{\mathbf{r}} \cdot \hat{\mathbf{r}} = 1$, $\hat{\mathbf{r}} \cdot \hat{\boldsymbol{\theta}} = 0$.	1. Basis vectors have an inner product given by the metric, $\mathbf{e}_i \cdot \mathbf{e}_j = g_{ij}$; see Eq. (1.3).
2. Components are such that the length squared is the sum of squares; for example, $v^2 = v_r^2 + v_\theta^2 + v_z^2$.	2. Components are such that the length squared is obtained by summing with the metric; for example, $v^2 = g_{ij} v^i v^j = g_{ij} \dot{x}^i \dot{x}^j$.

A vector function or operator is almost always given in books or tables in terms of physical components. Yet to obtain these expressions it is often more straightforward to use coordinate components. Indeed, it is important to be at home with both viewpoints since they complement each other and both find extensive application. It is likely that the coordinate viewpoint is less familiar and hence it here receives somewhat more discussion.

Spherical Coordinates

These coordinates are denoted by (r, θ, ϕ) or $x^i (i = 1, 2, 3)$ where $0 \leqslant r < \infty$, $-\pi < \theta < \pi$, $0 \leqslant \phi < 2\pi$. In terms of these coordinates the usual cartesian coordinates are given by $x = r \sin\theta \cos\phi$, $y = r \sin\theta \sin\phi$, $z = r \cos\theta$. Thus

$$ds^2 = dr^2 + r^2\, d\theta^2 + r^2 \sin^2\theta\, d\phi^2. \tag{1.14}$$

From (1.14) we can identify the elements of the metric tensor and then by the methods discussed previously for cylindrical coordinates we find $v_r = \dot{r}$, $v_\theta = r\dot{\theta}$, $v_\phi = r \sin\theta \dot{\phi}$ and $v^1 = \dot{r}$, $v^2 = \dot{\theta}$, $v^3 = \dot{\phi}$. Both sets give, of course, the same result for $v^2 = \mathbf{v} \cdot \mathbf{v}$.

Acceleration

The first thing that must be pointed out before discussing acceleration is the difference between the position vector $\mathbf{x}$ and the velocity $\mathbf{v}$. After fixing an origin in our coordinate system we can draw an arrow for each of these quantities. However, if we change our minds and pick a new fixed origin, a new arrow for **x** must be drawn. In contrast, no change is required for **v**. The quantity **v** is a *geometric object*, which has intrinsic meaning independent of any coordinate system. This is quite different from the case for **x**, as illustrated in Fig. 1.2. Clearly **x** requires an origin for it to have meaning, whereas **v** is independent of any coordinate system whatever and has intrinsic geometric meaning. (This is not true of course for the components of **v** whose numerical values at a point depend on the coordinate system chosen.) The vector **v**(**x**) resides in what is called the *tangent space* at the point **x**, denoted $T_{\mathbf{x}}$. We return to a more complete discussion of the tangent space and its definition in Section 6.1. For the present it suffices to say that vectors which are geometric objects are elements of a tangent space and not of the space of position vectors at all. This concept is perhaps best illustrated by considering motion on the surface of a sphere. Clearly the tangent vector to any trajectory does not lie in the spherical surface itself but rather in a two-dimensional tangent plane at the point in question. See Fig. 1.3.

The acceleration vector is also a geometric object since it is obtained as the limit of the difference of two velocity vectors. It lies also in the tangent plane. We noted earlier that the acceleration must account for the possible change in the basis vectors from one point to the next. It should be no surprise that the differentiation of a geometric object (velocity) to obtain the acceleration will be different from the differentiation of a nongeometric object (position coordinates) to obtain the velocity [cf. (1.5) and (1.6)].

To obtain the coordinate expressions for the acceleration we proceed as follows. Since $\mathbf{v} = v^i \mathbf{e}_i$, we have

$$\mathbf{a} = \frac{d\mathbf{v}}{dt} = \dot{v}^i \mathbf{e}_i + v^i \frac{d\mathbf{e}_i}{dt}. \tag{1.15}$$

We must now consider how the coordinate basis vectors change in time, that is, with position as the position vector changes in time. We note that

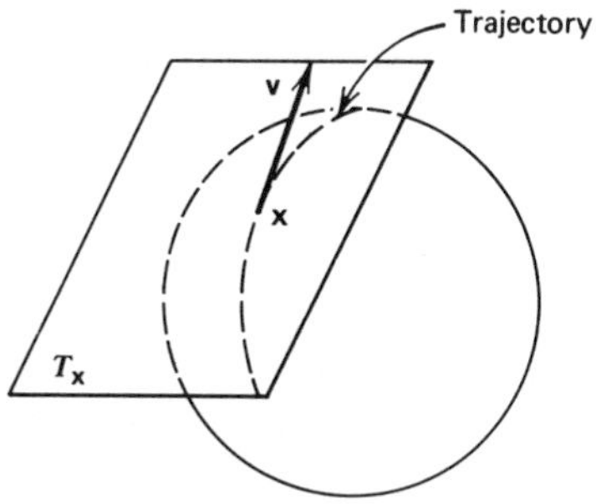

FIGURE 1.3. Trajectory on the surface of a sphere with the velocity at a point on the trajectory lying in the tangent space to the sphere at that point.

the first-order change in the basis vectors should be proportional to the changes in the dx^i for an infinitesimal displacement. The proportionality we make explicit by writing $d\mathbf{e}_i = \Gamma^k_{ij}\, dx^j\, \mathbf{e}_k$(note sums). The quantities Γ^k_{ij} are called *connection coefficients* and they contain the information about how the coordinate basis vectors twist and turn in going from one point to the next. Substitution into (1.15) gives

$$\mathbf{a} = \left[\dot{v}^i + v^j \Gamma^i_{jk} v^k\right]\mathbf{e}_i. \tag{1.16}$$

Thus we can identify the contravariant coordinate components of $\mathbf{a}$,

$$a^i = \dot{v}^i + \Gamma^i_{jk} v^j v^k. \tag{1.17}$$

Note that only that part of the connection coefficient that is symmetric under interchange of j and k is needed, and we will henceforth take the connection coefficients to be symmetric under interchange of the two lower indices. For (1.17) to be useful we must of course have the Γ^i_{jk}. But before proceeding to obtain a formula for them we define the *covariant derivative*. By the chain rule of differentiation we have

$$\dot{v}^i = \frac{\partial v^i}{\partial x^j}\frac{dx^j}{dt} = \frac{\partial v^i}{\partial x^j} v^j \equiv v^i{}_{,j} v^j, \tag{1.18}$$

where in the last expression we have defined the comma notation to denote ordinary partial differentiation. Substitution into (1.17) then gives

$$a^i = \left(v^i{}_{,j} + v^k \Gamma^i_{kj}\right) v^j, \tag{1.19}$$

where the quantity in parentheses is denoted by

$$v^i{}_{;j} \equiv v^i{}_{,j} + v^k \Gamma^i_{kj} \tag{1.20}$$

and is called the covariant derivative of the contravariant vector component v^i. A covariant derivative is denoted by a semicolon. Inserting (1.20) in (1.19) we have

$$a^i = v^i{}_{;j} v^j. \tag{1.21}$$

Now let us obtain the connection coefficients. We use the expression for the metric tensor in terms of the coordinate basis vectors. We have assumed the usual product rule for differentiation of a vector inner product and now consider the change in each side of $g_{ij} = \mathbf{e}_i \cdot \mathbf{e}_j$ for an infinitesimal change of position characterized by the coordinate differential dx^k.

$$g_{ij,k}\, dx^k = d\mathbf{e}_i \cdot \mathbf{e}_j + \mathbf{e}_i \cdot d\mathbf{e}_j. \tag{1.22}$$

We substitute for $d\mathbf{e}_i = \Gamma^k_{ij}\, dx^j\, \mathbf{e}_k$ to obtain

$$g_{ij,k} = g_{mj}\Gamma^m_{ki} + g_{im}\Gamma^m_{kj}. \tag{1.23}$$

By the index interchanges $i \leftrightarrow k$ and $j \leftrightarrow k$ we obtain two additional equations. Subtracting (1.23) from the sum of these two equations, and noting the symmetries for index interchange on the g_{ij} and $\Gamma^i{}_{jk}$, we obtain the result $2g_{km}\Gamma^m_{ij} = g_{kj,i} + g_{ki,j} - g_{ij,k}$. Multiplying by g^{kl} and summing over k we find

$$\Gamma^l_{ij} = \tfrac{1}{2} g^{lk}[g_{kj,i} + g_{ki,j} - g_{ij,k}]. \tag{1.24}$$

This equation then allows computation of the connection coefficients from a knowledge of the metric. Differential geometers have more efficient ways but (1.24) will suffice. We give the connection coefficients for cylindrical and spherical coordinates but leave their computation as an exercise.

For cylindrical coordinates $(r, \theta, z) = (x^1, x^2, x^3)$

$$\Gamma^1_{22} = -r, \quad \Gamma^2_{12} = \Gamma^2_{21} = \frac{1}{r}.$$

All others are zero.

For spherical coordinates $(r, \theta, \phi) = (x^1, x^2, x^3)$

$$\Gamma^1_{22} = -r, \quad \Gamma^1_{33} = -r\sin^2\theta, \quad \Gamma^2_{12} = \Gamma^2_{21} = \Gamma^3_{13} = \Gamma^3_{31} = \frac{1}{r},$$

$$\Gamma^2_{33} = -\sin\theta\cos\theta, \quad \Gamma^3_{23} = \Gamma^3_{32} = \cot\theta.$$

All others are zero.

Let us now use these connection coefficients and (1.17) to obtain the coordinate components of the acceleration vector.

For cylindrical coordinates,

$$a^1 = \dot{v}^1 + \Gamma^1_{jk} v^j v^k = \ddot{r} - r\dot{\theta}^2.$$

$$a^2 = \dot{v}^2 + \Gamma^2_{jk} v^j v^k = \ddot{\theta} + \frac{2}{r}\dot{r}\dot{\theta}. \tag{1.25}$$

$$a^3 = \dot{v}^3 + \Gamma^3_{jk} v^j v^k = \ddot{z}.$$

The physical components we identify from the expression

$$\mathbf{a}\cdot\mathbf{a} = g_{ij}a^i a^j = (\ddot{r} - r\dot{\theta}^2)^2 + r^2\left(\ddot{\theta} + \frac{2}{r}\dot{r}\dot{\theta}\right)^2 + \ddot{z}^2.$$

This gives

$$\begin{aligned} a_r &= \ddot{r} - r\dot{\theta}^2. \\ a_\theta &= r\ddot{\theta} + 2\dot{r}\dot{\theta}. \\ a_z &= \ddot{z}. \end{aligned} \tag{1.26}$$

For spherical coordinates we obtain the following results in similar fashion.

$$\begin{aligned} a^1 &= \ddot{r} - r\dot{\theta}^2 - r\sin^2\theta\dot{\phi}^2. \\ a^2 &= \ddot{\theta} + \frac{2}{r}\dot{r}\dot{\theta} - \sin\theta\cos\theta\dot{\phi}^2. \\ a^3 &= \ddot{\phi} + \frac{2}{r}\dot{r}\dot{\phi} + 2\cot\theta\dot{\phi}\dot{\theta}. \end{aligned} \tag{1.27}$$

The physical components are obtained as usual.

$$\begin{aligned} a_r &= \ddot{r} - r\dot{\theta}^2 - r\sin^2\theta\dot{\phi}^2. \\ a_\theta &= r\ddot{\theta} + 2\dot{r}\dot{\theta} - r\sin\theta\cos\theta\dot{\phi}^2. \\ a_\phi &= r\sin\theta\ddot{\phi} + 2\sin\theta\dot{r}\dot{\phi} + 2r\cos\theta\dot{\phi}\dot{\theta}. \end{aligned} \tag{1.28}$$

In concluding this section on velocity and acceleration in general coordinate systems, we briefly consider an application of these results. We establish a relation between the familiar kinetic energy and Newton's law, topics that are considered more fully in the next chapter.

We let F_i denote the covariant component of the force. (F_i is sometimes referred to as a "generalized force.") Then Newton's law is written in the form

$$ma_i = F_i.$$

The kinetic energy in general coordinates for a particle of mass m is given by

$$T = \tfrac{1}{2}m\mathbf{v}\cdot\mathbf{v} = \tfrac{1}{2}mg_{ij}v^i v^j,$$

where the last of the foregoing alternate expressions is the most useful for our present purposes. The metric g_{ij} is a function of the coordinates $\{x^i\}$ but not a function of the velocity components. We treat the coordinates and velocity components as independent variables and compute

$$\frac{\partial T}{\partial v^i} = mg_{ij}v^j \qquad \text{and} \qquad \frac{\partial T}{\partial x^i} = \frac{m}{2}v^k v^j g_{kj,i}.$$

$$\frac{d}{dt}\left(\frac{\partial T}{\partial v^i}\right) = mg_{ij}\dot{v}^j + mg_{ij,k}v^j v^k = mg_{ij}\dot{v}^j + \frac{m}{2}(g_{ij,k} + g_{ik,j})v^j v^k,$$

where we have explicitly written the symmetric piece of $g_{ij,k}$ under the interchange $j \leftrightarrow k$, which the sum with $v^j v^k$ picks out. Combining these results, we find

$$\frac{d}{dt}\left(\frac{\partial T}{\partial v^i}\right) - \frac{\partial T}{\partial x^i} = mg_{ij}\dot{v}^j + \frac{m}{2}(g_{ij,k} + g_{ik,j} - g_{jk,i})v^j v^k.$$

Comparing with (1.24) and (1.17) we see that this may be written in the form

$$\frac{d}{dt}\left(\frac{\partial T}{\partial v^i}\right) - \frac{\partial T}{\partial x^i} = mg_{ij}\dot{v}^j + mg_{il}\Gamma^l_{jk}v^j v^k$$

$$= mg_{il}\left(\dot{v}^l + \Gamma^l_{jk}v^j v^k\right) = mg_{il}a^l = ma_i. \tag{1.29}$$

From (1.29) we glean two results. The first is that Newton's law may be written in the form

$$\frac{d}{dt}\left(\frac{\partial T}{\partial v^i}\right) - \frac{\partial T}{\partial x^i} = F_i. \tag{1.30}$$

When the force can be computed from a potential function, $F_i = U_{,i}$, then it is customary to let $L = T - U$. The function L is called the *Lagrangian* function and (1.30) can be written as

$$\frac{d}{dt}\left(\frac{\partial L}{\partial v^i}\right) - \frac{\partial L}{\partial x^i} = 0.$$

These equations are called the Euler–Lagrange differential equations of Lagrangian mechanics and we consider them extensively in Chapter 3.

The second point we wish to make is that (1.29) provides an efficient means for computing the connection coefficients or the covariant components of the acceleration. Generally it is a simple matter to write down the kinetic energy in general coordinates. Once the metric is known from the infinitesimal line element $ds^2 = g_{ij}\,dx^i\,dx^j$, or otherwise,

$$T = \frac{m}{2}\frac{d\mathbf{s}}{dt}\cdot\frac{d\mathbf{s}}{dt} = \frac{m}{2}g_{ij}\frac{dx^i}{dt}\frac{dx^j}{dt}.$$

Computing the required derivatives and substituting into (1.29) gives immediately the covariant components of the acceleration **a**. From (1.17) we can identify immediately the connection coefficients, and by computing $\mathbf{a}\cdot\mathbf{a}$ we can identify the physical components of **a** as before.

1.3. MOVING COORDINATE SYSTEMS

In Sections 1.1 and 1.2 we concentrated on obtaining expressions for velocity and acceleration in frequently used coordinate systems: intrinsic, cylindrical, and spherical. Very often it is necessary to consider vectors in a vector space S when those vectors have been given or defined in a second vector space M. The coordinates in either of these reference frames might be cartesian, cylindrical, spherical, and so on. Usually we will think of S as (S)tationary, an inertial frame, and M as (M)oving. There is some general transformation T relating the vectors in M to those in S, $T\colon M \to S$. This transformation is one-to-one and all the vectors in S are related to some vector in M.* We do not consider the most general relationship possible but will confine our attention to coordinate systems S and M that have the same orientation (i.e., they are both right-handed coordinate systems) and that preserve the infinitesimal distance between neighboring points. Thus transformations T between S and M can be characterized as orientation and metric preserving. If such a transformation $T\colon M \to S$ in addition maps the zero vector in M into the zero vector in S, then the transformation preserves inner products ($T\mathbf{A}\cdot T\mathbf{B} = \mathbf{A}\cdot\mathbf{B}$), is linear, and hence is a rotation.

At this point we draw attention to a change in emphasis from the previous sections of this chapter. Heretofore we viewed velocity, accelera-

*Such a mapping is termed *bijective* by mathematicians.

tion, and so on, as geometric objects with an intrinsic existence independent of any reference frame. In this spirit we would view M and S as defined simply by two different choices for basis vectors in the underlying Euclidean vector space. We refer to this as the intrinsic view.

The mapping point of view we now adopt is different! Physical quantities of interest such as velocity can be expressed, or can be said to have a *representative*, in one frame or the other. One gets from the representations of the vectors in M to the representations in S by a transformation T. The vector spaces M and S may indeed be copies of one another but are still considered to be distinct. It is our opinion that, when discussing the relationships between the vector representations of physical quantities as given in two different reference frames, it is helpful to have the mapping as a visible component of the formalism. It has been our experience that distinguishing between the vector representatives of physical quantities in the frames M and S (i.e., the mapping viewpoint) adds precision, prevents confusion, and inhibits mistakes. Nevertheless, we hasten to add that the intrinsic view, treated carefully, is equivalent and is preferred by some authors.

We follow the convention that vectors denoted by capitals belong to M and lowercase vectors belong to S. For example, $\mathbf{X} \in M$ and $\mathbf{x} \in S$. In the spirit of our mapping viewpoint, we treat the vectors $\mathbf{x} \in S$ and $\mathbf{X} \in M$ as two different things and the transformation T, $\mathbf{x} = T\mathbf{X}$, takes vectors in M into vectors in S. This view of the transformation T is termed the *active view*, and is natural in the context of the mapping viewpoint.

In the context of the intrinsic view, where M and S simply represent different choices of bases for the underlying vector space, and when the transformation is a rotation R, a second view called the *passive view* is natural. In the passive view the vectors are not rotated at all but rather just the bases. The rotation R maps the basis in M into the basis in S. The active and passive views are contrasted in Figs. 1.4 and 1.5. The distinctions between the active and passive views are further clarified as we review the properties of rotations in detail.

A rotation R: $M \to S$ relates the components of vectors according to the equation

$$x^i = R^i_{\ j} X^j. \tag{1.31}$$

Equation (1.31) is the active view of the transformation that gives the components of $\mathbf{x} \in S$ in terms of the components of $\mathbf{X} \in M$. If the components are represented as a column, then (1.31) corresponds to the usual matrix multiplication for the linear operator R.

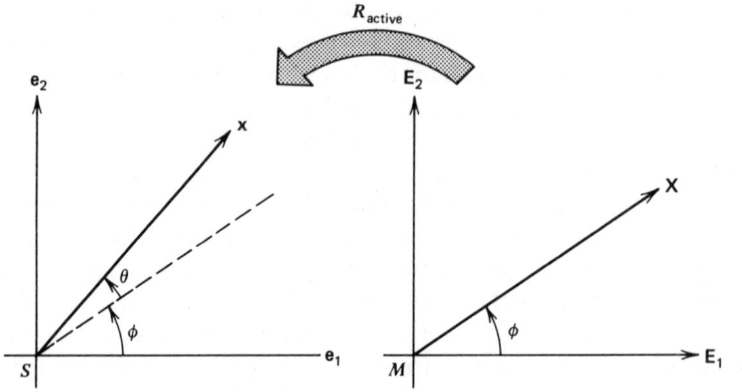

FIGURE 1.4. The active view of a positive rotation R that acts on the vectors and gives $\mathbf{x} = R\mathbf{X}$. In terms of components this rotation is given by $x^1 = X^1\cos\theta - X^2\sin\theta$ and $x^2 = X^1\sin\theta + X^2\cos\theta$. The rotation matrix $R = \begin{bmatrix} \cos\theta & -\sin\theta \\ \sin\theta & \cos\theta \end{bmatrix}$.

Rotations have the properties that they preserve inner products and the orientation of the axes, which imply respectively that

$$g_{ij} R^i{}_k R^j{}_l = g_{kl}. \tag{1.32}$$

$$+1 = [ijk] R^1{}_i R^2{}_j R^3{}_k. \tag{1.33}$$

We have made use of the following convenient expression for the determinant of an $(n \times n)$ matrix:

$$[i_1, i_2, \ldots, i_n]\det R = [j_1, j_2, \ldots, j_n] R^{i_1}{}_{j_1} R^{i_2}{}_{j_2} \ldots R^{i_n}{}_{j_n}. \tag{1.34}$$

We sum on the repeated indices in (1.34). The permutation symbol $[i_1, \ldots, i_n]$ has the following definition.

$$[i_1, \ldots, i_n] = \begin{cases} +1, & \text{if } i_1, \ldots, i_n \text{ is an even permutation of } 1, \ldots, n. \\ -1, & \text{if } i_1, \ldots, i_n \text{ is an odd permutation of } 1, \ldots, n. \\ 0, & \text{otherwise} \end{cases} \tag{1.35}$$

In matrix language (1.32) says that $R^t = R^{-1}$ where R^t denotes the transpose or adjoint operator corresponding to R, and R^{-1} denotes the inverse. Equation (1.33) is a statement that $\det R = +1$.

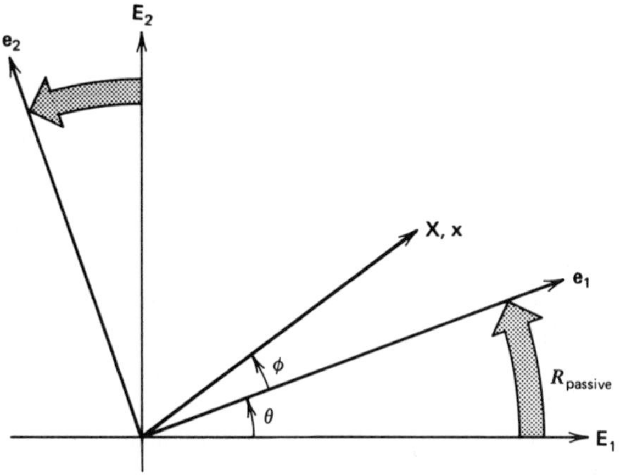

FIGURE 1.5. The passive view of a positive rotation R that acts on the basis vectors to give $\mathbf{x} = R\mathbf{X}$. The basis vectors are related by $\mathbf{e}_1 = \mathbf{E}_1\cos\theta + \mathbf{E}_2\sin\theta$ and $\mathbf{e}_2 = -\mathbf{E}_1\sin\theta + \mathbf{E}_2\cos\theta$. We note that the rotation matrix relating the components (x^1, x^2) to the components (X^1, X^2) is the rotation matrix in the active view through an angle $-\theta$.

We denote the bases of S and M with lower- and uppercase symbols, respectively, that is, $\{\mathbf{e}_i\}$ and $\{\mathbf{E}_i\}$, and consider the passive view. A rotation gives

$$\mathbf{x} = x^i\mathbf{e}_i = R^i{}_j X^j\mathbf{e}_i = X^j\mathbf{E}_j,$$

and thus

$$\mathbf{E}_j = \mathbf{e}_i R^i{}_j \quad \text{or } \mathbf{e}_j = \mathbf{E}_i (R^t)^i{}_j. \tag{1.36}$$

Often it is more convenient to construct a rotation matrix that is the product of several rotations about different axes by using (1.36) and mentally rotating the basis of S into the basis of M. We will see an example of this in Chapter 3 when we consider the Euler angle parameterization of an arbitrary rotation. Since both the mapping viewpoint and the intrinsic view, as well as active and passive transformations, all find application throughout the physics literature, it is important to understand all approaches. In this book we use both viewpoints and both active and passive transformations, but always attempt to make clear the approach being adopted. For the remainder of this section we consider mappings as active transformations.

A known mapping T: $M \to S$, which is an orientation and metric preserving transformation, can be uniquely decomposed into a translation and a rotation, t: $S \to S$ and R: $M \to S$, respectively, which we now proceed to define. Let $\mathbf{0}_M$ denote the zero vector in M and we let the vector $T\mathbf{0}_M$ in S be denoted by $\mathbf{a}$. Then the translation mapping t: $S \to S$ is defined by simple vector addition to $\mathbf{a}$. That is to say, if $\mathbf{x}$ is an arbitrary vector in S, then translated by t, $\mathbf{x}$ becomes $\mathbf{x} + \mathbf{a}$, that is, $t\mathbf{x} = \mathbf{x} + \mathbf{a}$. The inverse to this mapping t^{-1} is obviously given by $t^{-1}\mathbf{x} = \mathbf{x} - \mathbf{a}$. The rotation mapping R: $M \to S$ is composed of T followed by t^{-1}, that is, $t^{-1}T = R$. To check that R is indeed a rotation we consider its application to the zero vector $\mathbf{0}_M$ in M. We find $R\mathbf{0}_M = t^{-1}T\mathbf{0}_M = t^{-1}(T\mathbf{0}_M) = t^{-1}\mathbf{a} = \mathbf{a} - \mathbf{a} = \mathbf{0}_S$, which is the zero vector in S. Thus for $\mathbf{x} = T\mathbf{X}$ we have $\mathbf{x} = T\mathbf{X} = tR\mathbf{X} = R\mathbf{X} + \mathbf{a}$, where $\mathbf{X} \in M$ and $\mathbf{x} \in S$. The relationship between these vectors is sketched in Fig. 1.6. It may be helpful to view $\mathbf{a}$ as the vector in S to the origin of M. This is sketched in Fig. 1.7, where the vectors are the same as in Fig. 1.6.

Velocity

We obtain the velocity by differentation of $\mathbf{x} = R\mathbf{X} + \mathbf{a}$,

$$\dot{\mathbf{x}} = \dot{R}\mathbf{X} + R\dot{\mathbf{X}} + \dot{\mathbf{a}}. \tag{1.37}$$

We emphasize again that the vectors $R\mathbf{X}$ and $\mathbf{X}$ should not be confused. The first is in S, the second in M. The meaning of the various terms in (1.37) should also be clearly understood. The first term is a contribution to the absolute velocity $\dot{\mathbf{x}}$ due to the changing rotation relating S and M.

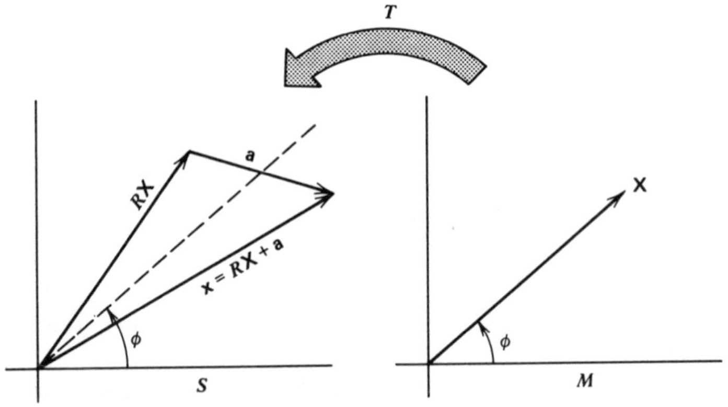

FIGURE 1.6. The relationship between the vectors involved in the transformation T.

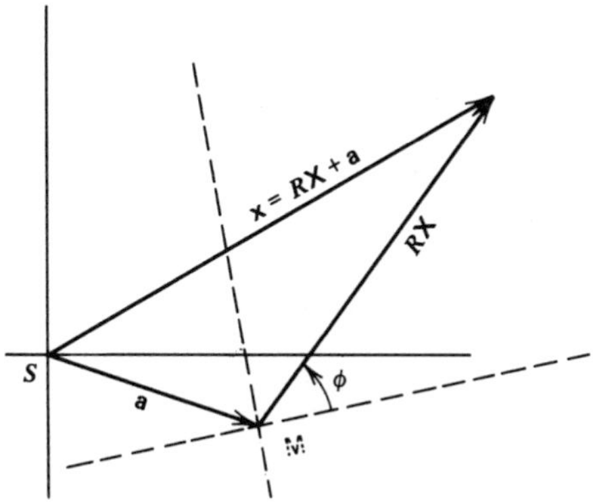

FIGURE 1.7. The same vector relationships as in Fig. 1.6 but with frame M positioned in S.

The second term is the image in S of the velocity $\dot{\mathbf{X}} \in M$. The third term is the velocity of the system M in S.

To further clarify the terms in (1.37) we will analyze some special cases.

Example 1. The case of purely translational motion, that is, motion where $\dot{R} = 0$, is immediate:

$$\dot{\mathbf{x}} = \mathbf{v} + \mathbf{v}_0, \tag{1.38}$$

where $\dot{\mathbf{x}}$ is the absolute velocity, $\mathbf{v} = R\dot{\mathbf{X}}$ is the relative velocity, and $\mathbf{v}_0 = \dot{\mathbf{a}}$ is the velocity of M in S.

Example 2. Consider the case where $\dot{\mathbf{x}}$ arises solely from the rotation of M, that is, $\dot{\mathbf{X}} = 0$ and $\dot{\mathbf{a}} = 0$. Then we have that

$$\dot{\mathbf{x}} = \dot{R}\mathbf{X}. \tag{1.39}$$

Note that $\mathbf{X} = R^{-1}\mathbf{x}$ and hence (1.39) leads to

$$\dot{\mathbf{x}} = \dot{R}R^{-1}\mathbf{x} = \dot{R}R^t\mathbf{x}. \tag{1.40}$$

If I is the identity matrix, then $(dI/dt) = d(RR^t)/dt = \dot{R}R^t + R\dot{R}^t = \dot{R}R^t + (\dot{R}R^t)^t = 0$. Thus we see that $\dot{R}R^t$ is skew-symmetric.* We show below that any skew-symmetric operator on $\mathbf{x}$ can be written as a vector product,

*An operator A is skew-symmetric if $A + A^t = 0$, that is, $A^t = -A$.

that is, in the form

$$\dot{\mathbf{x}} = \boldsymbol{\omega} \times \mathbf{x}. \tag{1.41}$$

Before relating (1.41) to (1.40) and $\boldsymbol{\omega}$ to $\dot{R}R^t$ we give some useful formulas that use the *Levi–Civita tensor*. This tensor generalizes to arbitrary coordinate systems the three-index permutation symbol defined in (1.35). We let $g \equiv \det(g_{ij})$ and define the Levi–Civita tensor by

$$\varepsilon_{ijk} = \sqrt{(g)}\,[ijk]. \tag{1.42}$$

The contravariant form of this tensor is

$$\varepsilon^{ijk} = \left(\sqrt{(g)}\right)^{-1}[ijk]. \tag{1.43}$$

(The Levi–Civita tensor is studied in Problems 1.4 and 1.5.)

The components of the cross product of two vectors is conveniently expressed in terms of the Levi–Civita tensor.

$$(\mathbf{A} \times \mathbf{B})_i = \varepsilon_{ijk} A^j B^k = [ijk] A^j B^k, \tag{1.44}$$

where this last form is explicitly for a cartesian system, that is, $g = 1$.

Now let $A = \dot{R}R^t$ and define a vector $\boldsymbol{\omega}$ by the formula $A_{ij} = -\varepsilon_{ijk}\omega^k$. Substitution into (1.40) gives

$$\dot{x}^i = A^i{}_j x^j = g^{im} A_{mj} x^j = -g^{im}\varepsilon_{mjk}\omega^k x^j = g^{im}(\boldsymbol{\omega} \times \mathbf{x})_m,$$

which gives (1.41). Figure 1.8 shows how these vectors relate to each other.

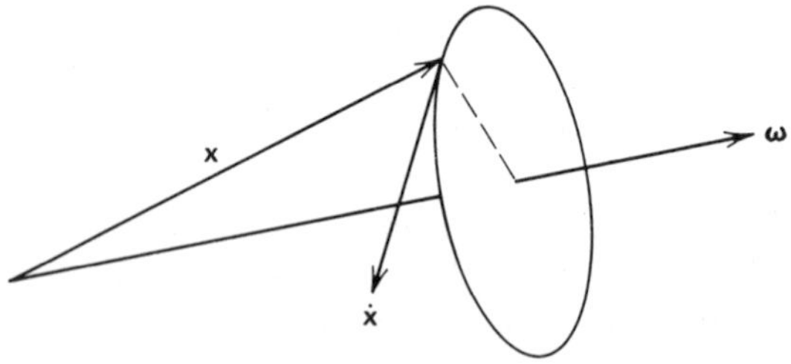

FIGURE 1.8. The relationship of position and velocity vectors to the instantaneous rotation vector $\boldsymbol{\omega}$.

An important and useful implication of (1.41) is that $\boldsymbol{\omega}$ may be interpreted as an instantaneous angular velocity vector. If a body rotates around a stationary point of the space S so that $\dot{\mathbf{x}}$ arises solely from the rotation of M, then at every instant there exists an instantaneous axis of rotation. The velocity of the points along this axis of rotation is zero. The velocity of every other point lies in a plane perpendicular to this axis and is proportional to the perpendicular distance to it. See Fig. 1.8.

Example 3. We turn now to the general relationship for velocity vectors as given in (1.37). We note that $\mathbf{X} = R^t(\mathbf{x} - \mathbf{a})$ and then

$$\dot{\mathbf{x}} = \dot{R}R^t(\mathbf{x} - \mathbf{a}) + R\dot{\mathbf{X}} + \dot{\mathbf{a}} = \boldsymbol{\omega} \times (\mathbf{x} - \mathbf{a}) + \mathbf{v} + \mathbf{v}_0. \tag{1.45}$$

Acceleration

We consider first the simple case where systems S and M are related by a translation only (R = identity operator).

Consider S to be an inertial coordinate system in which, by Newton's law, force = (mass) × (acceleration) or $\mathbf{f} = m\ddot{\mathbf{x}}$. Then since $\mathbf{x} = \mathbf{X} + \mathbf{a}$ we have

$$m\ddot{\mathbf{X}} = m\ddot{\mathbf{x}} - m\ddot{\mathbf{a}} = f - m\ddot{\mathbf{a}}. \tag{1.46}$$

Thus we see that in M, $-m\ddot{\mathbf{a}}$ plays the role of an "extra" force. Such forces are called *inertial forces* since they owe their existence to the nonuniform motion of the coordinate system. Clearly an appropriate motion of the coordinate system can make $m\ddot{\mathbf{X}}$ be zero. The example of so-called weightlessness is familiar. In a coordinate system centered on a freely falling astronaut, his just-released tool will have zero acceleration.

For coordinate systems related by both a rotation and a translation we have that $\mathbf{x} = R\mathbf{X} + \mathbf{a}$, and from (1.37)

$$\dot{\mathbf{x}} = \dot{R}\mathbf{X} + R\dot{\mathbf{X}} + \dot{\mathbf{a}} = R(R^t\dot{R}\mathbf{X} + \dot{\mathbf{X}}) + \dot{\mathbf{a}}. \tag{1.47}$$

Denote $\boldsymbol{\Omega}$ as the angular velocity in the rotating frame, that is, $\boldsymbol{\omega} = R\boldsymbol{\Omega}$. From Eqs. (1.40) and (1.41) we have

$$\dot{R}\mathbf{X} = \dot{R}R^t(\mathbf{x} - \mathbf{a}) = \boldsymbol{\omega} \times (\mathbf{x} - \mathbf{a}), \tag{1.48}$$

or

$$\dot{R}\mathbf{X} = R\boldsymbol{\Omega} \times R\mathbf{X} = R(\boldsymbol{\Omega} \times \mathbf{X}). \tag{1.49}$$

$$R^t\dot{R}\mathbf{X} = (\boldsymbol{\Omega} \times \mathbf{X}). \tag{1.50}$$

We have left the demonstration of (1.49) to Problem 1.8. Thus substituting (1.50) in (1.47) gives

$$\dot{\mathbf{x}} = R(\dot{\mathbf{X}} + [\boldsymbol{\Omega} \times \mathbf{X}]) + \dot{\mathbf{a}}. \tag{1.51}$$

Now differentiate once again

$$\ddot{\mathbf{x}} = \dot{R}(\dot{\mathbf{X}} + [\boldsymbol{\Omega} \times \mathbf{X}]) + R(\ddot{\mathbf{X}} + [\dot{\boldsymbol{\Omega}} \times \mathbf{X}] + [\boldsymbol{\Omega} \times \dot{\mathbf{X}}]) + \ddot{\mathbf{a}}. \tag{1.52}$$

Multiply through in the preceding equation by R^t and use (1.50) on the first two terms on the right-hand side. We define the vector $\ddot{\mathbf{A}}$ in M by the relation $R\ddot{\mathbf{A}} = \ddot{\mathbf{a}}$ and finally multiply by the scalar mass m to find

$$m\ddot{\mathbf{X}} = \mathbf{F} - m(\dot{\boldsymbol{\Omega}} \times \mathbf{X}) - 2m(\boldsymbol{\Omega} \times \dot{\mathbf{X}}) - m\boldsymbol{\Omega} \times (\boldsymbol{\Omega} \times \mathbf{X}) - m\ddot{\mathbf{A}}, \tag{1.53}$$

where

$$\mathbf{f} = m\ddot{\mathbf{x}} = R\mathbf{F}. \tag{1.54}$$

The last four terms in (1.53) play the role of forces and arise because of the noninertial nature of the coordinate system M. The first of these is present only when the rotation is not uniform. The next two are present even for uniform rotation. $-2m(\boldsymbol{\Omega} \times \dot{\mathbf{X}})$ is called the *Coriolis* force and $-m\boldsymbol{\Omega} \times (\boldsymbol{\Omega} \times \mathbf{X})$ is called the *centrifugal* force. See Fig. 1.9 for the vector relationships in the centrifugal force. The Coriolis force acts only on objects

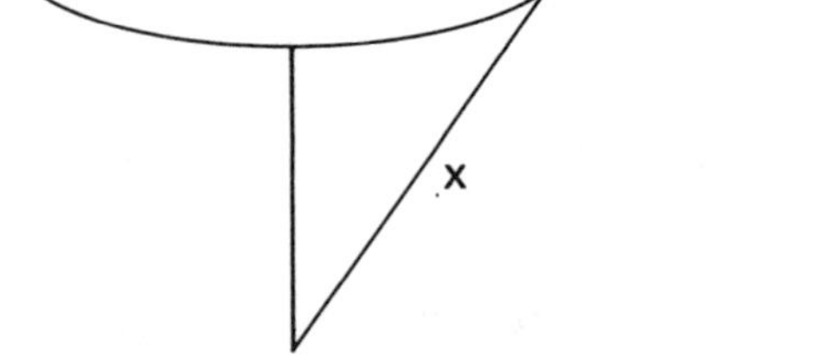

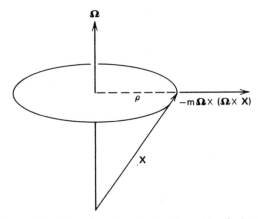

FIGURE 1.9. The centrifugal force lies along the line formed by the intersection of the two planes perpendicular to $\boldsymbol{\Omega}$ and $\boldsymbol{\Omega} \times \mathbf{X}$, respectively. It has magnitude $m\Omega^2\rho$.

with a velocity $\dot{\mathbf{X}}$ in the rotating frame. The centrifugal force acts on all objects. In the Northern (Southern) Hemisphere the Coriolis force always deflects objects to the right (left) that are moving along the surface of the earth. The last term results from the linear acceleration of the frame M in S.

This completes our discussion of the relationship between vectors of velocity and acceleration as seen in different reference frames or coordinate systems. To find the motion for a mechanical system it is always necessary to set up some frame of reference and a coordinate system therein. The results obtained in this chapter show how to pass from one frame or coordinate system in which it may be easy to set up the problem, to a second frame or coordinate system in which we may wish the answer. The issues are strictly kinematic and we have touched only briefly on questions of dynamics. The following chapters are all devoted in one way or another to solving the dynamical problem and for kinematic results we will often refer back to this chapter.

PROBLEMS

1.1. Show $|\tau| = 1$ and that $d\tau/dt = (v/R)\hat{\mathbf{n}}$, where $\tau = d\mathbf{x}/ds$ is the tangent to a trajectory and $\hat{\mathbf{n}}$ is the principal normal with R the radius of curvature of the trajectory at the point in question.

1.2. Compute the connection coefficients Γ^k_{ij} for cylindrical and spherical coordinates and verify the expressions given after (1.24).

1.3. Verify Eqs. (1.26) and (1.28).

1.4. Verify Eq. (1.44) for cylindrical coordinates. The Levi–Civita tensor also gives a convenient index notation for the familiar curl. Verify that $(\nabla \times \boldsymbol{\omega})^k = \varepsilon^{kij}\omega_{j;\,i} = \varepsilon^{kij}\omega_{j,\,i}$ for cylindrical coordinates.

1.5. Show

(a) $\varepsilon_{ijk} = (\sqrt{(g)})[ijk]$.

(b) $\varepsilon^{ijk} = (\sqrt{(g)})^{-1}[ijk]$,

(c) $\varepsilon^{ijk}\varepsilon_{ilm} = [\delta^j_l\,\delta^k_m - \delta^j_m\,\delta^k_l]$

where g is the det g_{ij}. Hint: Transform from cartesian to curvilinear coordinates and use $\det(AB) = (\det A)(\det B)$.

1.6. Find both physical and coordinate components of the velocity vector $\mathbf{v}$ and the acceleration $\mathbf{a}$ in parabolic coordinates μ, ν, ψ where $x = \mu\nu\cos\psi$, $y = \mu\nu\sin\psi$, $z = (\mu^2 - \nu^2)/2$. Hint: Show first that $ds^2 = (\mu^2 + \nu^2)(d\mu^2 + d\nu^2) + \mu^2\nu^2\,d\psi^2$.

1.7. Find the angular velocity vector $\boldsymbol{\omega}$ in terms of the components of the rotation R.

1.8. Show that $R\boldsymbol{\Omega} \times R\mathbf{X} = R(\boldsymbol{\Omega} \times \mathbf{X})$ from the definition of the vector product.

1.9. Find the angular velocity of a standard $33\frac{1}{3}$ rpm phonograph.

(a) Relative to the Earth.

(b) Relative to an inertial system.

1.10. Find the position, as seen by an observer on Earth, of a satellite in a circular orbit passing above the poles. State the position vector in a coordinate system attached to the Earth and with the center of the Earth as origin (Saletan and Cromer, 1971).

1.11. A stationary satellite is one placed in a circular orbit above a point on the equator and given a velocity just right to keep it exactly over that point constantly. Suppose a satellite were placed in orbit over a point on the equator with the correct speed (magnitude of velocity) to remain stationary, but the direction of the velocity vector was slightly out of the equatorial plane. Describe its subsequent motion as seen from Earth (Saletan and Cromer, 1971).

1.12. Consider the situation of a small spider of mass m attached to the rim of a wheel on an automobile. The wheel is rolling without slipping at constant speed V in a circle of radius R. The spider is at a distance a from the center of the wheel. Show that the time-averaged force that the spider exerts on the wheel in order to hang on has component $2V^2/R$ perpendicular to the plane of the wheel and $V^2[1 + a^2/(2R^2)]/a$ in the plane.

1.13. A small disk is attached to a large platform with its center a distance R from the vertical axis about which the large platform rotates with angular velocity Ω. The small disk also rotates about a vertical axis through its center with angular velocity ω. Find the force a small bug of mass m exerts on the disk if it maintains its position without slipping, a distance a from the center of the disk.

1.14. An iceboat sails due east in the Northern Hemisphere. If the coefficient of friction preventing lateral motion is μ, find the velocity above which the boat will slide sideways. Is this velocity different for the same iceboat sailing due west at the same latitude? Ignore terms of order ω^2 where ω is the angular velocity of the earth's rotation.

CHAPTER TWO

NEWTONIAN DYNAMICS

Nature and Nature's Laws lay hid in Night, God said, Let Newton be, and all was Light.

POPE

In this chapter we recall briefly the formulation of dynamics given by Newton. Our approach here is mainly one of review. We first focus on dynamics of a single particle, including forces of constraint. We then generalize to many-particle systems. We consider one-dimensional systems in detail and introduce the notions of phase space and phase curves. Phase flow is defined along with the concept of Liapunov stability. We conclude this chapter by considering open systems, that is, systems that may exchange mass or momentum with the surroundings.

2.1. SINGLE-PARTICLE DYNAMICS

Newton's law of motion is

$$\frac{dm\mathbf{v}}{dt} = \mathbf{F}(t, \mathbf{x}, \mathbf{v}). \tag{2.1}$$

As simple as this law may look, and as familiar as it may seem, a few comments are in order. The mass m of an object is defined by comparing the acceleration resulting from a known force with that of an object whose mass has been chosen as a standard. Clearly, if the mass is constant, the left-hand side of (2.1) is just $m\ddot{\mathbf{x}}$. The force function in (2.1) is assumed to depend only on the instantaneous values of $t, \mathbf{x}, \mathbf{v}$. In other words, if the initial position and velocity are known, then the object's subsequent motion is completely determined.

The determination of the force function, which properly characterizes the interaction, becomes the fundamental experimental problem. From the mathematical point of view, a specification of $\mathbf{F}(t, \mathbf{x}, \mathbf{v})$ constitutes a definition of the system. Objects for which the force function $\mathbf{F}(t, \mathbf{x}, \mathbf{v})$ vanishes are said to be "free particles" and define inertial coordinate systems. Inertial coordinate systems are just those systems in which free particles have zero acceleration. This is sometimes referred to as Newton's first law of motion. Newton's third law of motion states that the forces on two particles due to their mutual interaction are equal in magnitude and oppositely directed.

Care must be exercised in the application of (2.1) if the system under study is an open system. We return to this issue at the end of this chapter.

Unless specifically stated otherwise we shall consider the mass of the object under study to be constant. In this more usual case (2.1) can then be written as

$$m\frac{d\mathbf{v}}{dt} = \mathbf{F}(t, \mathbf{x}, \mathbf{v}). \tag{2.2}$$

It is also often true that the vector $\mathbf{F}$ can be written as the gradient of a scalar function.

$$\mathbf{F} = -\nabla U(\mathbf{x}). \tag{2.3}$$

In such situations there is an important constant of the motion called the energy. Using (2.3) in (2.2) we obtain

$$m\frac{d\mathbf{v}}{dt} + \nabla U = 0. \tag{2.4}$$

We take the inner product of this equation with $\mathbf{v}$ to find

$$\frac{d}{dt}\left(\frac{m}{2}\mathbf{v}\cdot\mathbf{v}\right) + \mathbf{v}\cdot\nabla U = 0 \tag{2.5}$$

If the potential function U does not depend explicitly on the time, then

$(d/dt)U = \nabla U \cdot \mathbf{v}$ and we find

$$\frac{d}{dt}\left(\frac{m}{2}v^2 + U\right) = 0. \tag{2.6}$$

Thus we have that the quantity

$$E = \tfrac{1}{2}mv^2 + U \tag{2.7}$$

is conserved, that is, is constant in time. We call this quantity the *total energy*. The first term on the right of (2.7) is called the *kinetic energy* and the second is called the *potential energy*. Clearly, any constant could be added to U without affecting the results and so the energy is defined only to within an arbitrary constant. Note again that the existence of an energy constant in this case depends on the potential U being a function only of the spatial position.

Let us now compute the change in kinetic energy of a system, whether or not a potential function exists. Using again (2.2) and taking the inner product of both sides with $\mathbf{v}$, we find that

$$\mathbf{v} \cdot m\frac{d\mathbf{v}}{dt} = \frac{d}{dt}\left(\frac{m}{2}v^2\right) = \mathbf{v} \cdot \mathbf{F}. \tag{2.8}$$

Integrating this equation from t_0 to t we find

$$\tfrac{1}{2}mv^2|_t - \tfrac{1}{2}mv^2|_{t_0} = \int_{t_0}^{t} \mathbf{F} \cdot \mathbf{v}\, dt. \tag{2.9}$$

The term on the right-hand side of (2.9) represents the work done in changing the kinetic energy of the particle from its initial to its final value.

Having defined the energy constant we discuss angular momentum. With respect to the origin, the *angular momentum* of an object with mass m and momentum $\mathbf{p} = m\mathbf{v}$, is given by

$$\mathbf{L} \equiv \mathbf{x} \times \mathbf{p}, \tag{2.10}$$

where $\mathbf{x}$ is the position vector of the object in question. Note that the angular momentum of a particle must be defined with respect to some fixed point in the coordinate system. Since it is natural and convenient, we simply use the origin of an inertial coordinate system. Differentiating (2.10) with respect to time, we find

$$\frac{d\mathbf{L}}{dt} = \frac{d\mathbf{x}}{dt} \times \mathbf{p} + \mathbf{x} \times \frac{d\mathbf{p}}{dt} = \mathbf{x} \times \mathbf{F} = \mathbf{T}. \tag{2.11}$$

T is the torque of the force **F** about the origin.

To summarize our single-particle results, we find that a force changes the particle's momentum and angular momentum according to Eqs. (2.1) and (2.11), respectively. If the force vanishes, then these two quantities are constant in time. The kinetic energy of the particle changes according to (2.9). If a potential U exists satisfying (2.3), then we speak of a potential energy of the particle and find its total energy to be conserved in such a field.

It often happens that the forces constituting **F** in (2.1) can conveniently be classified into external forces and forces of constraint. An example is furnished by a weight on the end of a string. The gravitational force is viewed as the external force and the force exerted on the weight by the string is the force of constraint. Some constraints are easier to deal with than others and it is a matter of convenience which forces are labeled "forces of constraint" and which ones are viewed as "external forces." Generally speaking, forces of constraint are contact forces (ultimately electromagnetic in origin) and arise almost automatically in the statement of the problem. Consideration of some examples and a little experience will for most cases remove any uncertainty in the reader's mind as to what should be viewed as a constraint force. Ultimately it is a question of judgment, but for most problems the choices are rather obvious.

Many constraints can be quantified by requiring the position and velocity to satisfy some constraining equations, which may change with time. For example, $f_\alpha(\mathbf{x}, \mathbf{v}, t) = 0$, $\alpha = 1, 2$. Sometimes constraints take the form of an inequality rather than an equality, or perhaps appear in the form of a relation among differentials. Constraints are broadly divided into two classes: holonomic and nonholonomic. Inequality constraints and nonintegrable differential constraints are termed *nonholonomic.* Otherwise they are termed *holonomic*.

To illustrate the techniques of solution we shall limit our attention to holonomic constraints that do not depend on the velocity. In this case we write the equations of constraint in the form

$$f_\alpha(\mathbf{x}, t) = \text{const.} \tag{2.12}$$

We may think of these equations as confining the motion of the particle to a surface or curve that may be changing with time. We write the equation of motion in the form

$$m\frac{d\mathbf{v}}{dt} = \mathbf{F}_{\text{ext}} + \mathbf{F}_c, \tag{2.13}$$

where $\mathbf{F}_c$ represents the forces of constraint. As before we assume a knowledge of $\mathbf{F}_{\text{ext}}$ but since the constraint forces are generally unknown we cannot solve (2.13) directly for the motion.

Forces of constraint that are normal to the surfaces defined by (2.12) are called *smooth* and in particular preclude dissipative forces of friction. In the case of smooth forces of constraint we may proceed as follows. The vectors ∇f_α are normal to the surfaces and indeed form a basis for all vectors that are normal. Thus we may expand the forces of constraint $\mathbf{F}_c$ in terms of this basis, that is,

$$\mathbf{F}_c = \lambda^\alpha(t)\nabla f_\alpha. \tag{2.14}$$

(Note sum on α!)

We also have that

$$\left(m\frac{d\mathbf{v}}{dt} - \mathbf{F}_{\text{ext}}\right)_\perp = 0, \tag{2.15}$$

where $\perp$ denotes that part orthogonal to the vector subspace of the ∇f_α. This gives one or two equations (depending on whether there are two or one equations of constraint) for the three unknown functions of position. The remaining equations required for a solution may be obtained by taking the total derivative of $f_\alpha(\mathbf{x}(t), t) =$ const. This gives

$$\frac{df_\alpha}{dt} = \frac{\partial f_\alpha}{\partial t} + \frac{\partial f_\alpha}{\partial x^i}\frac{dx^i}{dt} = \frac{\partial f_\alpha}{\partial t} + \mathbf{v}\cdot\nabla f_\alpha = 0, \tag{2.16}$$

which gives enough equations for a solution. Once the motion has been found, we can then substitute back into (2.13) and solve for $\mathbf{F}_c$. Assuming these steps have been carried out, we can then solve for the scalar function(s) $\lambda^\alpha(t)$ in (2.14) relating the constraint forces and the surfaces of constraint.

Substituting (2.14) for $\mathbf{F}_c$ in (2.13), taking the scalar product with $\mathbf{v}$ and using (2.16), we find the equation telling how the total energy changes with time.

$$\frac{d}{dt}\left(\frac{1}{2}mv^2 + U\right) = -\lambda^\alpha(t)\frac{\partial f_\alpha}{\partial t}, \tag{2.17}$$

where we have also assumed that the external forces can be obtained from a time-independent potential.

We note from (2.17) that if the surfaces do not move in time, then the energy is conserved. Otherwise the energy changes with time because of the constraint forces, despite the fact that the constraint forces are here assumed to act normal to their respective surfaces. This is easily understood for a moving surface since the net particle velocity is not entirely within the surface and $\mathbf{v} \cdot \mathbf{F}_c$ is not necessarily zero.

As an example consider a bead constrained to move on a circular hoop of radius R that rotates at constant angular velocity ω about the vertical direction. See Fig. 2.1. There is also a constant gravitational field in the vertical direction as well. The equations of constraint are

$$f_1(r, \theta, \phi) = r - R = 0,$$

$$f_2(r, \theta, \phi, t) = \phi - \omega t = 0.$$

It is appropriate to use spherical coordinates to describe the position of the bead. Using the gradient operator in spherical coordinates, we find $\nabla f_1 = \hat{\mathbf{r}}$, and $\nabla f_2 = \hat{\boldsymbol{\phi}}/r \sin\theta$. Thus we know that for the orthogonal complement to the vector space generated by $\nabla f_1 \propto \hat{\mathbf{r}}$ and $\nabla f_2 \propto \hat{\boldsymbol{\phi}}$ we may use $\hat{\boldsymbol{\theta}}$ as a basis vector. Now we use Eqs. (1.28) of Chapter 1 to write $\ddot{\mathbf{x}}$ in terms of spherical coordinates. We also note for the gravitational force $\mathbf{F}_{\text{ext}}$ that

$$\mathbf{F}_{\text{ext}} = -mg\hat{\mathbf{z}} = -mg(\hat{\mathbf{r}} \cos\theta - \hat{\boldsymbol{\theta}} \sin\theta).$$

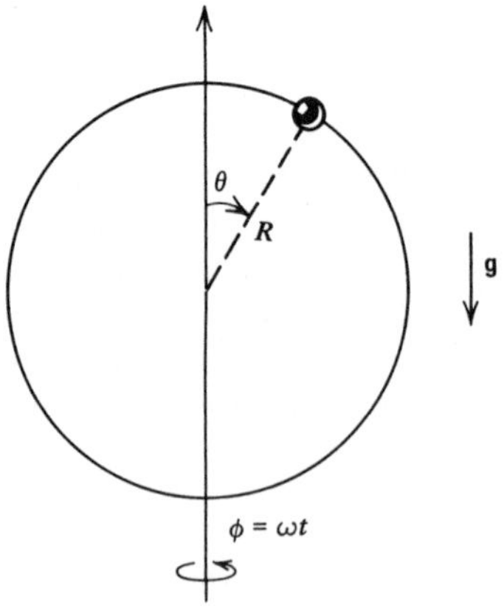

FIGURE 2.1. Bead on a uniformly rotating circular hoop in a uniform gravitational field.

Thus we find that

$$m\ddot{\mathbf{x}} - \mathbf{F}_{\text{ext}} = m\ddot{\mathbf{x}} + mg(\hat{\mathbf{r}}\cos\theta - \hat{\boldsymbol{\theta}}\sin\theta)$$

$$= m\{\hat{\mathbf{r}}(\ddot{r} - r\dot{\theta}^2 - r\sin^2\theta\dot{\phi}^2 + g\cos\theta)$$

$$+\hat{\boldsymbol{\theta}}(r\ddot{\theta} + 2\dot{r}\dot{\theta} - r\dot{\phi}^2\sin\theta\cos\theta - g\sin\theta)$$

$$+\hat{\boldsymbol{\phi}}(r\sin\theta\ddot{\phi} + 2\sin\theta\dot{r}\dot{\phi} + 2r\cos\theta\dot{\phi}\dot{\theta})\}.$$

Now make use of the equations of constraint and substitute in the above $r = R$, $\dot{r} = 0$, $\ddot{r} = 0$, $\dot{\phi} = \omega$, $\ddot{\phi} = 0$. We obtain

$$m\ddot{\mathbf{x}} - \mathbf{F}_{\text{ext}} = m\{\hat{\mathbf{r}}(-R\dot{\theta}^2 - R\omega^2\sin^2\theta + g\cos\theta)$$

$$+\hat{\boldsymbol{\theta}}(R\ddot{\theta} - R\omega^2\sin\theta\cos\theta - g\sin\theta) + \hat{\boldsymbol{\phi}}(2R\omega\dot{\theta}\cos\theta)\}. \tag{2.18}$$

To get the $\perp$ part we just take the scalar product with $\hat{\boldsymbol{\theta}}$. (This is equivalent to taking the scalar product with $\nabla f_1 \times \nabla f_2$, which would of course always give the $\perp$ piece.) We find for the differential equation of motion

$$R\ddot{\theta} - R\omega^2\sin\theta\cos\theta - g\sin\theta = 0.$$

We use the components of the velocity $\mathbf{v}$ in spherical coordinates that follow (1.14) in Chapter 1. This velocity inserted into (2.16) results in the following equations:

$$\dot{r} = 0 \quad \text{and} \quad \dot{\phi} - \omega = 0.$$

From (2.18) we find the constraint force, that is, that part orthogonal to $\hat{\boldsymbol{\theta}}$, and substitute into (2.14).

$$\mathbf{F}_c = m\{\hat{\mathbf{r}}(-R\dot{\theta}^2 - R\omega^2\sin^2\theta + g\cos\theta) + \hat{\boldsymbol{\phi}}(2R\omega\dot{\theta}\cos\theta)\}$$

$$= \lambda^1\nabla f_1 + \lambda^2\nabla f_2.$$

and so we obtain by examination

$$\lambda^1 = -m(R\dot{\theta}^2 + R\omega^2\sin^2\theta - g\cos\theta)$$

and

$$\lambda^2 = mR^2\omega\dot{\theta}\sin 2\theta.$$

The change in the energy from (2.17) is given by

$$\frac{dE}{dt} = mR^2\omega^2\dot{\theta}\cos 2\theta.$$

2.2. MANY-PARTICLE SYSTEMS

To treat a system with many individual particles interacting requires only that we carefully apply the laws we have discussed for a single particle. Consequently, we omit obvious details.

We use the index α to label the individual particles and note that sums over particles will be denoted explicitly. The equations of motion become

$$m_\alpha \ddot{\mathbf{x}}_\alpha = \mathbf{F}_\alpha \qquad (\alpha = 1, 2, \ldots) \tag{2.19}$$

The momentum of a system of particles is the sum of the individual momenta:

$$\mathbf{P} = \sum_\alpha m_\alpha \dot{\mathbf{x}}_\alpha \tag{2.20}$$

Using (2.19) we compute the time derivative of this equation. The force on the αth particle is due to interactions with other particles in the system under study and forces external to this system.

$$\mathbf{F}_\alpha = \sum_\beta \mathbf{F}_{\alpha\beta} + \mathbf{F}_\alpha^{\text{ext}}. \tag{2.21}$$

$\mathbf{F}_{\alpha\beta}$ is the force on the αth particle due to the βth particle. $\mathbf{F}_{\alpha\alpha} = 0$ for all α. By Newton's third law of motion $\mathbf{F}_{\alpha\beta} = -\mathbf{F}_{\beta\alpha}$. Thus we find that

$$\sum_\alpha \mathbf{F}_\alpha = \sum_\alpha \mathbf{F}_\alpha^{\text{ext}}, \tag{2.22}$$

and

$$\frac{d\mathbf{P}}{dt} = \sum_\alpha m_\alpha \ddot{\mathbf{x}}_\alpha = \sum_\alpha \mathbf{F}_\alpha^{\text{ext}}. \tag{2.23}$$

Immediately we recognize that if the external forces on a system are zero (the system is closed), then the total momentum is a constant in time.

Let $M \equiv \Sigma_\alpha m_\alpha$ and define the *center-of-mass* of a system according to the relation

$$M\mathbf{X}_{cm} \equiv \sum_\alpha m_\alpha \mathbf{x}_\alpha. \tag{2.24}$$

Then

$$\mathbf{P} = \sum_\alpha m_\alpha \dot{\mathbf{x}}_\alpha = M\dot{\mathbf{X}}_{cm} \tag{2.25}$$

and

$$\frac{d\mathbf{P}}{dt} = \sum_\alpha m_\alpha \ddot{\mathbf{x}}_\alpha = M\ddot{\mathbf{X}}_{cm} = \sum_\alpha \mathbf{F}_\alpha^{\text{ext}}. \tag{2.26}$$

The above result is called the *center-of-mass theorem* and shows that the center-of-mass moves as if all the mass of the system were concentrated there and all the external forces acted there.

We frequently use the symbol T to denote the kinetic energy. The kinetic energy of a system of particles is the sum of the kinetic energies of the individual particles.

$$T = \sum_\alpha \tfrac{1}{2} m_\alpha \mathbf{v}_\alpha^2 = \sum_\alpha \tfrac{1}{2} m_\alpha |\dot{\mathbf{x}}_\alpha|^2. \tag{2.27}$$

Let $\mathbf{y}_\alpha = \mathbf{x}_\alpha - \mathbf{X}_{cm}$ denote the position of the αth particle with respect to the center-of-mass. We then find that (details are left to Problem 2.1)

$$T = \tfrac{1}{2} \sum_\alpha m_\alpha |\dot{\mathbf{y}}_\alpha|^2 + \tfrac{1}{2} M |\dot{\mathbf{X}}_{cm}|^2. \tag{2.28}$$

In other words, the total kinetic energy is the sum of the kinetic energy associated with the center-of-mass motion, plus the kinetic energy of the individual particles relative to the center-of-mass. We also find that the rate of change in the total kinetic energy of the system is given by

$$\frac{dT}{dt} = \sum_\alpha m_\alpha \dot{\mathbf{x}}_\alpha \cdot \ddot{\mathbf{x}}_\alpha = \sum_\alpha \dot{\mathbf{x}}_\alpha \cdot \mathbf{F}_\alpha. \tag{2.29}$$

That is to say, the change in the total kinetic energy of the system is given by the sum of the work done on the individual particles in the system.

Consider the case where the forces $\mathbf{F}_\alpha$ are derived from a time-independent potential. The force on the αth particle depends on its position in the potential field according to

$$\mathbf{F}_\alpha = -\nabla_\alpha U, \tag{2.30}$$

where ∇_α denotes the gradient with respect to the coordinates of the αth particle. Then we have that

$$\frac{dT}{dt} = -\sum_\alpha \dot{\mathbf{x}}_\alpha \cdot \nabla_\alpha U. \tag{2.31}$$

Since $U = U(\mathbf{x}_1, \mathbf{x}_2, \ldots)$ does not depend explicitly on time we find once again that

$$\frac{d}{dt}(T + U) = 0. \tag{2.32}$$

The total energy is again a constant of the motion.

Consider the total angular momentum for a system of particles.

$$\mathbf{L} = \sum_\alpha \mathbf{x}_\alpha \times m_\alpha \dot{\mathbf{x}}_\alpha = \sum_\alpha \mathbf{y}_\alpha \times m_\alpha \dot{\mathbf{y}}_\alpha + M\mathbf{X}_{cm} \times \dot{\mathbf{X}}_{cm}. \tag{2.33}$$

$$\mathbf{L} = \mathbf{L}_{cm} + \mathbf{X}_{cm} \times M\dot{\mathbf{X}}_{cm}. \tag{2.34}$$

We see that the angular momentum of a system of particles about the coordinate origin is given by the angular momentum of the particles about the center-of-mass, plus the angular momentum of the center-of-mass, about the coordinate origin.

Consider how the torque due to the external forces acting on the system affects the angular momentum. Let $\mathbf{Z}$ denote an arbitrary vector in an inertial coordinate system and we compute the torque about the point $\mathbf{Z}$.

$$\mathbf{T}_\mathbf{Z} = \sum_\alpha \mathbf{z}_\alpha \times \mathbf{F}_\alpha = \sum_\alpha m_\alpha \mathbf{z}_\alpha \times \ddot{\mathbf{x}}_\alpha, \tag{2.35}$$

where $\mathbf{x}_\alpha = \mathbf{z}_\alpha + \mathbf{Z}$. We find that

$$\mathbf{T}_\mathbf{Z} = \frac{d}{dt}\mathbf{L}_\mathbf{Z} + M(\mathbf{X}_{cm} - \mathbf{Z}) \times \ddot{\mathbf{Z}}. \tag{2.36}$$

Subject to the conditions that either $\mathbf{Z}$ is the center-of-mass or $\ddot{\mathbf{Z}} = 0$, the time rate of change of the angular momentum about the point marked by **Z** is given by the total torque about **Z**. Otherwise there is an additional contribution given by the second term on the right of (2.36).

2.3. MOTION IN ONE DIMENSION

For motion that takes place in one dimension we can obtain many, very specific results. We also remark that the results to follow have broader application than one might expect because many problems formulated in more than one dimension reduce to one-dimensional problems. This can result from constraints, as in the example of the bead on the hoop in Section 2.1. Reduction to a one-dimensional problem can also result from the discovery of constants of the motion such as energy, which lead to a partial solution. Many examples of the reduction to one dimension occur throughout the book.

Equation (2.2) in this special case becomes

$$m\ddot{x} = F \tag{2.37}$$

and we can always find a potential function, which is not the case for motion in more than one dimension:

$$U(x, t) = -\int_{x_0}^{x} F(t, y)\, dy. \tag{2.38}$$

If we have a system where the force does not depend explicitly on time,* then, as shown before, the energy is a constant of the motion.

$$E = \tfrac{1}{2}m\dot{x}^2 + U(x) = \text{const.} \tag{2.39}$$

Thus with one constant of the motion, in one dimension, the solution of the problem is reduced to an integration

$$t - t_0 = \int_{x_0}^{x} \frac{dx}{\sqrt{2/m[E - U(x)]}}. \tag{2.40}$$

The reduction of the problem of motion to an integration such as in (2.40) is called "reduction to quadratures."

*Such systems are termed *autonomous*.

If the motion has turning points, that is, points where $\dot{x}$ vanishes, then these points are easily found from (2.39) and depend only on the form of the potential function $U(x)$. Points where the force in (2.37) vanishes are called *equilibrium points*, and from (2.38) we see that equilibrium points correspond to those points that satisfy $dU/dx = 0$. Thus many important features of motion in one dimension are obtained immediately by an examination of the potential without solving explicitly for the motion. A qualitative analysis of a system, by examining the temporal and spatial dependence of the potential function, is frequently more instructive in regard to general system behavior than is an explicit solution.

2.4. PHASE CURVES

In addition to an examination of the potential for a qualitative analysis of motion, there exists a second important technique that we now introduce. Equation (2.2) may be reduced to a system of first-order, differential equations through the introduction of another variable. Let $\mathbf{y} = \dot{\mathbf{x}}$ and then (2.2) becomes the system

$$\dot{\mathbf{y}} = \mathbf{f}(t, \mathbf{x}, \mathbf{y}),$$

$$\dot{\mathbf{x}} = \mathbf{y}, \tag{2.41}$$

where we let $\mathbf{f} = \mathbf{F}/m$. Many properties of the system can be deduced by looking at the curves in what is called phase space. *Phase space* is the space of points $(\mathbf{y}, \mathbf{x})$, whereas we will refer to the space of points $(\mathbf{x})$ as *configuration space*. In the case of a single dimension for the motion, phase space is two dimensional and the pattern obviously generalizes to higher dimensions: phase space has twice the number of dimensions of the configuration space for the motion.

As shown in the previous section, the energy integral leads to a complete solution to the problem of motion for one dimension. For one dimension Eqs. (2.41) become

$$\dot{y} = f(x), \qquad \dot{x} = y, \tag{2.42}$$

where for the present discussion we consider only forces that do not depend on the velocity y nor on the time explicitly. In this case the system (2.42) has a solution of the form $x = \phi(t)$, $y = \dot{\phi}(t)$, and (2.39) (the law of conservation of energy) allows us to find the phase curves easily, since for a given E, Eq. (2.39) is just a functional relation for y and x that defines the phase curves.

We note that a phase curve may consist of a single point that would correspond to an equilibrium point of the system. Such points have $\mathbf{y} = 0$ and $\mathbf{f} = 0$ in (2.41) and thus $\nabla U = 0$ at equilibrium points. Furthermore we note that a fundamental theorem of ordinary differential equations tells us that for given initial conditions, solutions to Eqs. (2.41) exist and are unique. Thus through every phase point there is only one phase curve; that is, phase curves do not intersect for (2.41). We clarify these ideas further by considering some examples.

Example 1. Consider the familiar problem of one-dimensional motion for a ball moving in a uniform gravitational field. $\ddot{x} = g$. Then for the phase space variables (y, x) we have

$$\dot{y} = g, \qquad y = \dot{x}, \tag{2.43}$$

with the familiar solution

$$y(t) = y_0 + gt, \qquad x(t) = x_0 + y_0 t + \tfrac{1}{2}gt^2. \tag{2.44}$$

The equation for phase curves from (2.39) becomes

$$y^2 = 2g(x - x_0) + y_0^2. \tag{2.45}$$

See Fig. 2.2 for a sketch of the phase curves.

Example 2. Let us consider the equation of motion for a simple pendulum of length L:

$$\ddot{x} + \frac{g}{L}\sin x = 0. \tag{2.46}$$

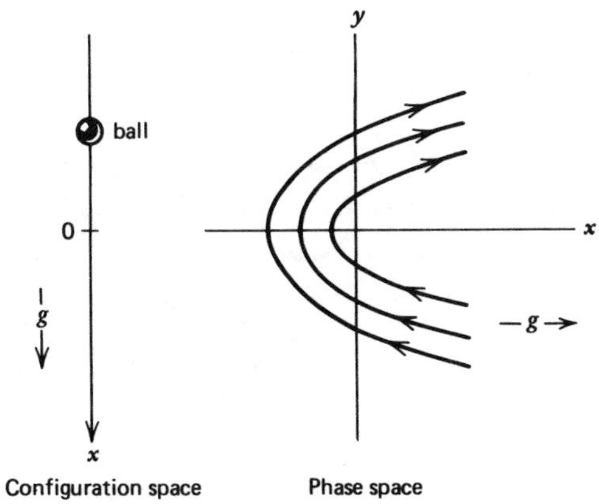

FIGURE 2.2. The phase curves for a ball moving in one dimension in a uniform gravitational field consist of a sequence of nonintersecting parabolas that all open toward positive x.

We first define a dimensionless time parameter τ according to $\tau = t\sqrt{g/L}$. Then (2.46), with differentiation with respect to τ, becomes simply

$$\ddot{x} + \sin x = 0. \tag{2.47}$$

To follow our standard procedure for obtaining the phase curves we write

$$\dot{y} = -\sin x, \qquad \dot{x} = y, \tag{2.48}$$

which coupled with (2.39) gives for phase curves

$$\frac{y^2}{2} = (E - 1) + \cos x, \tag{2.49}$$

which are sketched in Fig. 2.3. We have normalized the energy so that $x = 0$, $y = 0$ corresponds to $E = 0$. Note that for $E \ll 1$, x must be small and (2.49) just becomes $x^2 + y^2 = 2E$. Thus the phase curves near the origin are circles and correspond to a simple harmonic oscillator with differential equation of motion: $\ddot{x} + x = 0$. The origin in Fig. 2.3 is an example of an "O" (elliptic) point and physically corresponds to the pendulum bob at rest at its lowest point. The "X" (hyperbolic) points in Fig. 2.3 look like phase curves crossing but this is not the case as one can see by an examination of the flow directions with time, which are indicated by

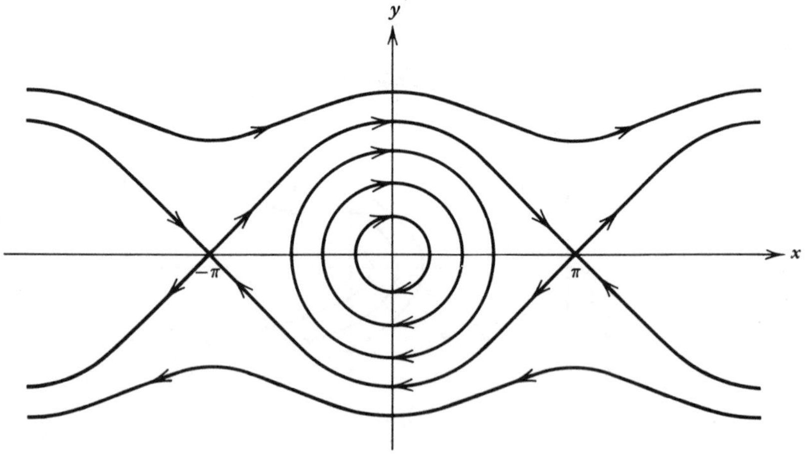

FIGURE 2.3. Phase curves for a simple pendulum.

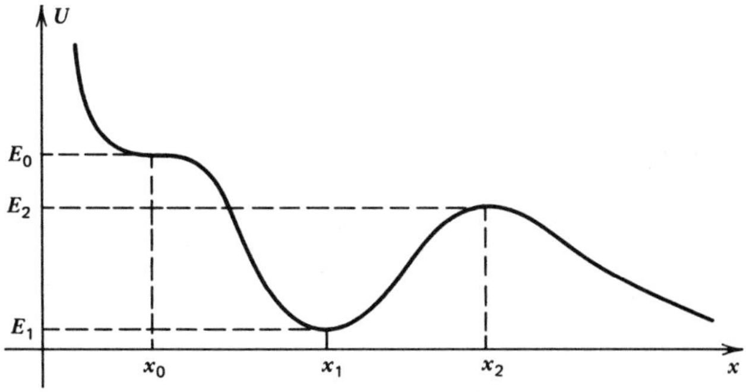

FIGURE 2.4. An arbitrary potential function with an inflection point at x_0, a local minimum at x_1, and a local maximum at x_2.

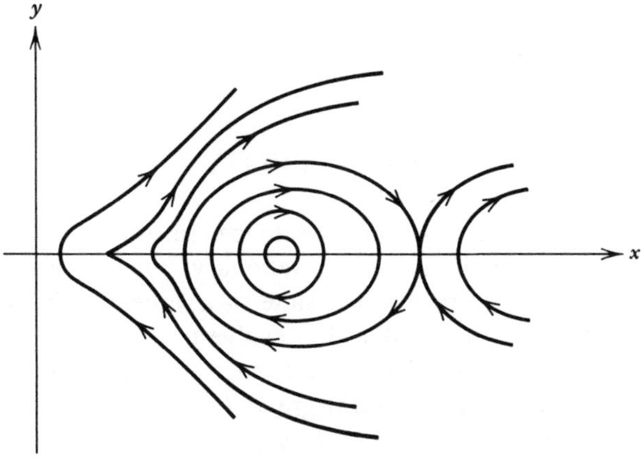

FIGURE 2.5. Phase curves corresponding to the potential function given in Fig. 2.4. The points on the x-axis for these two figures are aligned. The point x_0 is an isolated (unstable) equilibrium point and gives a cusp in the phase plot.

the arrows. The "X" points correspond to phase points of the system for departure or arrival which are obtained asymptotically, that is, as $t \to \pm\infty$. They correspond to the pendulum being vertical, with its bob at the highest point (unstable).

Example 3. We consider a potential $U(x)$, which is pictured in Fig. 2.4. To make the phase plot of Fig. 2.5 corresponding to this potential function we study the level curves of the equation

$$y^2 = 2[E - U(x)] \tag{2.50}$$

for various choices for the energy parameter E.

As discussed at the end of Section 2.3, examination of the potential gives general qualitative information about the motions. For the potential of Fig. 2.4, near the point x_1 the motion is periodic, and near the points x_0 and x_2 we find the system to be unstable. That is to say, the system will not stay near these points if slightly disturbed. The behavior of the solutions near these special points is explored in Problem 2.6.

We emphasize that in one sense these qualitative methods for analyzing the behavior of solutions are more instructive than solving the equations of motion directly. This is because considerations of the phase curves and the potential give information about all solutions and their dependence and behavior on the constants of the motion.

2.5. STABILITY

One of the most interesting questions to be asked about any dynamical system concerns the stability of its stationary (time-independent) states. Generally speaking, we say that the state of a system is *stable* if under a small perturbation the system does not evolve to a new, radically different state, but instead returns in time to the former one. For example, a mass on the end of a rigid rod in a uniform gravitational field has two stationary states, both with the rod vertical. One has the mass on top and one has the mass at the bottom. A simple experiment shows one of these states is stable and one is not. This definition of stability strictly applies only to systems wherein there is some form of dissipation that eventually damps out the perturbations. Consider again the physical pendulum of a mass at the bottom of a rigid rod but this time postulate the complete absence of dissipative forces. Then after a small perturbation the pendulum just goes into a new oscillatory state and does not tend to return to the bottom again. Yet we still regard the bottom position of the idealized system as "stable" in comparison to the state with the mass at the top.

The following definition is given in terms of phase space and is applicable to any system, whether dissipation is present or not. The definition considers the motion of the system phase point in phase space and not just the motion in configuration space. The definition we give is called *Liapunov stability* and allows us to deal in a precise way with situations such as that considered earlier for the rigid pendulum.

The differential equations of motion (2.41) or (2.2) can be written in even more compact form. We let $\boldsymbol{\xi}$ denote the "position" vector in phase space. The vector $\boldsymbol{\xi}$ in terms of the variables defined for (2.41) is given by $\boldsymbol{\xi} = (\mathbf{y}, \mathbf{x})$ and $\dot{\boldsymbol{\xi}} = (\dot{\mathbf{y}}, \dot{\mathbf{x}})$. Then (2.41) becomes

$$\dot{\boldsymbol{\xi}} = \mathbf{v}(\boldsymbol{\xi}, t), \tag{2.51}$$

and in terms of the variables of (2.41), the phase space velocity $\mathbf{v}$ is given by $\mathbf{v}(\boldsymbol{\xi}, t) = (\mathbf{f}(t, \mathbf{x}, \mathbf{y}), \mathbf{y})$. Let $\boldsymbol{\phi}(t)$ denote a solution to the differential equation with initial position $\boldsymbol{\phi}_0$. The solution $\boldsymbol{\phi}(t)$ is said to be *Liapunov stable* if given any $\varepsilon > 0$, there exists $\delta > 0$ (depending only on ε and not on t) such that for every solution $\boldsymbol{\xi}(t)$ with $|\boldsymbol{\xi}(0) - \boldsymbol{\phi}_0| < \delta$ the inequality $|\boldsymbol{\xi}(t) - \boldsymbol{\phi}(t)| < \varepsilon$ is satisfied for all $t > 0$. If $\boldsymbol{\phi}_0 = \boldsymbol{\phi}(0)$ is an equilibrium point, that is, $\boldsymbol{\phi}(t) = \boldsymbol{\phi}_0$ for all t, then we say that the equilibrium point $\boldsymbol{\phi}_0$ is Liapunov stable. See Fig. 2.6. We say that $\boldsymbol{\phi}(t)$ is *asymptotically stable* if $\lim_{t \to +\infty} |\boldsymbol{\xi}(t) - \boldsymbol{\phi}(t)| = 0$. This is the expected behavior near a stable equilibrium point when some form of dissipation is present.

Let us look at two examples.

Example 1. $\dot{x} = x$, $\dot{y} = -y$, where $\boldsymbol{\xi} = \begin{pmatrix} x \\ y \end{pmatrix}$. Clearly $\boldsymbol{\xi} = \begin{pmatrix} 0 \\ 0 \end{pmatrix}$ is an equilibrium point. Is it Liapunov stable? In this case we can solve for the motion easily to find that $x = x_0 e^t$ and $y = y_0 e^{-t}$. Then we have $|\boldsymbol{\xi}(t)|^2 = x_0^2 e^{2t} + y_0^2 e^{-2t}$ and because of the exponential increase with t for the first term, no matter how small we make $|\boldsymbol{\xi}_0|^2 = x_0^2 + y_0^2$ there will always come a time at which $|\boldsymbol{\xi}(t)|$ will exceed any arbitrarily chosen ε. Hence the equilibrium point $\boldsymbol{\xi} = \begin{pmatrix} 0 \\ 0 \end{pmatrix}$ is Liapunov unstable.

Example 2. Consider Example 2 of Section 2.4: $\dot{x} = y$, $\dot{y} = -\sin x$ (simple pendulum). We first write the solutions for the phase curves in the form $y^2/2 = E + (\cos x - 1)$. Clearly $x = 0$, $y = 0$ is an equilibrium point for this system and the constants in the solution for the phase curves are chosen so that $E = 0$ labels the equilibrium point. The equilibrium point is Liapunov stable since given $\varepsilon > 0$ we can always choose the energy small enough that

$$|\boldsymbol{\xi}(t)|^2 = 2E + 2(\cos x - 1) + x^2 < 2E + x_{\max}^2 < \varepsilon \qquad \text{for all } t > 0.$$

$x_{\max}$ is given by $\cos x_{\max} = 1 - E$, where for $E \ll 1$ we have $x_{\max} \simeq (2E)^{1/2}$.

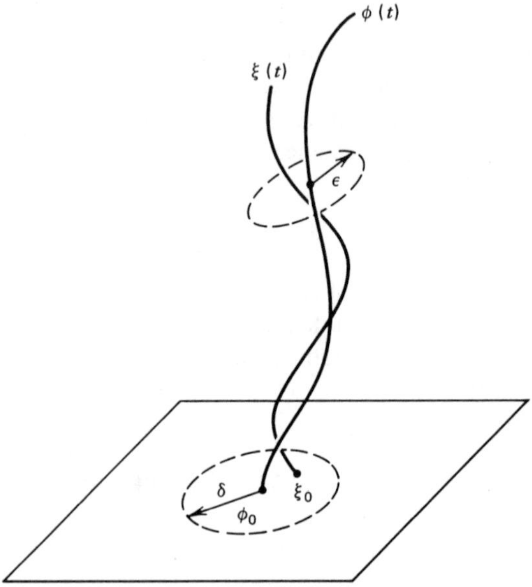

FIGURE 2.6. A trajectory $\boldsymbol{\phi}(t)$ that is Liapunov stable since solutions $\boldsymbol{\xi}(t)$ which start near $\boldsymbol{\phi}(0)$ always remain nearby.

So just choose $E < \varepsilon/4$ to be certain that the equilibrium point satisfies the conditions for Liapunov stability. It is part of Problem 2.8 to show that the other phase points are not Liapunov stable.

2.6. PHASE FLOW

The phase point representing the state of a mechanical system will evolve along a phase curve in phase space as the motion proceeds. Choose an initial point in phase space; then the motion of the system will "push" this point to a neighboring one and thereby generate a sequence of connected points, which represents the "flow" of the system from one phase point to the next along a phase curve of the system. This notion of phase flow can be very helpful in visualizing many dynamical results having geometrical interpretation in phase space. Consequently, it is worth some effort to make the definition of phase flow precise. For this purpose a number of definitions are useful, which quantify differentiability of mappings.

Either configuration space or phase space may be given the structure of a manifold, but for the time being we avoid being mathematically precise in the definition of a manifold. We satisfy ourselves with the following intuitive notions. We think of the manifold of interest as a set of connected points. In the neighborhood surrounding each point we can erect an Euclidean coordinate system. When coordinate systems overlap for nearby points, they do so in a smooth way. Examples of manifolds are: the linear space $\mathbb{R}^n$ or any open subset U of $\mathbb{R}^n$, the set of points on a spherical shell called S^2, the circle S^1, the torus $T^2 = (S^1 \times S^1)$.* Let U be an open domain of the manifold of interest, which may be configuration space, phase space, or one of the other useful manifolds we shall encounter in the future.

A *differentiable function* $f: U \to \mathbb{R}$ (read "f maps points in U into points in $\mathbb{R}$") defined on a domain U with coordinates $x^1, \ldots, x^n$ is a function $f(x^1, \ldots, x^n)$ continuously differentiable r times where $1 \leqslant r < \infty$. By a *differentiable mapping* $f: U \to V$ of a domain U with coordinates $x^1, \ldots, x^n$ into a domain V with coordinates $y^1, \ldots, y^m$ we mean a mapping given by differentiable functions $y^i = f^i(x^1, \ldots, x^n)$. This means that if $y^i: V \to \mathbb{R}$ are the coordinates in V, then $(y^i \circ f): U \to \mathbb{R}$ are differentiable functions in U $(i = 1, \ldots, m)$.† By a *diffeomorphism* we mean a one-to-one mapping $f: U \to V$ such that both f and the inverse map $f^{-1}: V \to U$ are differentiable mappings.

Now we consider our fundamental differential system

$$\dot{\boldsymbol{\xi}} = \mathbf{v}(\boldsymbol{\xi}) \tag{2.52}$$

to have solutions that can be extended to arbitrary values of t. We look at the solutions to system (2.52) whose initial conditions at time $t = 0$ are represented by the point $\boldsymbol{\xi}_0$ in phase space. The solution at any subsequent time depends on $\boldsymbol{\xi}_0$ and we denote it by

$$\boldsymbol{\xi}(t) = g^t \boldsymbol{\xi}_0. \tag{2.53}$$

In this way we have defined a mapping of the phase space M into itself:

$$g^t: M \to M. \tag{2.54}$$

The mapping g^t is a diffeomorphism and the set of diffeomorphisms $\{g^t: t \in \mathbb{R}\}$ form a group. $g^t \circ g^s = g^{t+s}$, the mapping g^0 is the identity, and g^{-t} is the inverse of g^t. Also the mapping formed by $g: \mathbb{R} \times M \to M$ where $g(t, \boldsymbol{\xi}) = g^t \boldsymbol{\xi}$ is differentiable. We sum up all these properties by saying that

*The set of points $A \times B$ (cartesian product of two sets) denotes the set of pairs (a, b) where $a \in A$ and $b \in B$.

†The notation $(g \circ f)(x)$ means $g(f(x))$.

the transformations g^t form a *one-parameter group of diffeomorphisms* of phase space. This group is what we call the *phase flow*.

An interesting example is furnished again by the familiar one-dimensional harmonic oscillator, $\ddot{x} + x = 0$. In this case the phase flow is just the group of rotations of the phase plane through the angle t about the origin. The mapping g^t is a rotation about the origin. See Fig. 2.3, where the interior circles correspond to phase curves for the harmonic oscillator.

As a further example let us consider a system such that $U = -x^4$. Then we have that $\dot{y} = 4x^3$, $y = \dot{x}$, $E = y^2/2 - x^4$. Some phase curves of the flow of this system are sketched in Fig. 2.7.

In the case of more than one dimension for configuration space, the constant energy surfaces are of higher dimension than the simple curves in a plane. Equations (2.52) and (2.53) with their associated discussions are still applicable but a visualization of the phase curves and the constant energy surfaces may be difficult. As in the case for one dimension the phase flow takes place on the energy surfaces (as long as energy is indeed a constant of the motion).

We may consider the simple example of the two-dimensional harmonic oscillator:

$$\dot{x}_1 = y_1, \quad \dot{y}_1 = -x_1, \quad \dot{x}_2 = y_2, \quad \dot{y}_2 = -x_2. \tag{2.55}$$

In the form of (2.52), the vector $\boldsymbol{\xi}$ has components given by $\boldsymbol{\xi} = (y_1, y_2, x_1, x_2)$ and $\dot{\boldsymbol{\xi}} = (-x_1, -x_2, y_1, y_2)$. The conservation of energy theorem gives $E = (y_1^2 + y_2^2)/2 + (x_1^2 + x_2^2)/2$ and the energy surfaces are spheres in 4-space. The projection of the phase curves from the energy surfaces into configuration space (x_1, x_2) gives the orbits, which would again be circles about the origin. Note, however, that for general systems,

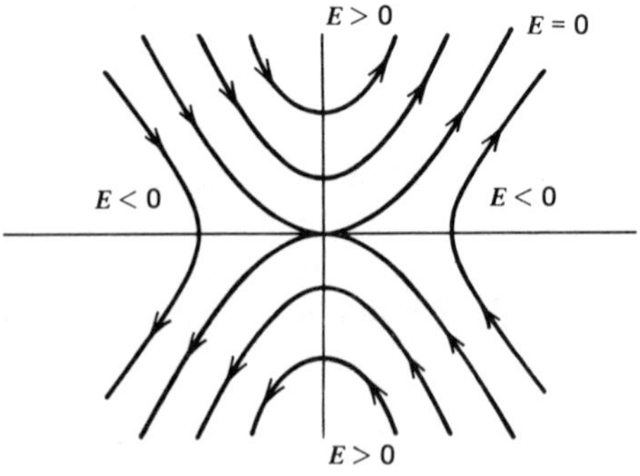

FIGURE 2.7. The phase flow for $U = -x^4$.

such projections of phase curves on the energy surface into configuration space, may lead to apparent crossings. It is only in phase space that we know such phase curves do not cross.

2.7. OPEN SYSTEMS

Many times it is necessary to consider mechanical systems that are not closed. These systems may not only have external forces on them, but there may also be a change in the number of particles or mass in the system. For such situations we must generalize (2.23). We emphasize that if one does indeed keep track of each and every particle that participates in a dynamical problem, then no error is made. However, sometimes it is not practical to keep track of all particles for all time, or even desirable in those cases where the system of interest may be very much smaller.

We think of a fixed volume V and let the subscript α label the particles that are in V. The sum involved in (2.23) is over *all* particles involved in the interaction, whether in V or not. We choose to split the sums of (2.23) into two parts: the sum over particles in V with label α, and a sum over particles participating in the interaction but not in V. The second part we simply denote as $\mathbf{Q}$ and identify it as the momentum flux out of the volume V. The number of particles in V may of course change with time. With this splitting (2.23) becomes

$$\frac{d\mathbf{P}}{dt} = \mathbf{F}_{\text{ext}} - \mathbf{Q}, \tag{2.56}$$

where $\mathbf{P} = \sum_\alpha m_\alpha \mathbf{v}_\alpha$ is the total momentum of the particles in V, $\mathbf{F}_{\text{ext}} = \sum_\alpha \mathbf{F}_\alpha^{\text{ext}}$ is the total external force on the particles in V, and $\mathbf{Q}$ denotes the momentum flux out of the volume V, that is, the net rate at which positive momentum flows out of V. The key to correct application of (2.56) lies in a careful and judicious choice of the volume V. We illustrate by considering some traditional examples that are not difficult but that indicate the range of possibilities.

Example 1. We first consider a chain that is coiled up near a hole in a table and falls through. The problem is to describe the motion of the lower end of the chain. The volume we focus on is indicated in Fig. 2.8 by the dashed line. Each link, as it is added to the chain, enters the system with zero velocity and thus there is no momentum flux into this volume. The total force on the system is just $\rho y g$ where ρ is the constant mass per unit length

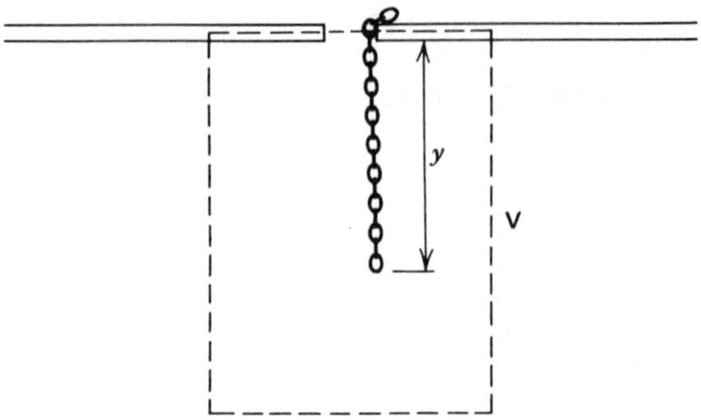

FIGURE 2.8. A chain falling through a hole in a table.

of the chain. Thus, for this example (2.56) just becomes

$$\frac{d}{dt}(\rho y\dot{y}) = \rho g y. \tag{2.57}$$

If we let $z = y\dot{y}$, then this equation may be written in the form

$$z\frac{dz}{dy} = gy^2, \tag{2.58}$$

which may be easily integrated to give $\dot{y} = (2gy/3)^{1/2}$ and the energy as $E = -(1/6)\rho g y^2$. The table top is the reference level for the potential energy.

Example 2. We consider as our second example a rocket projected vertically upward in a uniform gravitational field. See Fig. 2.9. We let j denote the constant rate of mass ejection out through the nozzle of the rocket motor. Let u denote the velocity with which these exhaust gases leave relative to the nozzle. Then we have that $\mathbf{P} = m\dot{z}\hat{z}$, $\mathbf{F} = -mg\hat{z}$, $\mathbf{Q} = -j(u - \dot{z})\hat{z}$. Substitution into (2.56) and an integration yields the result

$$\dot{z} = -u\ln\left(\frac{m}{m_0}\right) - gt, \tag{2.59}$$

where $j = -(dm/dt)$ and m_0 denotes the initial mass. Taking values of $j = m_0/100$ sec and $u = 10{,}000$ ft/sec we find that to reach escape velocity of 7 mi/sec, the ratio of initial to final mass of the rocket is $\simeq 60$.

In this chapter we have reviewed the basic equations of Newtonian mechanics, both for single and many particle systems. We considered in

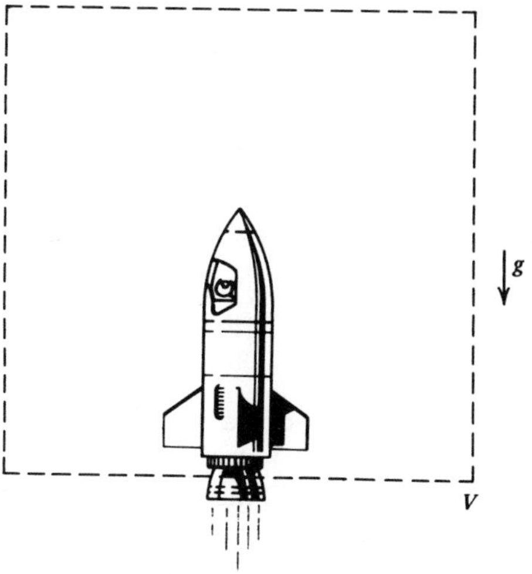

FIGURE 2.9. Schematic of rocket and the associated fixed volume.

detail holonomic constraints and ways in which the solutions to constrained problems may be calculated. The concept of phase flow was discussed and we'll see that this forms one of the primary tools for obtaining a qualitative understanding of the dynamics of mechanical systems. Liapunov stability is one example where the phase trajectories are of primary interest. The chapter is concluded by consideration of open systems which can exchange mass and momentum with their surroundings.

PROBLEMS

2.1. Derive Eqs. (2.28).

2.2. Show that if there are no external forces, and the forces of interaction lie along the line joining the particles and depends only on the separation between them, then a potential such as that postulated for (2.30) exists.

2.3. Show that the motion for the vector $\mathbf{r} = \mathbf{r}_2 - \mathbf{r}_1$ for the interaction of two bodies where $U(\mathbf{r}_1, \mathbf{r}_2) = U(|\mathbf{r}|)$, is the same as a mass point with mass $m_1 m_2/(m_1 + m_2)$.

2.4. If the interaction forces between particles of an N body system lie along the line joining their centers, then show that such forces exert no net torque on the system.

2.5. A small mass m hangs on a string of length L, vertically and at rest in a uniform gravitational field. The upper end of the string is suddenly subjected to a constant acceleration **a** making a small angle θ with the vertical. Describe the subsequent motion of the mass.

2.6. Find the motion of the one-dimensional system with potential depicted in Fig. 2.4 in the neighborhood of the points x_0, x_1, x_2.

2.7. Show that the tangents of the phase curve with $E = U(x_2)$ near the point x_2 are given by $y = (|d^2U/dx^2|_{x_2})^{1/2}(x - x_2)$. $U(x)$ is the potential of Fig. 2.4.

2.8. Consider the phase curves of a system with one degree of freedom that is periodic. Let $A(E)$ denote the area that the phase curve corresponding to energy E encloses. Show that the period of the motion around this curve is $T = (dA/dE)$. Use this result to show that periodic motion on a closed phase curve for a simple pendulum is not Liapunov stable.

2.9. Consider the system defined by the equation $\ddot{x} = x$. Sketch the image of a circle $x^2 + (y - 1)^2 < 1/4$ under the action of the phase flow and justify your result (Arnold, 1978).

2.10. Consider a cable of length L and mass per unit length λ on a horizontal surface being pulled along a horizontal direction with a constant force K. Find the tension in the cable at a distance x from the point of application of the force K. Now assume that, although the surface is frictionless, the cable accumulates dust uniformly along its length at the constant rate J. Find the tension as before. As a rather more interesting case, consider the rate of mass accumulation per unit length to be given by $j = 2J_0(L - x)/L^2$. Find the tension at the point x as a function of time when the cable starts at rest.

2.11. A force law for particle motion in one dimension is $F = -ax + b/x^3$, where a and b are constants.

(a) Find a potential and sketch it.

(b) Find the positions for equilibrium and check their stability. Are they Liapunov stable?

(c) Find the period for small oscillations and for a given energy find the turning points of the motion.

(d) Using the energy E and the turning points, show that the period for arbitrary periodic motion is the same as that found in part (c).

(e) Sketch the phase curves.

2.12. A chain of length L and mass M is stretched vertically its full length above a table with the lower end just touching the top of the table. The chain is then dropped. Find the force exerted by the chain on the table as a function of time. What is the maximum value of this force? What energy does the chain impart to the table? Verify that this is consistent with the force found.

2.13. Suppose that you are to design a roller coaster that is to start from rest and move under the influence of gravity along a frictionless track described by the equation $y = f(x)$, where $f(0) = 0$. Find a necessary condition on $f(x)$ and its derivatives so that the roller coaster never leaves the track.

2.14. Consider an open-topped freight car of mass m initially coasting on smooth, level rails with a speed v_0. It is raining, with the rain falling vertically, and the rain accumulates in the initially empty freight car.

(a) Find the speed v of the car after it has collected a mass m' of rainwater.

(b) Now assume that the car has a small vertical drainpipe in the floor that allows the water to leak out as fast as it comes in. Find the speed v of the car after it has collected a mass m' of rainwater.

CHAPTER THREE

LAGRANGIAN DYNAMICS

Mechanics in becoming geometry remains nontheless mechanics. The partition between mechanics and geometry has broken down and the nature of each of them has diffused through the whole....

EDDINGTON

One of the most convenient ways of formulating the laws of motion is by means of a variational principle. A principle is rather different from a force law. To contrast them we recall, on one hand, Newton's second law: $\mathbf{F} = m\mathbf{a}$ for a particle of mass m, where $\mathbf{F}$ and $\mathbf{a}$ are the vector force function and the vector acceleration, respectively. This force law is a vector equation relating the three components of the acceleration to the three force components. With appropriate initial conditions, these three equations determine completely the subsequent motion of the particle.

On the other hand, the Principle of Least Action requires a particle to move so that a certain quantity called the "action" is an extremum. This "action" is determined by a single scalar function called the Lagrangian, which is discussed fully in this chapter. It is indeed remarkable that a single scalar function, plus a correct application of the principle also leads to a solution for the problem of motion. In this chapter we formulate the Principle of Least Action and explore its implications.

It should be pointed out that the set of situations where Newton's laws are applicable does not entirely overlap with those cases where the Principle of Least Action can be appropriately applied. However, there is a great deal of overlap, and in these situations the methods mutually clarify one another.

3.1. FUNCTIONAL DERIVATIVES

In order to deal appropriately with the Principle of Least Action we must define the derivative of a function whose domain may be other functions. Think of $\phi: E \to F$ as a mapping from a Banach space E to a Banach space F. A *Banach* space is a complete, normed vector space, and most of the familiar and useful vector spaces of physics are Banach spaces, for example, $\mathbb{R}^3$, Hilbert spaces of quantum mechanics, and so on. (For a more complete discussion of Banach spaces in the context of mathematical physics see, for example, Choquet-Bruhat and DeWitt, 1982.) We give some examples of such Banach spaces in mappings. (1) $E = \mathbb{R}$ (the real numbers) and $F = \mathbb{R}$. This is the situation in elementary calculus. (2) $E = \mathbb{R}^3$, and $F = \mathbb{R}^3$. We would call ϕ a vector function of position. (3) $E =$ {the space of infinitely differentiable functions $g: \mathbb{R} \to \mathbb{R}$}, and $F = \mathbb{R}$. Such a mapping $\phi: E \to F$ is called a *functional*, denoting a function of a function.

As an explicit example of a functional we consider the length of a curve $y(t)$ between the points t_0 and t_1. The infinitesimal distance ds is given by $ds = (dt^2 + dy^2)^{1/2} = dt(1 + (dy/dt)^2)^{1/2} = dt(1 + (\dot{y})^2)^{1/2}$. Then the length of the curve $y(t)$ is given by $S = \int_{t_0}^{t_1} dt(1 + (\dot{y})^2)^{1/2}$. The mapping $S[y(t)]$ is clearly a map on the space of such curves into $\mathbb{R}$, that is, S is a functional.

We now give a definition of the differential of ϕ that works for all such examples above and is consistent with the definition of a differentiable function given in Section 2.6. Let E and F be two Banach spaces and let U be an open set of E. We denote the norm on the Banach space E as $\| \ \|_E$ and similarly the norm on F as $\| \ \|_F$. Let $h \in E$ and x_0 and $(x_0 + h) \in U$. Let ϕ: $U \to F$ be a mapping where $\phi(x_0)$ and $\phi(x_0 + h) \in F$. The mapping ϕ: $U \to F$ is said to be *differentiable at* x_0 iff* there exists a continuous, linear mapping $D\phi_{x_0}$ of E into F such that $\phi(x_0 + h) - \phi(x_0) = D\phi_{x_0}(h) + R(h)$ where $R(h) \in F$ satisfies

$$\lim_{\|h\|_E \to 0} \frac{\|R(h)\|_F}{\|h\|_E} = 0$$

$D\phi_{x_0}$ is called the *differential of* ϕ *at the point* x_0 and if ϕ is differentiable at

*"iff" = "if and only if".

each point of U we simply write $D\phi$ and call it the *differential of* ϕ. We note without proof that if the differential exists it is unique. In the case that the points of E are functions, that is, ϕ is a functional, $D\phi$ is often referred to as the functional derivative or first variation and is frequently denoted by $\delta\phi$. In the physics literature this latter notation is customary for functional derivatives and we adopt it.

To become familiar with $D\phi$ let us look at some examples:

Example 1. Let $E = \mathbb{R}$, $F = \mathbb{R}$, and $\phi(x) = x^2$.

$$x^2\colon \mathbb{R} \to \mathbb{R}, (x_0 + h)^2 = x_0^2 + 2x_0 h + h^2.$$

$$(x_0 + h)^2 - x_0^2 = 2x_0 h + h^2, \; Dx^2_{x_0} = 2x_0, \; R(h) = h^2.$$

Example 2. For this example $E = \mathbb{R}^3$, $F = \mathbb{R}$. $\phi\colon \mathbb{R}^3 \to \mathbb{R}$; ϕ is a scalar function of position on Euclidean 3-space and the vector $\mathbf{x} \in \mathbb{R}^3$ maps to $\phi(\mathbf{x}) \in \mathbb{R}$.

$$\phi(\mathbf{x}_0 + \mathbf{h}) = \phi(\mathbf{x}_0) + \left.\frac{\partial \phi}{\partial x^i}\right|_{\mathbf{x}_0} h^i + 0(h^2),$$

$$D\phi_{\mathbf{x}_0}(\mathbf{h}) = \left.\frac{\partial \phi}{\partial x^i}\right|_{\mathbf{x}_0} h^i \quad \text{and so } D\phi = \nabla\phi.$$

This last example clearly illustrates that, in general, the differential is not just a number. It is indeed a linear mapping (operator) from E into F. In Example (2) the differential of ϕ is $\nabla\phi$ and $\nabla\phi$ is a linear operator of $\mathbb{R}^3$ into $\mathbb{R}$, since $[\nabla\phi(\mathbf{x}_0)\cdot\mathbf{h}] \in \mathbb{R}$ for all $\mathbf{h} \in \mathbb{R}^3$. Note that in our usage here $\nabla\phi$ is the operator under discussion, not ∇.

If we were to consider a function from $\mathbb{R}^3$ to $\mathbb{R}^3$ then $D\phi_{\mathbf{x}_0}$ would be a matrix because the linear mappings from $\mathbb{R}^3$ into $\mathbb{R}^3$ consist of 3×3 matrices.

Example 3. As a final example consider the following: let $L(x, y, z)$ be a real-valued, differentiable function of three variables. Let a one-dimensional motion be given by the curve $x(t)$ with the associated velocity $\dot{x}(t)$. Then $\Phi(x) = \int_{t_0}^{t_1} L(\dot{x}(t), x(t), t)\,dt$ is a functional on such motions and we compute the differential of Φ at $x(t)$. The function $h(t)$ is another motion.

$$\Phi(x + h) - \Phi(x) = \int_{t_0}^{t_1} [L(\dot{x} + \dot{h}, x + h, t) - L(\dot{x}, x, t)]\,dt$$

$$= \int_{t_0}^{t_1} \left\{ \frac{\partial L}{\partial \dot{x}} \dot{h} + \frac{\partial L}{\partial x} h \right\} dt + 0(h^2).$$

Then it follows that

$$\delta\Phi_x(h) = \int_{t_0}^{t_1} \left\{ \frac{\partial L}{\partial \dot{x}} \dot{h} + \frac{\partial L}{\partial x} h \right\} dt$$

$$= \int_{t_0}^{t_1} \left\{ -\frac{d}{dt}\left(\frac{\partial L}{\partial \dot{x}}\right) + \frac{\partial L}{\partial x} \right\} h \, dt + \left(h \frac{\partial L}{\partial \dot{x}} \right)\Bigg|_{t_0}^{t_1},$$

where the last equation is obtained through integration by parts.

The crosshatched area between the curves in Fig. 3.1 represents the numerical difference between $\Phi(x + h)$ and $\Phi(x)$, which $\delta\Phi_x(h)$ approximates to first order in h. Naturally, the smaller h is, then the more accurately $\delta\Phi_x(h)$ approximates the actual difference.

To be even more explicit in this example, consider again the functional giving the length of a curve. For Φ to represent the length of a curve $y(t)$ between the points t_0 and t_1, $L = (1 + \dot{y}^2)^{1/2}$. This specific example for a functional also invites one to consider what curve would have the minimum length between the fixed points $y(t_0) = y_0$ and $y(t_1) = y_1$.

Here again the notions of elementary calculus can be extended. The point x_0 in the Banach space E is a *stationary point* of the function ϕ if $D\phi_{x_0}(h) = 0$ for all h. In the language of functionals, we would say that the curve $x(t)$ is a stationary point of the functional Φ, if $\delta\Phi_x(h) = 0$ for all curves h.

These ideas generalize in a natural way for higher dimensions. A curve $\mathbf{x}(t) \in \mathbb{R}^n$ is an extremal (stationary point) of the functional Φ if $\delta\Phi_{\mathbf{x}}(\mathbf{h}) = 0$ for all curves $\mathbf{h}(t)$. In applying this to mechanics problems L is called the *Lagrange function* or the *Lagrangian*. The Lagrangian $L(\dot{\mathbf{x}}, \mathbf{x}, t)$ denotes a

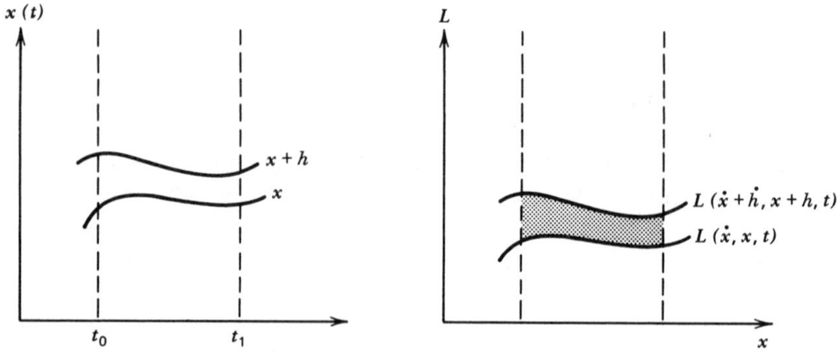

FIGURE 3.1. Relationship between variation of a function and variation of a functional.

mapping $L: \mathbb{R}^n \times \mathbb{R}^n \times \mathbb{R} \to \mathbb{R}$. The functional of interest is

$$\Phi(\mathbf{x}) = \int_{t_1}^{t_2} L(\dot{\mathbf{x}}, \mathbf{x}, t)\, dt. \tag{3.1}$$

The differential of Φ or the first variation of Φ is given by

$$\delta\Phi_{\mathbf{x}}(\mathbf{h}) = \int_{t_1}^{t_2} \left(-\frac{d}{dt}\frac{\partial L}{\partial \dot{\mathbf{x}}} + \frac{\partial L}{\partial \mathbf{x}} \right) \cdot \mathbf{h}\, dt + \left(\mathbf{h} \cdot \frac{\partial L}{\partial \dot{\mathbf{x}}} \right) \Bigg|_{t_1}^{t_2}. \tag{3.2}$$

From Eq. (3.2) we obtain the very important result. The functional Φ on the space of curves joining $(\mathbf{x}_1, t_1)$ and $(\mathbf{x}_2, t_2)$ is extremal for the curve $\mathbf{x}(t)$, that is, $\delta\Phi_{\mathbf{x}}(\mathbf{h}) = 0$ for all $\mathbf{h}(t)$ where $\mathbf{h}(t_1) = \mathbf{h}(t_2) = 0$, iff

$$\frac{d}{dt}\left(\frac{\partial L}{\partial \dot{\mathbf{x}}} \right) - \frac{\partial L}{\partial \mathbf{x}} = 0 \tag{3.3}$$

all along the curve $\mathbf{x}(t)$.

Equations (3.3) are called the *Euler–Lagrange* equations and constitute the fundamental differential system for Lagrangian mechanics. Note that there is one equation in (3.3) for each component of $\mathbf{x}$. We emphasize that (3.3) is not an equation to determine L. The mapping L is presumed known! It is the motion $\mathbf{x}(t)$ that is to be determined from (3.3).

We also wish to remind the reader of the connection with mechanics first made in Section 1.2 and Eq. (1.30). We showed the equivalence of Newton's laws with the Euler–Lagrange equations, in those instances where a potential function exists for the force. The following section considers this connection in more detail.

3.2. HAMILTON'S PRINCIPLE OF LEAST ACTION

We introduce Hamilton's Principle of Least Action and demonstrate that this principle leads to the correct equations of motion in those situations where a potential function exists.

We let N denote the number of particles in our mechanical system and let $\mathbf{r}_1, \mathbf{r}_2, \ldots, \mathbf{r}_N$ denote the position vectors of the N particles, scaled by the mass according to $\mathbf{r}_i = (m_i)^{1/2}\mathbf{x}_i$. We define the vector $\mathbf{r} = (\mathbf{r}_1, r_2, \ldots, \mathbf{r}_N)$ in $\mathbb{R}^{3N}$ with a cartesian metric. Similarly, for the scaled force vector: $\mathbf{f} = (\mathbf{f}_1, \mathbf{f}_2, \ldots, \mathbf{f}_N) = -(\nabla_1 U, \nabla_2 U, \ldots, \nabla_N U) \equiv -\nabla U$, where ∇_i denotes the gradient with respect to $\mathbf{r}_i$. In this convenient notation we may write down Newton's law in the form $\ddot{\mathbf{r}} = -\nabla U$. We also can write conveniently the kinetic energy function $T = (1/2)\,\dot{\mathbf{r}} \cdot \dot{\mathbf{r}}$. Then we let the Lagrangian L be $L = T - U$. The Euler–Lagrange equations for this function are

$$\frac{d}{dt}\left(\frac{\partial L}{\partial \dot{\mathbf{r}}}\right) - \frac{\partial L}{\partial \mathbf{r}} = \frac{d}{dt}\left(\frac{\partial T}{\partial \dot{\mathbf{r}}}\right) + \frac{\partial U}{\partial \mathbf{r}}. \tag{3.4}$$

For the given kinetic energy we see that (3.4) are identical to Newton's equations. Therefore we may state *Hamilton's Principle of Least Action* as follows. The motion of a conservative mechanical system (one for which a potential function exists) is so as to extremize the functional

$$S = \int_{t_1}^{t_2} L(\dot{\mathbf{x}}, \mathbf{x}, t)\, dt, \tag{3.5}$$

where $L = T - U$. The functional S is called the *action*.

It is important to recognize that the Lagrangian L is not unique. An arbitrary total derivative of time, $d\psi/dt$, can be added to L in (3.5) and $\delta S_{\mathbf{x}}(\mathbf{h})$ does not change at all. The function L and $L + (d\psi/dt)$ give the same equations of motion.

Conversely, if two Lagrangians L and L' give the same equations of motion, then their difference must be a total time derivative. We give a proof for one dimension. Let $L'(\dot{x}, x, t) = L(\dot{x}, x, t) + \Lambda(\dot{x}, x, t)$, where both L' and L satisfy the Euler–Lagrange Eqs. (3.3). We emphasize that the Euler–Lagrange equations are not a set of differential equations for the function L, but rather (3.3) determine $\mathbf{x}(t)$. L is assumed to be a known function of $\dot{\mathbf{x}}$, $\mathbf{x}$, and t. Since by assumption the motion is determined by (3.3), where for L we may use either L or L', it must be the case that

$$\frac{d}{dt}\left(\frac{\partial \Lambda}{\partial \dot{x}}\right) - \frac{\partial \Lambda}{\partial x} = \ddot{x}\frac{\partial^2 \Lambda}{\partial \dot{x}^2} + \dot{x}\frac{\partial^2 \Lambda}{\partial x\, \partial \dot{x}} + \frac{\partial^2 \Lambda}{\partial t\, \partial \dot{x}} - \frac{\partial \Lambda}{\partial x} \equiv 0. \tag{3.6}$$

This equation must be true for all $\ddot{x}$, since $\ddot{x}$ is determined by the equations of motion, as discussed earlier. Also since L and L' depend only on $(\dot{x}, x, t)$, so must $\Lambda = L - L'$. Hence none of the partial derivatives of Λ in (3.6) can involve $\ddot{x}$. Since $\ddot{x}$ is an arbitrary function insofar as (3.6) is concerned, we must have that the coefficient of $\ddot{x}$ vanishes identically. Thus $(\partial^2\Lambda)/(\partial \dot{x}^2) = 0$, which may be integrated to yield $\Lambda = \dot{x}\alpha(x, t) + \beta(x, t)$. We then substitute this result back into (3.6) and find that $(\partial\alpha/\partial t) - (\partial\beta/\partial x) = 0$, which is satisfied for $\alpha = (\partial\psi/\partial x)$ and $\beta = (\partial\psi/\partial t)$. This gives $\Lambda = (d\psi/dt)$.

We have shown in one dimension that two Lagrangians give the same equations of motion if and only if they differ by a total time derivative. The theorem is also true for motion in arbitrary dimensions and we leave the proof to Problem 3.2.

3.3. LAGRANGE'S EQUATIONS AND GENERALIZED COORDINATES

One of the most useful features of Lagrangians is that often a Lagrangian will involve a coordinate velocity component, but not that coordinate itself. For such a coordinate (3.3) gives $(d(\partial L/\partial \dot{x}^i)/dt) = 0$ and we have immediately that $\partial L/\partial \dot{x}^i =$ constant. This constitutes a partial integration of the system (3.3). A coordinate that is absent from the Lagrangian in this way is called *cyclic*. We will see numerous examples throughout the remainder of this book. It must be noted, however, that this fortunate circumstance of having a cyclic coordinate does not "just happen." This occurs when there is a symmetry in the system, that is, the Lagrangian for the system is invariant under some transformation such as a rotation. To make the symmetry manifest by having a cyclic coordinate, it is sometimes necessary to make a judicious choice for the coordinates. To see how to choose coordinates for Lagrangians, look once again at constraints.

As in Chapter 2, consider constrained motion where we have k holonomic equations of constraint:

$$f_\alpha(\mathbf{r}_1, \mathbf{r}_2, \ldots, \mathbf{r}_N, t) = f_\alpha(\mathbf{r}, t) = 0, \qquad \alpha = 1, \ldots, k. \tag{3.7}$$

By an appropriate choice of generalized coordinates, we can, in many cases, completely eliminate from consideration the forces of constraint implied by (3.7). Insofar as it is possible actually to carry out the elimination of the constraint equations, we have one of the major economies introduced by the Lagrangian formulation of mechanics!

We consider a transformation from $(\mathbf{r}_1, \mathbf{r}_2, \ldots, \mathbf{r}_N)$ to a set of linearly independent functions called *generalized coordinates* $q^i(i = 1, \ldots, 3N)$. The number of degrees of freedom for our constrained system is $n = 3N - k$, and we consider first a transformation for the last k generalized coordinates. In particular we choose the last k of these generalized coordinates to be given as follows:

$$q^{n+\beta} = Q^\beta(f_1(\mathbf{r}, t), f_2(\mathbf{r}, t), \ldots, f_k(\mathbf{r}, t)), \qquad \beta = 1, \ldots, k, \tag{3.8}$$

where the $Q^\beta(y_1, y_2, \ldots, y_k)$ are a set of k functions, linearly independent in the k arguments. Often for the Q^β we just use the constraint functions themselves, but this is a matter of choice and convenience. In any case the

Jacobian of the Q^β with respect to the f_α must be nonzero so that the transformation will be one-to-one and the equations may be inverted to find

$$f_\alpha\left(q^{n+1}, q^{n+2}, \ldots, q^N, t\right) = 0. \tag{3.9}$$

The constraint equations, if not of the form (3.7), can preclude the construction of such an invertable transformation. Usually this is possible, however, and the importance of such a choice lies in the observation that the constraint Eqs. (3.7) give

$$q^{n+\beta} = Q^\beta(0, \ldots, 0), \qquad \beta = 1, \ldots, k. \tag{3.10}$$

In words, the last k coordinates of the set of generalized coordinates are constant, independent of time.

Along with (3.8) we have some set of transformation equations that relate the remaining generalized coordinates q^i and $\mathbf{r} = (\mathbf{r}_1, \mathbf{r}_2, \ldots, \mathbf{r}_N)$. Including all transformation equations at once, we simply write

$$q^i = q^i(\mathbf{r}, t), \qquad i = 1, \ldots, 3N, \tag{3.11}$$

where we remember that for $n < i \leqslant 3N$ the transformation is given in terms of the constraint equations through (3.8). The Jacobian of this transformation in (3.11) must also be nonzero.

To obtain the equations of motion in these generalized coordinates we compute as follows. From (3.11) we find $\dot{q}^j = (\partial q^j/\partial r^k)\dot{r}^k + (\partial q^j/\partial t)$ and thus $(\partial \dot{q}^j/\partial \dot{r}^i) = (\partial q^j/\partial r^i)$, which depends on having holonomic constraints. Using this result, we obtain

$$\frac{\partial L}{\partial \dot{r}^i} = \frac{\partial L}{\partial \dot{q}^j}\frac{\partial \dot{q}^j}{\partial \dot{r}^i} = \frac{\partial L}{\partial \dot{q}_j}\frac{\partial q^j}{\partial r^i}.$$

We find

$$\frac{d}{dt}\left(\frac{\partial q^j}{\partial r^i}\right) = \frac{\partial^2 q^j}{\partial r^k\,\partial r^i}\dot{r}^k + \frac{\partial^2 q^j}{\partial t\,\partial r^i}$$

$$= \frac{\partial}{\partial r^i}\left(\frac{\partial q^j}{\partial r^k}\dot{r}^k + \frac{\partial q^j}{\partial t}\right) = \frac{\partial \dot{q}^j}{\partial r^i}.$$

Using the foregoing results,

$$\frac{d}{dt}\left(\frac{\partial L}{\partial \dot{r}^i}\right) = \frac{d}{dt}\left(\frac{\partial L}{\partial \dot{q}^j}\right)\frac{\partial q^j}{\partial r^i} + \frac{\partial L}{\partial \dot{q}^j}\frac{\partial \dot{q}^j}{\partial r^i}, \tag{3.12}$$

and

$$\frac{\partial L}{\partial r^i} = \frac{\partial L}{\partial \dot{q}^j}\frac{\partial \dot{q}^j}{\partial r^i} + \frac{\partial L}{\partial q^j}\frac{\partial q^j}{\partial r^i}. \tag{3.13}$$

Subtracting (3.13) from (3.12) to obtain the Euler–Lagrange equations, we find

$$\left\{\frac{d}{dt}\left(\frac{\partial L}{\partial \dot{q}^j}\right) - \frac{\partial L}{\partial q^j}\right\}\frac{\partial q^j}{\partial r^i} = 0.$$

Since by assumption the Jacobian does not vanish, the matrix $\partial q^j/\partial r^i$ is invertible and we find that

$$\frac{d}{dt}\left(\frac{\partial L}{\partial \dot{q}^i}\right) - \frac{\partial L}{\partial q^i} = 0. \tag{3.14}$$

For $n < i \leqslant 3N$, L does not depend on the $\dot{q}^i$, since they equal zero; $(\partial L/\partial q^i) = 0$ and L does not depend on the q^i either. *Thus in (3.14) we need only interpret the range of the index i to be over 1,..., n!* The Euler–Lagrange equations have been reduced to a set in terms of generalized coordinates whose number corresponds to the number of degrees of freedom in the system. In practice it is almost always possible to choose generalized coordinates from the outset, and Eqs. (3.8) are almost never written down.

As an example, we consider a mass m on a string of fixed length L in a uniform gravitational field. The upper end of the string is attached to a mass M which can move in the horizontal direction only in the plane of the pendulum. See Fig. 3.2. We write down the Lagrangian for this system. The position of mass M is $(x, 0)$; the position of mass m is $(x + L\sin\theta, -L\cos\theta)$. The velocity vectors for these two masses are found by differentiation to be $(\dot{x}, 0)$ and $(\dot{x} + \dot{\theta}L\cos\theta, L\dot{\theta}\sin\theta)$, respectively. The kinetic energy for the system is given by $T = \frac{1}{2}M\dot{x}^2 + \frac{1}{2}m(\dot{x}^2 + 2\dot{x}\dot{\theta}L\cos\theta + \dot{\theta}^2L^2)$. The potential energy of the system is $U = -mgL\cos\theta$. The Lagrangian is then

$$L = \tfrac{1}{2}(M + m)\dot{x}^2 + \tfrac{1}{2}m(L^2\dot{\theta}^2 + 2\dot{x}\dot{\theta}L\cos\theta) + mgL\cos\theta. \tag{3.15}$$

This Lagrangian contains only the variables that correspond to the unconstrained degrees of freedom. The variable x is cyclic and the resulting constant of the motion is the momentum in the x direction.

Sometimes it is not possible or convenient to eliminate constraints completely by a choice of generalized coordinates, even though they may be

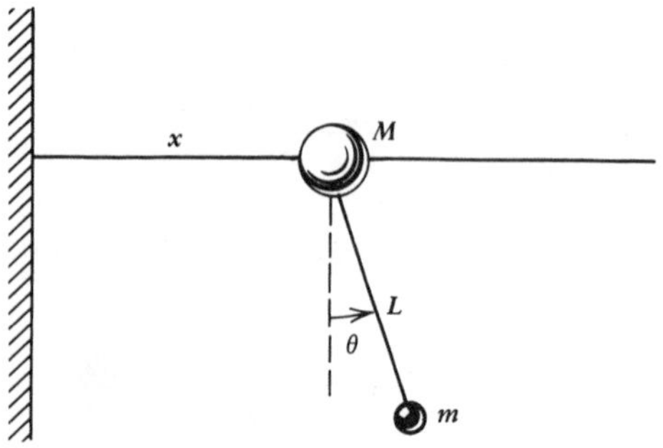

FIGURE 3.2. Planar pendulum suspended from a moving mass.

holonomic. Let a mechanical system have a Lagrangian $L(\dot{\mathbf{q}}, \mathbf{q}, t)$ subject to the constraint equations

$$f_\alpha(\mathbf{q}, t) = 0, \qquad \alpha = 1, \ldots, k. \tag{3.16}$$

We consider the functional derivative of the action and require

$$\delta S_{\mathbf{q}}(\mathbf{h}) = \int_{t_1}^{t_2} \left(-\frac{d}{dt}\left(\frac{\partial L}{\partial \dot{\mathbf{q}}}\right) + \frac{\partial L}{\partial \mathbf{q}} \right) \cdot \mathbf{h}\, dt + \left(\mathbf{h} \cdot \frac{\partial L}{\partial \dot{\mathbf{q}}} \right) \Bigg|_{t_1}^{t_2} = 0, \tag{3.17}$$

but subject to the constraint on $\mathbf{h}(t)$ that

$$\delta f_{\alpha\mathbf{q}}(\mathbf{h}) = \frac{\partial f_\alpha}{\partial \mathbf{q}} \cdot \mathbf{h} = 0. \tag{3.18}$$

These constraints on $\mathbf{h}(t)$, the k Eqs. (3.18), result from demanding that $\mathbf{q}(t) + \mathbf{h}(t)$ satisfy (3.16) as well as $\mathbf{q}(t)$. As before, we demand that $\mathbf{h}(t_1) = \mathbf{h}(t_2) = 0$, but no longer are the nearby motions $\mathbf{q}(t) + \mathbf{h}(t)$ totally arbitrary. The function $\mathbf{h}(t)$ must be such that (3.18) are satisfied. The content of (3.18) is that $\mathbf{h}$ must be orthogonal to each of the k vectors $(\partial f_\alpha / \partial \mathbf{q})$. Thus $\mathbf{h}$ is arbitrary except that it lies in the orthogonal complement to the space generated by the vectors $(\partial f_\alpha / \partial \mathbf{q})$. This is exactly the situation considered in Chapter 2 for smooth constraints, where we used the notation ∇f_α for $(\partial f_\alpha / \partial \mathbf{q})$. Thus to satisfy (3.17) it is sufficient that

$$-\frac{d}{dt}\left(\frac{\partial L}{\partial \dot{\mathbf{q}}}\right) + \frac{\partial L}{\partial \mathbf{q}} = \lambda^\alpha(t) \frac{\partial f_\alpha}{\partial \mathbf{q}}. \quad \text{(Note sum on } \alpha.) \tag{3.19}$$

The $\lambda^\alpha(t)$ are Lagrange multipliers. We recognize from our former discussion of smooth, holonomic constraints in Section 2.1 that Eqs. (3.19) are correct. For the case of nonholonomic constraints additional uncertainties are introduced as to whether the equations obtained by functional variation are correct or not. Without recourse to Newton's equations and forces, one cannot decide this issue, and thus the utility of the Lagrangian formulation with such constraints is reduced. In addition, the inclusion of the constraints into the equations of motion is to some extent arbitrary and is ultimately justified by recourse to an examination of the forces. Consequently, we give no further discussion of constraints here.*

3.4. CENTRAL FORCE FIELDS AND KEPLER'S PROBLEM

A central-force field is one that is invariant under an arbitrary rotation about a central point—the center of force. Consequently, the force may be written in the form $\mathbf{F}(\mathbf{r}) = f(r)\hat{\mathbf{r}}$ and there exists a potential $U(r)$, such that $\mathbf{F}(\mathbf{r}) = -\nabla U(r)$. In such a force field the angular momentum with respect to the center of force is conserved since the torque $\mathbf{T} = \mathbf{r} \times \mathbf{F}(\mathbf{r}) = 0$. Thus $\mathbf{L}$, computed about the center of force, is constant. We define the direction of $\mathbf{L}$ to be the direction of the z-axis in a cylindrical coordinate system and examine the motion of a particle of mass m in such a central field.

Since $\mathbf{L}$ remains constant, the motion will remain in the $z = 0$ plane. The Lagrangian for such a system is

$$L = \tfrac{1}{2}m(\dot{r}^2 + r^2\dot{\theta}^2) - U(r). \tag{3.20}$$

The coordinate θ is a cyclic coordinate and $(\partial L/\partial \theta) = 0$. Thus $(\partial L/\partial \dot{\theta}) = mr^2\dot{\theta}$ is a constant. This constant is the magnitude of the angular momentum $\mathbf{L}$.

Let us define the angular momentum per unit mass as $h \equiv r^2\dot{\theta}$. Then the total energy, which we know also to be constant, can be written in the form

$$E = \tfrac{1}{2}m\dot{r}^2 + \frac{1}{2}\frac{mh^2}{r^2} + U(r). \tag{3.21}$$

This is the same formula for the energy that we would obtain for motion of a particle in one dimension with a potential

$$U_{\text{eff}}(r) = U(r) + \frac{mh^2}{2r^2}. \tag{3.22}$$

*For alternative perspectives on the issue of nonholonomic constraints see Goldstein, 1980; Saletan and Cromer, 1970, 1971.

$U_{\text{eff}}(r)$ is called the *effective potential*. Thus the central force problem can be reduced to quadratures where

$$t = \int \left[\frac{2(E - U_{\text{eff}})}{m} \right]^{-1/2} dr + \text{const.} \tag{3.23}$$

In similar fashion, $(d\theta/dt) = (h/r^2) = (d\theta/dr)(dr/dt) = (d\theta/dr)[2(E - U_{\text{eff}})/m]^{1/2}$ leads to the orbit equation

$$\theta = \int \left[\frac{2(E - U_{\text{eff}})}{m} \right]^{-1/2} \frac{h}{r^2} dr + \text{const.} \tag{3.24}$$

The constants occurring in (3.23) and (3.24) are determined by the initial conditions. Since $(d\theta/dt) = (h/r^2) > 0$, θ is monotonic and continually increasing.

The detailed behavior of the orbits obtained from (3.24) depends on the form of the potential $U(r)$. All orbits, however, must lie in that region of the $r \geqslant 0$ half-line where $U_{\text{eff}}(r) \leqslant E$. At the boundaries of this region $E = U_{\text{eff}}$, in which case (3.21) implies that $\dot{r} = 0$. The values of r for which $U_{\text{eff}}(r) = E$ determine the turning points of the motion. We call these turning points $r_{\min}$ and $r_{\max}$, where $0 \leqslant r_{\min} \leqslant r_{\max} \leqslant \infty$. If $0 \leqslant r_{\min} < r_{\max} < \infty$, then the motion is bounded and takes place in an annular region of the $z = 0$ plane. The angle θ varies monotonically as the coordinate r oscillates between $r_{\min}$ and $r_{\max}$. Only if the increase in the angle θ during the period of oscillation is a rational multiple of π will the orbit in the plane be closed. A point on the orbit where $r = r_{\min}$ is referred to as a *pericenter* and a point where $r = r_{\max}$ is an *apocenter*. Refer to Fig. 3.3. The suffix "center" is replaced by "gee," "helion," and "lune" where the center of force is the earth, sun, or moon, respectively (e.g., apogee). If the orbit is not closed in the annulus then it is dense in the annulus and it can be shown (Arnold, 1968) that the only central potentials for which the orbits are closed are those proportional to r^{-1} or r^2. These central potentials correspond, respectively, to the Newtonian gravitational potential and the spherical harmonic oscillator.

Kepler's problem refers to the analysis of the orbits in the Newtonian potential. In this case $U(r) = -\alpha/r$ with α a positive constant (often written as GMm). Then $U_{\text{eff}}(r) = -\alpha/r + (mh^2/2r^2)$. The behavior of this effective potential as a function of r is sketched in Fig. 3.4. Phase curves for this potential in the $(\dot{r}, r)$ phase plane are sketched in Fig. 3.5.

Equation (3.24) can be integrated directly to give

$$\theta = \cos^{-1}\left\{ \frac{mh/r - \alpha/h}{(2mE + \alpha^2/h^2)^{1/2}} \right\} + \text{const.} \tag{3.25}$$

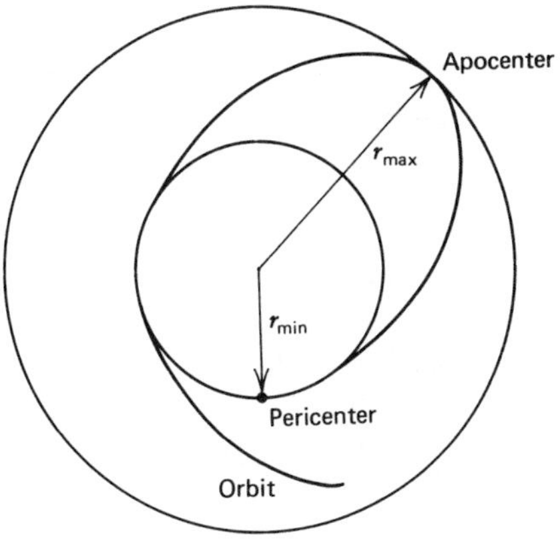

FIGURE 3.3. Section of a general orbit for a central force field.

We take $\theta = 0$ wherever necessary in order that the constant in (3.25) be zero. Then using the definitions

$$e = \left(1 + \frac{2mh^2E}{\alpha^2}\right)^{1/2}, \qquad p = mh^2/\alpha \tag{3.26}$$

we have from (3.25)

$$\frac{p}{r} = 1 + e\cos\theta. \tag{3.27}$$

It is straightforward to check that the choice of constant corresponds to choosing $\theta = 0$ when the particle is at a pericenter. Equation (3.27) is the equation of a conic section with one focus at the origin. The quantity e is called the *eccentricity* and p is called the *semilatus rectum*.

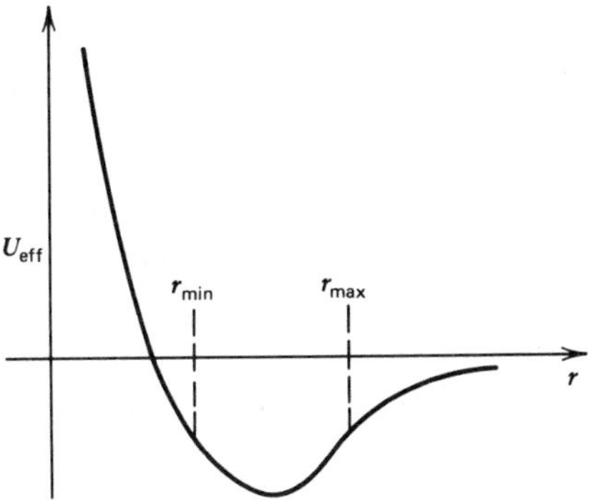

FIGURE 3.4. The effective potential for Kepler's problem.

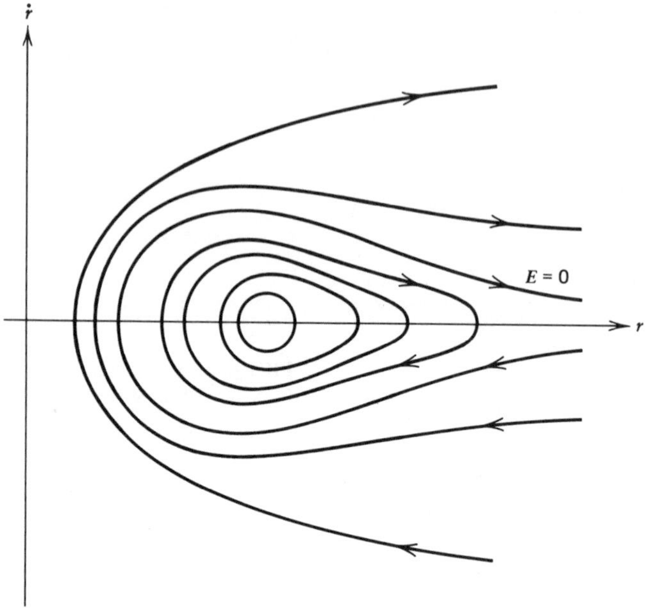

FIGURE 3.5. Phase curves for Kepler's problem in the $(\dot{r}, r)$ subspace.

If $E < 0$, then $e < 1$ and (3.27) is the equation of an ellipse. The usual formulas from analytical geometry give for the semimajor axis and the semiminor axis, respectively:

$$a = p(1 - e^2)^{-1} = \frac{-\alpha}{2E}; \qquad b = p(1 - e^2)^{-1/2} = h\left(\frac{-m}{2E}\right)^{1/2}. \tag{3.28}$$

For values of r_{min} and r_{max} we find

$$r_{\text{min}} = \frac{p}{1+e} = a(1 - e); \qquad r_{\text{max}} = \frac{p}{1-e} = a(1 + e). \tag{3.29}$$

For the geometrical relationships see Fig. 3.6. For a given value of the angular momentum per unit mass h, the minimum value that E may have is that value corresponding to the minimum in the effective potential as sketched in Fig. 3.4. This value $E_{\text{min}} = -\alpha^2/(2mh^2)$ gives an eccentricity equal to zero. This corresponds to a circular orbit with radius mh^2/α. One can show (Problem 3.6) that the sectorial velocity, the rate of change of the

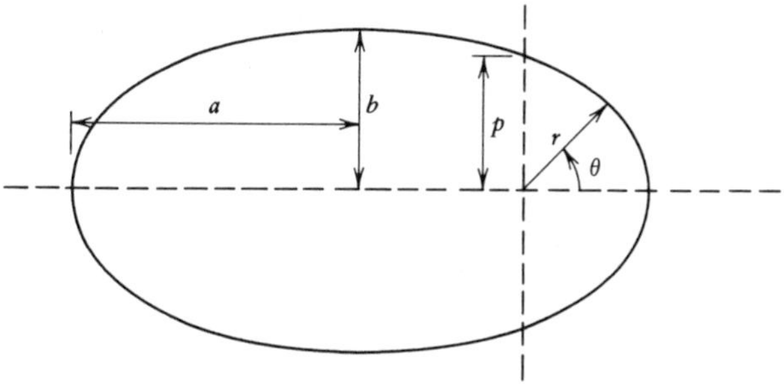

FIGURE 3.6. An elliptic orbit.

orbital area swept out by the radius vector, is $(dA/dt) = h/2$. Thus the area of the ellipse and the period of the motion are related by $A = \pi ab = hT/2$. But using (3.28) we find

$$T = 2\pi a^{3/2}\left(\frac{m}{\alpha}\right)^{1/2} = \pi\alpha\left(\frac{-m}{2E^3}\right)^{1/2}, \tag{3.30}$$

which is Kepler's third law relating the square of the period to the cube of the semimajor axis.

If $E \geqslant 0$, then the motion is infinite rather than being periodic. For $E = 0$, $e = 1$ and the orbit is a parabola with $r_{\min} = p/2$. This case requires that the particle start from infinity at rest. For $E > 0$, $e > 1$ and the orbit is a hyperbola. We find $r_{\min} = p/(1 + e) = a(e - 1)$, where $a = p/(e^2 - 1) = \alpha/2E$. Figure 3.7 shows the relationship among these quantities for a hyperbolic orbit.

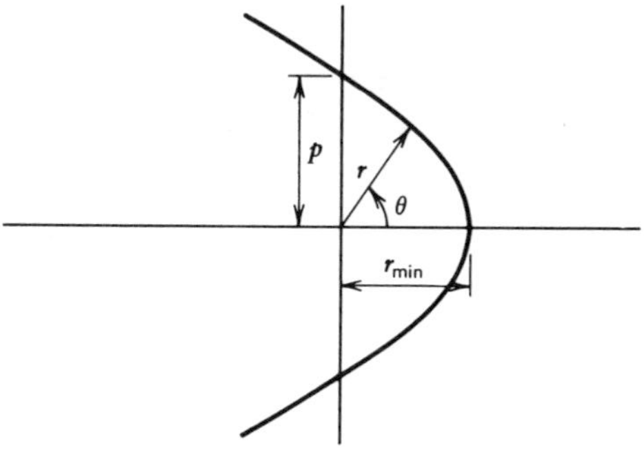

FIGURE 3.7. A hyperbolic orbit.

A convenient parametric representation exists for the time dependence of the orbit. Assume $E < 0$ and thus consider elliptic orbits. Substitution of the effective potential $U_{\text{eff}} = -\alpha/r + (mh^2/2r^2)$ into (3.23) with (3.26) and (3.27) leads to the result

$$t = \left(\frac{ma}{\alpha}\right)^{1/2} \int \left[a^2e^2 - (r-a)^2\right]^{-1/2} r\,dr + \text{const.} \tag{3.31}$$

Making the substitution $(r - a) = -ae\cos\xi$ and integrating gives

$$r = a(1 - e\cos\xi) \quad \text{and} \quad t = \left(\frac{ma^3}{\alpha}\right)^{1/2}(\xi - e\sin\xi). \tag{3.32}$$

The constants have been chosen so that $t = 0$ when $r = r_{\text{min}}$. Similar results are obtained in the case of hyperbolic orbits. (See Problem 3.8).

3.5. LAGRANGIAN FOR PARTICLE MOTION IN AN ELECTROMAGNETIC FIELD

Let us consider the traditional problem of a charged particle moving in a combined electric and magnetic field. The fields are assumed given and are not coupled dynamically to the particle motion. This system has important applications and will enable us to illustrate the construction of a Lagrangian for a mechanical system in which the correct potential energy is not obvious and is furthermore velocity dependent.

In Gaussian units the Lorentz force law for a particle with velocity **v** and charge e is

$$\mathbf{F} = e\left(\mathbf{E} + \frac{1}{c}\mathbf{v} \times \mathbf{B}\right), \tag{3.33}$$

where $\mathbf{E}(\mathbf{x}, t)$ and $\mathbf{B}(\mathbf{x}, t)$ are, respectively, the electric and magnetic fields and c is the speed of light. The usual Maxwell equations imply that these fields may be written in terms of a scalar potential Φ and a vector potential **A**.

$$\mathbf{E} = -\nabla\Phi - \frac{1}{c}\frac{\partial \mathbf{A}}{\partial t}. \tag{3.34}$$

$$\mathbf{B} = \nabla \times \mathbf{A}. \tag{3.35}$$

If we substitute for the fields in terms of these potentials we obtain the following expression for the force law.

$$m\ddot{\mathbf{x}} = \mathbf{F} = -e\nabla\Phi - \frac{e}{c}\frac{\partial \mathbf{A}}{\partial t} + \frac{e}{c}\dot{\mathbf{x}} \times (\nabla \times \mathbf{A}). \tag{3.36}$$

We see in (3.36) a velocity-dependent force and so anticipate a velocity-dependent potential function. It is most convenient to deal with this equation in component language. Thus we rewrite (3.36) in component form.

$$m\ddot{x}_i = -e\Phi_{,i} - \frac{e}{c}A_{i,t} + \frac{e}{c}\varepsilon_{ijk}\dot{x}^j\varepsilon^{kmn}A_{n,m}. \tag{3.37}$$

Probably the easiest way to proceed to a correct Lagrangian is to write down that part of the Lagrangian that is obvious and then add whatever is needed to get the correct equations of motion. Consequently, we let $L = \frac{1}{2}m\dot{x}^2 - e\Phi(\mathbf{x}, t) + f(\dot{\mathbf{x}}, \mathbf{x}, t)$. We substitute this expression into the Euler–Lagrange equations and compare with (3.37). This gives the following equation for f:

$$\ddot{x}^j\frac{\partial^2 f}{\partial \dot{x}^j\,\partial \dot{x}^i} + \dot{x}^j\frac{\partial^2 f}{\partial x^j\,\partial \dot{x}^i} + \frac{\partial^2 f}{\partial t\,\partial \dot{x}^i} - \frac{\partial f}{\partial x^i} = \frac{e}{c}\left(A_{i,t} - \varepsilon_{ijk}\varepsilon^{kmn}\dot{x}^jA_{n,m}\right). \tag{3.38}$$

Since there are no acceleration terms on the right-hand side of (3.38), the coefficient of $\ddot{x}^j$ must equal zero, that is, $(\partial^2 f/\partial \dot{x}^j\,\partial \dot{x}^i) = 0$. This may be integrated to find f in the form $f = \dot{x}^j\alpha_j(\mathbf{x}, t) + \beta(\mathbf{x}, t)$. Substituting back into (3.38) and comparing terms shows that $\beta = 0$ and $\boldsymbol{\alpha}(\mathbf{x}, t) = (e/c)\mathbf{A}(\mathbf{x}, t)$. Thus we have

$$L = \frac{1}{2}m\dot{x}^2 - e\Phi(\mathbf{x}, t) + \frac{e}{c}\dot{\mathbf{x}} \cdot \mathbf{A}(\mathbf{x}, t). \tag{3.39}$$

For $L = T - U$, then $U = e\Phi - (e/c)\mathbf{v} \cdot \mathbf{A}$.

This is an example of a system for which the Lagrangian is constructed in such a way that the Euler–Lagrange equations give the correct equations of motion. The force $\mathbf{F}$ is not obtained from ∇U. Some applications of (3.39) are considered in the problems.

3.6. CONFIGURATION MANIFOLDS

We have seen in Section 3.3 that generalized coordinates q^i play a prominent role in the Lagrangian formulation of mechanics and provide one major simplification insofar as they eliminate consideration of constraints.

In this section we see how they also lead to a useful visualization of the system motion.

As an example we consider once again the system depicted in Fig. 3.2 with Lagrangian (3.15). The generalized coordinates of this system are (x, θ), where $-\infty < x < \infty$ and $0 \leqslant \theta < 2\pi$. We can visualize this set of points (x, θ) as a cylindrical surface in $\mathbb{R}^3$. The set of points on a circle is often denoted as S^1, parameterized by the coordinate θ. Then the points given in terms of the generalized coordinates (x, θ) are the elements of the set $\mathbb{R}^1 \times S^1$.

We refer to the set of points determined by the generalized coordinates of a mechanical system as the *configuration manifold* for that system. The state of the system at any instant is represented by a point on the manifold. The time history, as the system evolves from state to state, is then represented by a connected set of points on the configuration manifold and we refer to this set of points as a *trajectory* or *motion* of the system.

In the preceding example $\mathbb{R}^1 \times S^1$ is the configuration manifold for the simple pendulum with moving point of support as shown in Fig. 3.2. The sequence of points $(x(t), \theta(t))$ that the system evolves through gives a trajectory on the manifold. A sketch of these quantities is given in Fig. 3.8.

We also note from Fig. 3.8 that, in a localized region of the configuration manifold, it is not necessary to use an angle coordinate, if for some reason another choice is preferred. For a local region we might use the (ξ, η) coordinates on the surface of the cylinder, as indicated in Fig. 3.8. It is clear that the point on the opposite side of the cylinder, that is, opposite from $(\xi, \eta) = (0, 0)$, will have an ambiguous η value, since it can be reached by going in both the positive and negative directions. Thus we don't want to try to extend the locally cartesian system (ξ, η) too far. Nevertheless, many ways of putting coordinates on a localized region are possible.

By way of anticipation we might also wonder just how the phase space of Chapter 2 can be extended to the case of generalized coordinates. The generalized velocity components $\dot{q}^i$ play almost a dual role in the Lagrangian formulation. In obtaining the Euler–Lagrange equations the $\dot{q}^i$ are treated as independent variables from the q^i, and so indeed it is the $(\dot{q}^i, q^i)$ that appear as the natural coordinates for Lagrangian systems. In Chapter 6 we consider in detail this "extended phase space" with coordinates $(\dot{q}^i, q^i)$.

Motivated by these portending generalizations and the obvious utility of configuration manifolds for visualizing the motions of systems in Lagrangian mechanics, we consider briefly the definition and description of differentiable manifolds.

Let M be a set of points on which open sets have been defined. For example on the real line $\mathbb{R}$, one usually takes the open intervals $a < x < b$ to be the open sets. Often the open sets in M will be defined using a distance function or metric. These open sets are assumed to be such that the union of

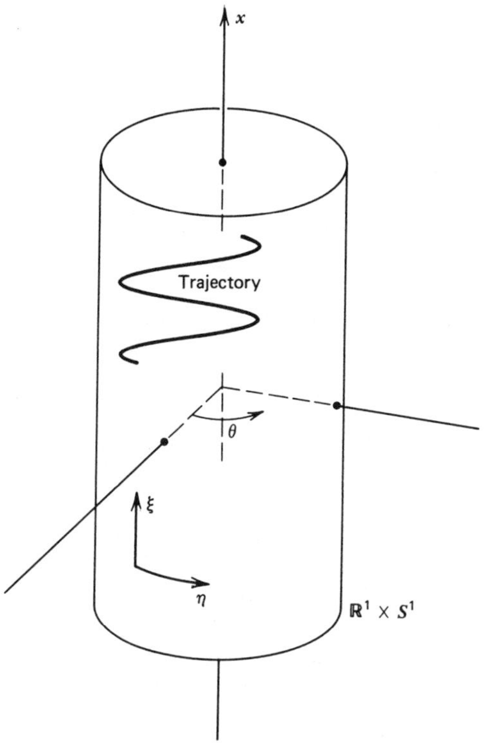

FIGURE 3.8. The configuration manifold for a simple pendulum with moving point of support. A segment of a possible trajectory is also drawn.

a countable collection of them contains M. Furthermore, we assume that any two points in M may be separated by finding disjoint open sets, each of which contains one point but not the other.*

A *chart* on M is a pair (U, ϕ), where U is an open subset of M and ϕ is a bijection (one-to-one, onto mapping) of U onto some open neighborhood V in $\mathbb{R}^n$; ϕ: $U \to V$. The coordinates $(x^1, \ldots, x^n)$ of the image $\phi(p) \in \mathbb{R}^n$ of the point $p \in M$ are called the *coordinates of p* in the chart (U, ϕ). A chart (U, ϕ) is sometimes called a *local coordinate system*. The set M is required to have a finite or countable collection of charts such that every point in M is in at least one chart.

We compose the mappings ϕ_j and ϕ_i^{-1} to form the mapping $\phi_j \circ \phi_i^{-1}$: $\phi_i(U_i \cap U_j) \to \phi_j(U_i \cap U_j)$, which is the mapping of an open set of $\mathbb{R}^n$ into another open set. See Fig. 3.9. An *atlas* on M is a set of charts $\{(U_i, \phi_i)\}$ of M

*The requirements on M in this paragraph are contained in the technical statement that M must be a Hausdorff topological space.

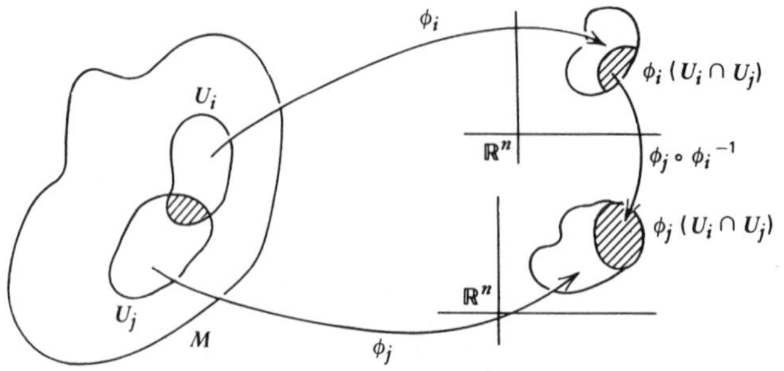

FIGURE 3.9. Diagram showing the relationship between the charts, including the composition mappings, and the intersection of the neighborhoods. The crosshatched regions denote the intersection $U_i \cap U_j$ and its images under ϕ_i and ϕ_j.

such that the open neighborhoods U_i cover M and the mappings $\phi_j \circ \phi_i^{-1}$ are differentiable as defined in Section 2.5. In terms of coordinates these composition mappings are of the form $x^i \to y^j = f^j(x^i)$, where the functions f^j must be differentiable. Unless stated otherwise, we will consider these composition mappings to be differentiable an arbitrary number of times and we refer to such overlap of charts as being *smooth*.

In qualitative terms an atlas on M may be thought of as a patchwork quilt. The "patches" on the "quilt" correspond to the chart neighborhoods that all stand in one-to-one correspondence with $\mathbb{R}^n$. These "patches" are all pieced together so that they overlap in a smooth way. Locally (i.e., within a patch), the manifold just looks like $\mathbb{R}^n$. This was the role of the (ξ, η) coordinate system (patch) on the configuration manifold $\mathbb{R}^1 \times S^1$ that we previously considered and sketched in Fig. 3.8.

Two atlases are equivalent if their union is also an atlas. A *differentiable manifold M* is a collection of equivalent atlases. We consider only connected manifolds (i.e., those that cannot be broken up into two disjoint sets), and thus the number n in $\mathbb{R}^n$ is the same for all charts and is called the *dimension of the manifold*.

The construction of the charts coincides with putting coordinate labels on the manifold points, except that sometimes the labels do not define open sets. Angular coordinates such as $0 \leqslant \theta < 2\pi$ do not range over open intervals; one end is closed and one end is open because 0 and 2π refer to the same point. In such cases we usually must take two open sets to cover the entire range. With this caveat in mind, however, labeling points in M is tantamount to constructing charts. We can talk about the configuration manifold Fig. 3.8 as $\mathbb{R}^1 \times S^1$, but as soon as we label the points with the labels (x, θ) or (ξ, η), we are in effect constructing a chart. The charts allow us in a specific way to talk about the points on a manifold. As an example of using the charts we say that the mapping $f\colon M \to \mathbb{R}$ is differentiable at

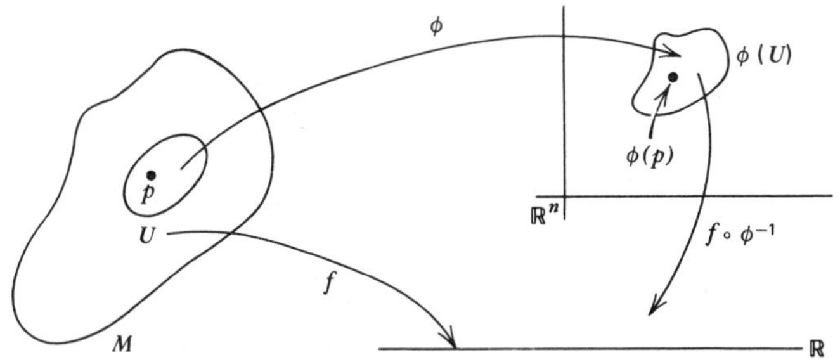

FIGURE 3.10. Schematic of the mapping $f \circ \phi^{-1} : \mathbb{R}^n \to \mathbb{R}$ induced by the real-valued function f on the manifold M and the chart bijection ϕ.

the point $p \in M$ if in a chart (U, ϕ), $f \circ \phi^{-1}$ is differentiable at the point $\phi(p)$. See Fig. 3.10.

Example 1. Consider some examples of manifolds and a choice of atlas. $\mathbb{R}^n$ itself can be viewed as a manifold with a single chart $(\mathbb{R}^n, \mathrm{id})$, where id denotes the identity mapping. The configuration manifold of a single particle in a potential $V(x, y, z)$ is $\mathbb{R}^3$.

Example 2. The circle $S^1 = \{(x, y) \in \mathbb{R}^2 \mid x^2 + y^2 = 1\}$ is a manifold. As open sets for the charts we may take the points $U_1 = \{\theta \mid (\pi/4) < \theta < (7\pi/4)\}$ and $U_2 = \{\theta \mid (-3\pi/4) < \theta < (3\pi/4)\}$. The bijections ϕ_1, ϕ_2 are just defined by $\phi_i(\theta) = \theta$, $i = 1, 2$. These neighborhoods taken together clearly cover all the points of S^1. For this manifold an atlas must consist of at least two charts. Any attempt to cover S^1 with a single open set invariably leaves out at least one point. For example, $0 < \theta < 2\pi$ leaves out the point with label 0 or 2π.

Example 3. The 2-sphere S^2, which is the set of points $S^2 = \{\mathbf{x} \in \mathbb{R}^3 \mid x^2 + y^2 + z^2 = 1\}$, is a manifold. For an atlas with two charts we might take the open sets $U_S = S^2 - \mathrm{SP}$ and $U_N = S^2 - \mathrm{NP}$, which consist of S^2 with the single point at the south pole (SP) removed and the single point at the north pole (NP) removed, respectively. The associated mappings are called stereographic projections and are most easily visualized geometrically. The map ϕ_S: $S^2 \to \mathbb{R}^2$ is obtained by drawing a straight line from the south pole, through the sphere, to the plane $\mathbb{R}^2$. The point of intersection on the sphere is mapped to the point on the plane that the line intersects. See Fig. 3.11. In this case $\phi_S(U_S) = \mathbb{R}^2$. The map ϕ_N is obtained by a similar construction as shown in Fig. 3.11. Other atlases that consist of different

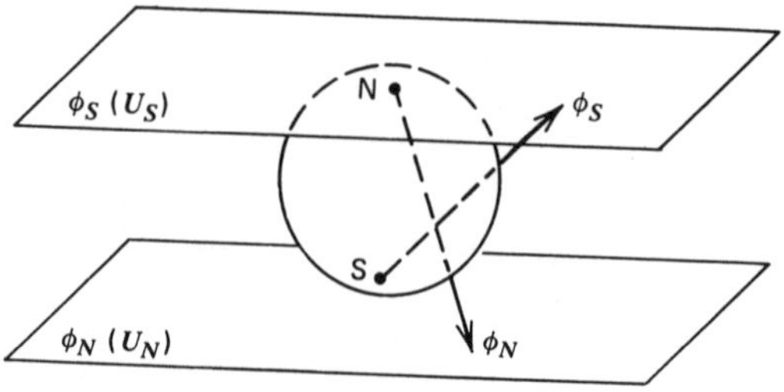

FIGURE 3.11. The bijections called stereographic projections that give an atlas for S^2.

charts may be constructed, but again at least two charts are needed. S^2 is a two-dimensional manifold, since it maps into $\mathbb{R}^2$ with the chart mappings.

Example 4. The next example is the configuration manifold for a double pendulum. The mechanical system is sketched in Fig. 3.12 as well as the configuration manifold that is the torus T^2. The manifold T^2 may be viewed as the cartesian product $S^1 \times S^1 = T^2$; consequently, as open sets for an atlas we may take the sets $U_1 \times U_1$, $U_1 \times U_2$, $U_2 \times U_1$, and $U_2 \times U_2$, where U_1 and U_2 are the open sets on S^1 as given in Example 2. The first element in these cartesian product pairs we denote with the angle θ and the second with the angle ψ, as is done in Fig. 3.12. The bijections $\phi_{11}, \phi_{12}, \phi_{21}, \phi_{22}$ are given by $\phi_{ij}(\theta, \psi) = (\phi_i(\theta), \phi_j(\psi)) \in \mathbb{R}^2$, $i, j = 1, 2$. T^2 is a two-dimensional manifold because the bijections map to $\mathbb{R}^2$.

Example 5. We note again the configuration manifold of the system sketched in Fig. 3.2. The manifold itself is sketched in Fig. 3.8. In this case

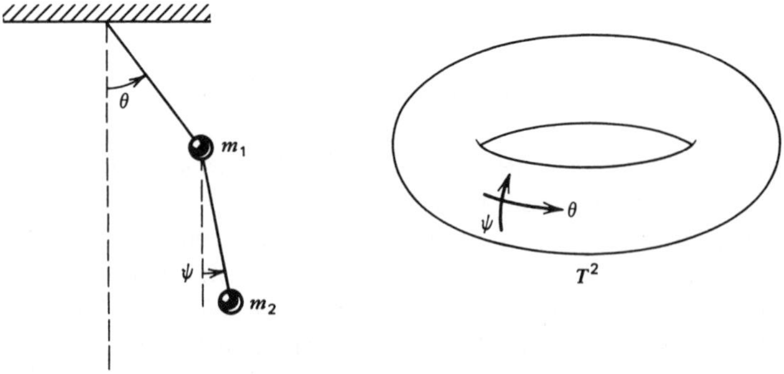

FIGURE 3.12. The mechanical system of a double pendulum and its corresponding configuration manifold T^2. A curve on the surface of the torus represents some motion of the mechanical system.

one might take an atlas consisting of the charts $\mathbb{R} \times U_1$ and $\mathbb{R} \times U_2$, where U_1 and U_2 are again the charts on S^1 in Example 2. The bijections θ_i: $\mathbb{R} \times U_i \to \mathbb{R}^2$ are the mappings defined by $\theta_i(x, \theta) = (\mathrm{id}(x), \phi_i(\theta)) = (x, \theta) \in \mathbb{R}^2$.

In Examples 4 and 5 the charts that were specifically selected consisted of sets that were cartesian products and the chart bijections were also constructed as "product" mappings. These examples are specfic instances of *product manifolds* that have their charts and mappings defined in a natural way. If M and N are two manifolds with atlases $\{(U_i, \phi_i)\}$ and $\{(V_i, \psi_i)\}$, respectively, then an atlas for the product manifold $M \times N$ consists of the charts $\{(W_{ij}, \phi_{ij})\}$, where the W_{ij} are of the form $U_i \times V_j$ and the bijections ϕ_{ij} are of the form (ϕ_i, ψ_j). Specifically, θ_{ij}: $W_{ij} \to \mathbb{R}^m \times \mathbb{R}^n$ is the map given by $\theta_{ij}(u, v) = (\phi_i(u), \psi_j(v))$, where $(u, v) \in U_i \times V_j$.

It is useful to recognize that each of the preceding manifolds can be viewed as a submanifold of $\mathbb{R}^n$ for some n. As an example, it is clear that the manifold of Fig. 3.8 is contained within $\mathbb{R}^3$; the cylindrical shell is just a surface in $\mathbb{R}^3$.

Assume M is a subset of $\mathbb{R}^n$ and let $\mathbf{x}_0$ be an arbitrary point in M. Let U be an arbitrary neighborhood of $\mathbf{x}_0$ in $\mathbb{R}^n$. If there exists $n - k$ functions f_α: $U \to \mathbb{R}$ $(\alpha = 1, \ldots, n - k)$ such that $U \cap M = \{\mathbf{x} \in U \mid f_\alpha(\mathbf{x}) = 0,\ \alpha = 1, \ldots, n - k\}$, and such that the vectors ∇f_α are linearly independent, then M is an *embedded submanifold* of $\mathbb{R}^n$ of dimension k.

Again we return to the configuration manifold of Fig. 3.8. We notice that the constraint equation is just that $r = L =$ constant. We have $f(r, \theta, z) = r - L$ and $\nabla f = \hat{\mathbf{r}}$. All those points in $M = \mathbb{R}^1 \times S^1$ are just those satisfying the constraint equations. The manifold S^1 itself is an embedded submanifold of $\mathbb{R}^2$ where the constraint function is $f(x, y) = x^2 + y^2 - 1$.

For N particles free to move and interact in three dimensions, we have a manifold of $\mathbb{R}^{3N}$. The constraints that are placed on their interactions and motions through (3.7) lead to the configuration manifold M of the system. If all constraints have been eliminated through the use of generalized coordinates, then the trajectories or motions on M are only limited by the initial conditions given by the system. If certain constraints still remain because of convenience or because they could not be eliminated through a choice of generalized coordinates, then motions on M will be restricted and all regions of the manifold will not be accessible to the motions of the system.

As a final example of a configuration manifold we consider the free rotations of a rigid body. A *rigid body* is an object for which the distance between all mass elements remains fixed. In the next chapter we study the dynamics of rigid bodies in detail, but here we examine only the configuration manifold. If we focus on some fiducial orientation to which we refer all subsequent orientations, then the orientation of a rigid body is specified by giving a rotation matrix. The configuration manifold for rigid bodies is a manifold of rotation matrices. This manifold is important not only because of the importance of rotations, which we have encountered since Chapter 1, but also because it serves as a good example of an embedded submanifold. This manifold is important from many viewpoints and warrants the effort to become fully familiar with its properties.

Consider the set of all real 3×3 matrices. These matrices may be thought of as a point in $\mathbb{R}^9$. Now select the subset consisting of orthogonal 3×3 matrices that represent rotations as was discussed in Chapter 1. The orthogonality condition (1.32) places six constraints on the elements of the 3×3 matrices, leaving three degrees of freedom. Among these remaining rotations we find two disjoint sets corresponding to the proper rotations and the improper rotations, that is, those with determinant equal to $+1$ or -1, respectively. Thus taking the orthogonal 3×3 matrices with determinant equal to $+1$, we obtain the set of matrices denoted $SO(3)$. We note that with determinant equal to $+1$ these matrices can be reached in a continuous fashion from the identity. [Aside from being an interesting manifold, $SO(3)$ is a three-parameter Lie group of transformations and is considered further from this perspective in Chapter 10.]

We need charts (local coordinates) and begin by considering the usual parameterization of $SO(3)$. Let $R \in SO(3)$ be a rotation matrix. Since R is orthogonal (unitary) all its eigenvalues are of unit magnitude (see Problem 3.13). Thus if we choose a basis for $\mathbb{R}^3$ in which R is diagonal, we find by taking the determinant of R that $\lambda_1\lambda_2\lambda_3 = +1$. Since we must have that $|\lambda_i| = 1$, $i = 1, 2, 3$, we know that only the following cases are possible. (1) All eigenvalues are real and positive; thus all are equal to $+1$. This is not a very interesting case, since R then is just the identity matrix. (2) All eigenvalues are real with two equal to -1 and one equal $+1$. We will see that (2) is a special case of (3) to follow. (3) In this case the usual characteristic equation that gives the eigenvalues $\det(R - \lambda I) = 0$ has two complex roots and one real root. In this case the eigenvalues may be given as $+1, e^{i\theta}, e^{-i\theta}$. Despite the fact that we have started with R as a rotation on a real vector space, it is not possible to diagonalize R over the reals and we must extend the basis to complex vectors in order to diagonalize R. Nevertheless we can find the components of R in a real basis as follows. We recall (1.32) from Chapter 1:

$$g_{ij} R^i{}_m R^j{}_k = g_{mk}. \tag{3.40}$$

This equation may be interpreted as giving the scalar product between two column vectors labeled with indices m and k. These column vectors are formed from the columns of the rotation matrix R. We use a cartesian coordinate system and thus if $m = k$ the scalar product of the columns equals 1. If $m \neq k$, then the column vectors are orthogonal. Thus the columns of the rotation matrix form a triple of orthonormal column vectors.

To obtain the entries in these columns, we consider once again the eigenvectors of R. In the space of vectors on which R operates, we can, without loss of generality, choose the eigenvector that corresponds to the eigenvalue $+1$ to be $\mathbf{e}_1$, the first basis vector. With this choice the first column of R must have the entries $(1, 0, 0)$. The next two columns must be of unit magnitude, orthogonal to this column, and orthogonal to each other. We choose the columns with entries $(0, \cos\theta, \sin\theta)$, $(0, -\sin\theta, \cos\theta)$. These clearly satisfy all the criteria mentioned and R can then be written in the form

$$R = \begin{bmatrix} 1 & 0 & 0 \\ 0 & \cos\theta & -\sin\theta \\ 0 & \sin\theta & \cos\theta \end{bmatrix} \tag{3.41}$$

With the right-hand-rule convention this rotation matrix corresponds to a positive rotation around the $\mathbf{e}_1$ vector through an angle θ. (R is viewed as an active transformation.) Note that $\theta = \pi$ gives case (2).

Since R was chosen to be an arbitrary rotation matrix, that is an arbitrary element of $SO(3)$, any rotation can be represented in the form of (3.41). Hence three parameters are needed to characterize an element of $SO(3)$ uniquely: two parameters to give the direction of $\mathbf{e}$ on the unit sphere and then one angle ψ to specify the rotation around this axis. This allows us also to form a mental image of the manifold $SO(3)$, which we know to be three-dimensional. We think of two parameters specifying a point on the unit sphere and a vector $\psi\mathbf{e}$ from the origin out to this point. The length ψ of this vector corresponds to the magnitude of the rotation angle around $\mathbf{e}$. The points corresponding to $\psi = 2\pi$ are to be viewed as identical with the point $\psi = 0$. Thus the manifold of $SO(3)$ can be imagined as a sphere that involutes back on itself at a radius of 2π through the origin.

The standard parameterization of this configuation manifold is in terms of the Euler angles (ϕ, θ, ψ), where the appropriate ranges are $0 \leqslant \phi \leqslant 2\pi$, $0 \leqslant \theta \leqslant \pi$, and $0 \leqslant \psi \leqslant 2\pi$. Figure 3.13 shows the relationship between the basis vectors $\{\hat{\mathbf{e}}_i\} = (\hat{\mathbf{x}}, \hat{\mathbf{y}}, \hat{\mathbf{z}})$ and the basis vectors $\{\hat{\mathbf{E}}_i\} = (\hat{\mathbf{X}}, \hat{\mathbf{Y}}, \hat{\mathbf{Z}})$ in the passive view. However, to correspond to a positive, active transformation, the basis $(\hat{\mathbf{X}}, \hat{\mathbf{Y}}, \hat{\mathbf{Z}})$ in Fig. 3.13 is rotated away from $(\hat{\mathbf{x}}, \hat{\mathbf{y}}, \hat{\mathbf{z}})$, in contrast to Fig. 1.5, which has the roles of these bases reversed for a positive, passive transformation. Equation (1.36) tells us how these basis vectors are related and we note that the row index is the summed index on the rotation matrix.

With the foregoing cautionary remarks in mind, we can construct an arbitrary rotation matrix $R \in SO(3)$ in the form of a product

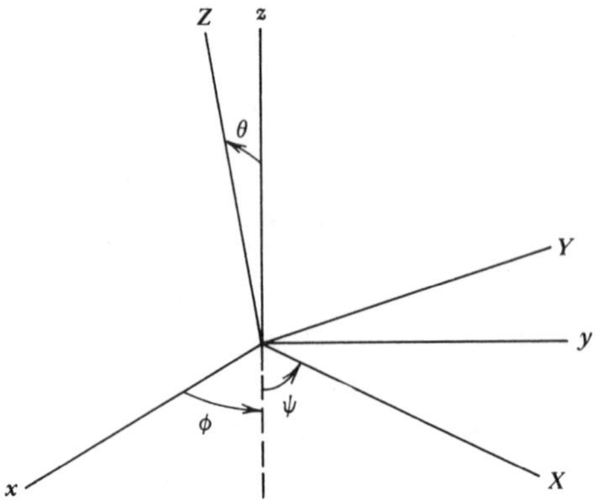

FIGURE 3.13. Euler angle parameterization for a rotation in $SO(3)$.

$R = R_z(\phi)R_x(\theta)R_z(\psi)$, where the subscript on the "R" denotes the axis about which a rotation takes place. The order of these rotations, from left to right, follows from (1.36) of Chapter 1. When applied to the basis $\{\hat{\mathbf{e}}_i\}$ to go to the basis $\{\hat{\mathbf{E}}_i\}$, the rotation applied first occurs on the extreme left of the product. For transformation of a position vector $\mathbf{X}$ into $\mathbf{x}$ the order is reversed.

For brevity we use the notation $\sin\alpha \equiv s_\alpha$ and $\cos\alpha \equiv c_\alpha$; then from (3.41), the rotations about each of the axes are

$$R_z(\alpha) = \begin{bmatrix} c_\alpha & -s_\alpha & 0 \\ s_\alpha & c_\alpha & 0 \\ 0 & 0 & 1 \end{bmatrix},$$

$$R_y(\alpha) = \begin{bmatrix} c_\alpha & 0 & s_\alpha \\ 0 & 1 & 0 \\ -s_\alpha & 0 & c_\alpha \end{bmatrix}, \tag{3.42}$$

$$R_x(\alpha) = \begin{bmatrix} 1 & 0 & 0 \\ 0 & c_\alpha & -s_\alpha \\ 0 & s_\alpha & c_\alpha \end{bmatrix}.$$

Thus we find for the generic point in $SO(3)$

$$R = \begin{bmatrix} c_\psi c_\phi - s_\psi c_\theta s_\phi & -c_\phi s_\psi - s_\phi c_\theta c_\psi & s_\phi s_\theta \\ s_\phi c_\psi + c_\phi c_\theta s_\psi & -s_\psi s_\phi + c_\psi c_\theta c_\phi & -c_\phi s_\theta \\ s_\theta s_\psi & s_\theta c_\psi & c_\theta \end{bmatrix} \tag{3.43}$$

In the context of Chapter 1, where a rotation relates the moving frame M to a stationary frame S, (3.43) gives a parameterization of the most general rotation relating these two reference frames.

In conclusion, we summarize briefly the picture of mechanics that has emerged in this chapter. No longer must one have the vector function of force $\mathbf{F}(\dot{\mathbf{x}}, \mathbf{x}, t)$ and apply Newton's second law. Rather, we must have a single scalar function $L(\dot{\mathbf{x}}, \mathbf{x}, t)$ and apply Hamilton's Principle of Least Action. It is of course an overstatement to claim that one no longer needs to consider forces. There are many systems, particularly those nonconservative systems involving dissipation of some sort, where Lagrangian mechanics is often helpless. Despite the foregoing caveat a large number of interesting mechanical systems can be fully and most easily understood by possessing a single function called the Lagrangian. Lagrangian mechanics, through its use of generalized coordinates, leads directly to constants of the motion when a coordinate is cyclic. These generalized coordinates also lead to a helpful visualization of system motion as it traces out trajectories on configuration manifolds. Usually the generalized coordinates can be chosen to eliminate constraints from consideration.

PROBLEMS

3.1. Verify that the extremals of a functional

$$\Phi(\mathbf{x}) = \int_{t_1}^{t_2} L(\dot{\mathbf{x}}, \mathbf{x}, t)\, dt$$

do not depend on the coordinate system chosen for $\mathbb{R}^n$.

3.2. Show that two Lagrangians $L(\dot{\mathbf{x}}, \mathbf{x}, t)$ and $L'(\dot{\mathbf{x}}, \mathbf{x}, t)$ that give the same equations of motion differ by a total time derivative.

3.3. Show that the Lagrangian for a simple pendulum of length L and mass m whose point of support moves on a vertical circle of radius R at the constant angular velocity ω is

$$L = \tfrac{1}{2} mL^2\dot{\theta}^2 + mRL\omega^2 \cos(\omega t - \theta) + mgL\cos\theta,$$

where θ is the angle the pendulum makes with the vertical and $t = 0$ when the point of support is at the bottom of the circle. Hint: Eliminate total time derivatives.

3.4. A mass m is fastened by a weightless string wrapped around a circular cylinder of radius a and moves in a plane perpendicular to the cylinder. The mass is given an initial velocity v_0 and the string wraps around the cylinder. Show that the angular velocity of the mass about the axis of the cylinder is given by $\omega = v_0(r^2 - a^2)^{1/2}/r^2$, where r is the radial distance from the axis of the cylinder to the mass. Is the angular momentum conserved?

3.5. Consider the simple one-dimensional Lagrangian $L = \frac{1}{2}m\dot{x}^2 - ax$. Obtain the motion of the system $x(t)$. Now construct some arbitrary function $y(t)$ that has the same values as $x(t)$ at some arbitrary times t_1 and t_2 but that differs in the interval between them. Verify explicitly that the action is larger for $y(t)$ than for the motion $x(t)$.

3.6. Prove Kepler's second law of planetary motion for a general central force field: The radius vector sweeps out equal areas in equal times.

3.7. For the Newtonian gravitational potential phase space is four-dimensional. The (r, θ) plots of the orbits, and the $(\dot{r}, r)$ plots as given in Fig. 3.5, show phase curves in two-dimensional sections of phase space. What do the phase curves look like in the other possible 2-sections? If phase curves do not intersect why is it that the curves in these planes do?

3.8. Obtain the parametric equations of an $E > 0$ orbit corresponding to those of (3.32) in the text.

3.9. A particle is constrained to move on the surface of a paraboloid of revolution opening in the $\hat{\mathbf{z}}$ direction. There is a uniform gravitational force in the $-\hat{\mathbf{z}}$ direction. Obtain the differential equations of motion and show that the motion takes place such that z is between z_{min} and z_{max}. Show that the projection of the radius vector in the (x, y) plane sweeps out equal areas in equal times.

3.10. Consider a particle of mass m that is constrained to move without friction, under the action of gravity, on the inner surface of a fixed hemispherical bowl of radius R.

(a) Construct a Lagrangian for this system.

(b) If the particle is moving in a stationary horizontal circular path, find its angular velocity.

(c) If the particle undergoes a slight perturbation away from this stationary orbit, describe its subsequent motion.

(d) Under what condition is the orbit of (c) closed?

3.11. Construct the configuration manifold for the system described in Problem 3.3. Sketch a trajectory of the system on this manifold. Do the same thing for the system of Problem 3.10.

3.12. Consider a simple pendulum of length L and mass m with its point of attachment fastened to a mass M that is on a vertical circle of radius R. The mass M is free to move on the circle without friction, subject only to a uniform gravitational force, as is the small mass m. Construct a Lagrangian for this system and describe the motion qualitatively. Construct a configuration manifold of appropriate dimension and sketch a system trajectory. How is this manifold embedded?

3.13. Show that the eigenvalues of a unitary operator all have unit magnitude and then show how this result relates to orthogonal matrices. Hint: A unitary operator is one for which the adjoint and inverse are equal. Use a standard Hermitian inner product.

3.14. Consider a charged particle moving in the electrostatic field between two concentric cylindrical conductors, the inner one with radius a and the outer one with radius b. The potential difference between them is V with the inner conductor at the higher potential.

(a) Reduce the problem of motion to quadratures.

(b) How does the motion differ for oppositely charged particles?

(c) Show that circular motion is possible with a radius inversely proportional to $(V \ln(b/a))^{1/2}$.

3.15. Consider a charged particle moving in exactly the same field as the previous problem except for the addition of a uniform magnetic field B_0 parallel to the axes of the cylinders.

(a) Find an effective potential and reduce the problem of motion to quadratures.

(b) Under what conditions on the constants of the motion will there be retrograde motion in the trajectory?

3.16. Two masses m_1 and m_2 are connected by a string through a small hole in a very smooth table. Assume that under the table m_2 is constrained to move vertically and m_1 is constrained to the surface of the table.

(a) Choose an appropriate set of coordinates and obtain the Lagrangian.

(b) Find the equation(s) of motion and reduce to quadratures.

(c) Discuss the motion in detail and explain the existence of any bounds for the distance of m_1 from the center of the hole. How do these depend on the parameters of the problem? Under what conditions does m_1 go in a circle?

3.17. A triangular block of mass M is free to slide on a frictionless horizontal surface. A uniform solid cylindrical log of mass m and radius r is constrained to roll without slipping on the inclined surface of the block.

(a) Choose an appropriate set of coordinates and obtain the equations of motion for the log and block.

(b) Obtain the forces of constraint on the log and on the block.

(c) Discuss the relative merits of the Lagrangian and Newtonian approaches to this problem.

3.18. A particle of mass m slides on a large, smooth, stationary spherical surface of radius a. At what angle from the vertical will the particle leave the sphere if it starts from rest at the top?

CHAPTER FOUR

THE DYNAMICS OF RIGID BODIES

The art of reasoning is nothing more than a language well arranged.

CONDILLAC

A macroscopic object that cannot be deformed by the forces that act on it is called a *rigid body*. As with "particles" or "nonviscous fluids," a rigid body is a mathematical idealization, which is never realized exactly. Nevertheless, the dynamics of a great many commonplace objects, in many diverse situations, is well modeled by a rigid-body description.

Aside from the usefulness of rigid-body dynamics as a physical model, a study of rigid bodies leads to many valuable insights, since it brings together nice examples of Lagrangians, constants of motion, and moving coordinate systems. Many topics discussed in previous chapters find application in a unified context as we study in detail the motion of rigid bodies.

We begin with some general considerations and then discuss the inertia operator and its role in rigid-body motion. Euler's equations for rigid-body motion are studied, both with and without external torques. The energy and angular momentum constants serve as a basis for a thorough study of the torque-free case.

4.1. GENERAL PROPERTIES

A rigid body may be viewed as a large collection of point particles, each and every one constrained to maintain a fixed distance from all the others. This constraint on the relative positions of all the particles makes it necessary to specify only two things in order to specify the configuration of the rigid body uniquely. First, one must specify the position in the chosen reference system of one selected particle or mass point. We refer to this particular point as the *body point* and note that often the center-of-mass is chosen for the body point. Second, one must specify the orientation of the rigid body with respect to a fixed set of axes. Once the position vector **X** of the body point and the orientation of the rigid body are specified, the configuration of the rigid body is uniquely determined. The configuration manifold of a rigid body is $\mathbb{R}^3 \times SO(3)$: the position vector **X** in $\mathbb{R}^3$ for the body point and the rotation matrix R in $SO(3)$ for the orientation.

For certain special rigid bodies in which all mass points lie along a line, the configuration manifold is of smaller dimension. The orientation of a straight wire is given simply by specifying two angles, corresponding to the place on the unit sphere through which the wire would pierce if oriented along a radial line. The configuration manifold for this wire is $\mathbb{R}^3 \times S^2$. If an arbitrary mass distribution for a rigid body has more than one linear dimension, then the full configuration manifold $\mathbb{R}^3 \times SO(3)$ is required. A distribution with less than one dimension is a "mass point" or "particle" and is treated in the preceding chapters. In this chapter we focus on objects requiring the full configuration manifold $\mathbb{R}^3 \times SO(3)$.

In many situations the rigid body may be considered free of external forces. As studied in Section 2.2, the center-of-mass of a rigid body has a uniform velocity in the absence of external forces. In such a case we choose the body point to be the center-of-mass and consider the motion of the rigid body in the inertial system in which the center-of-mass is at rest. In this reference system the motion of the rigid body consists of rotations about the center-of-mass, which is a stationary point.

In a second large class of problems, external forces are present, but some point in the rigid body is fixed in the inertial reference system. We choose this fixed point to be the body point and then once again the motion consists entirely of rotations around the body point. Hence we focus our analysis on the rotations of rigid bodies about a body point. This point is either the center-of-mass or a point fixed in the inertial reference system.

With this fixed body point as an origin, we select two reference systems. The first is an inertial system with a set of fixed axes. This first system is the (S)tationary system of previous discussions (cf. Section 1.3). The second reference system has a set of axes firmly attached to the rigid body. This second system is an example of the (M)oving coordinate systems previously encountered. This particular M system is special because it has the same origin as S. There are no translations relating M and S. We frequently refer to this M as the (B)ody system, since the axes are fixed in the rigid body and the origin is taken at the fixed body point.

The systems B and S are related by a rotation only. The configuration of the rigid body in S is completely specified by giving the rotation R that when applied to basis vectors (passive view) would rotate the basis of S into the basis of B. In terms of the usual Euler angles, an arbitrary rotation is given by (3.43). Thus the configuration manifold for all systems considered in this chapter is $SO(3)$.

The motion of a rigid body is consequently represented by a "trajectory" in $SO(3)$. When outside forces are absent, (2.31) shows that the kinetic

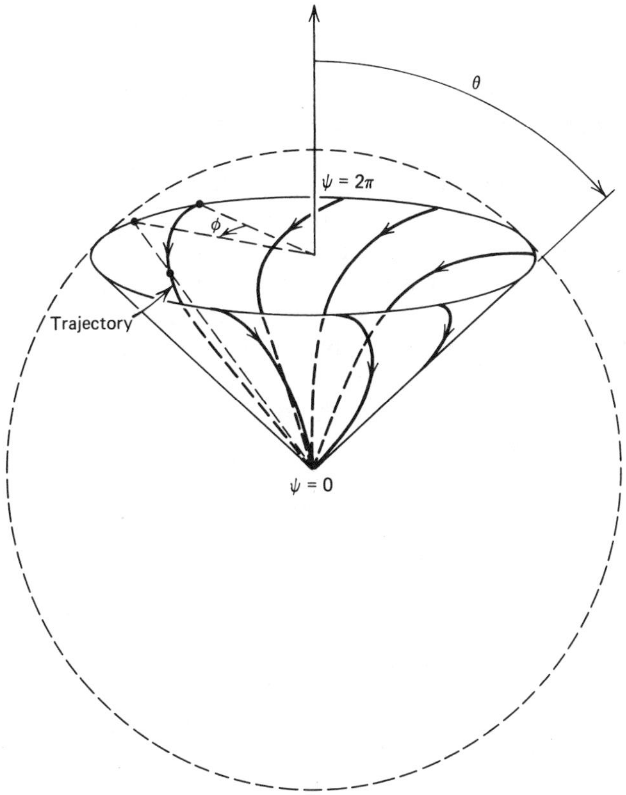

FIGURE 4.1. A trajectory in the manifold $SO(3)$. The trajectory is confined to a surface of constant θ. The Euler angles ϕ and ψ are related here by $d\psi = -8d\phi$. The sketch corresponds to a system trajectory for a cone rolling on a plane with the apex a fixed point. The ratio of the radius of the base of the rolling cone to the length of its side is $\frac{1}{8}$ for the trajectory sketched. This system is studied in Problem 4.5.

energy of the rigid body is constant. Likewise, the absence of external forces implies the absence of external torques, and (2.36) shows that the angular momentum with respect to the body point is also constant. Thus any "trajectory" in $SO(3)$, representing motion of a rigid body free of external forces, must conserve kinetic energy T and angular momentum $\mathbf{L}$.

Even in situations where the body point is simply a fixed point and external forces are acting on the system, the forces are usually such that one or more of the free-motion constants is conserved and the trajectory in $SO(3)$ is restricted. An example of a trajectory in $SO(3)$ is given in Fig. 4.1. The trajectory is for the system studied in Problem 4.5.

4.2. MOMENT OF INERTIA TENSOR

We again adopt the notation of Section 1.3: vectors in S are denoted with lower-case letters and vectors in B are denoted with upper-case letters. For example, the position vector for a mass point in S is denoted by $\mathbf{x}$ and the position vector for the same mass point in B is $\mathbf{X}$. In the active view, these are different vectors related by the rotation R, that is, $\mathbf{x} = R\mathbf{X}$. The vectors $\mathbf{x}$ and $\mathbf{X}$ are representatives of the same physical vector in two different reference systems. In the passive view $\mathbf{x}$ and $\mathbf{X}$ are the same vector but referred to two different basis sets related by a rotation.

When the velocity in S arises only from the rotation of B in S, we have from (1.41) that $\dot{\mathbf{x}} = \boldsymbol{\omega} \times \mathbf{x}$, where $\boldsymbol{\omega}$ is the angular velocity vector of the rigid body in S. With this velocity the angular momentum of a particle of mass m about the origin is

$$\mathbf{l} = \mathbf{x} \times m\dot{\mathbf{x}} = m\mathbf{x} \times (\boldsymbol{\omega} \times \mathbf{x}). \tag{4.1}$$

We can sum this equation for the many particles that make up the rigid body

$$\mathbf{l} = \sum_a \mathbf{l}_{(a)} = \sum_a m_{(a)}\mathbf{x}_{(a)} \times (\boldsymbol{\omega} \times \mathbf{x}_{(a)}) \tag{4.2}$$

Converting (4.2) to index form and using (1.44) along with part (c) of Problem 1.5, we obtain

$$l^i = \sum_a m_{(a)}\varepsilon^{ijk}\varepsilon_{klm}x_{(a)j}\omega^l x^m_{(a)} = \sum_a m_{(a)}\left[x^2_{(a)}\omega^i - x^i_{(a)}x_{(a)j}\omega^j\right]$$

$$l^i = \left(\sum_a m_{(a)}\left[x^2_{(a)}\delta^i_j - x_{(a)j}x^i_{(a)}\right]\right)\omega^j. \tag{4.3}$$

We identify the quantity

$$k^i{}_j = \sum_a m_{(a)}\left[x_{(a)}^2\delta_j^i - x_{(a)j}x_{(a)}^i\right] \tag{4.4}$$

as the *inertia operator* or *inertia tensor*. If we view our rigid body as a continuous distribution of matter, then the sum in (4.4) is more appropriately written as an integral.

$$k^i{}_j = \int_{\text{body}} \rho(\mathbf{x})\left[x^2\delta_j^i - x_j x^i\right] d^3x. \tag{4.5}$$

Equation (4.2) can then be written in the form

$$\mathbf{l} = k\boldsymbol{\omega}. \tag{4.6}$$

In words, the foregoing equation says that the linear operator k maps the angular velocity vectors $\boldsymbol{\omega}$ into the angular momentum vectors $\mathbf{l}$. Recall that lower-case letters denote quantities expressed in the S system.

All quantities in (4.6) may be expressed in the body system B by using the rotation operator. We have $\mathbf{l} = R\mathbf{L}$, $\boldsymbol{\omega} = R\boldsymbol{\Omega}$, and for the inertia operator we define $K = R^t k R$. Equation (4.6) implies

$$\mathbf{L} = K\boldsymbol{\Omega}. \tag{4.7}$$

Since the rotation operator R satisfies (1.32) and (1.33), we obtain for K the following integral expression:

$$K^i{}_j = \int_{\text{body}} \rho(\mathbf{X})\left[X^2\delta_j^i - X^i X_j\right] d^3X. \tag{4.8}$$

This gives the inertia operator in B the frame attached to the rigid body. This is the frame in which the inertia tensor is most easily calculated because in the frame fixed to the rigid body the density at a point does not change. From the passive point of view, (4.6) and (4.7) are the same. In the passive view the pairs $(\mathbf{l}, \mathbf{L})$, $(\boldsymbol{\omega}, \boldsymbol{\Omega})$, and (k, K) denote the same quantities with different components in the two reference systems S and B.

The kinetic energy of the rigid body may now be compactly written:

$$\begin{aligned} T &= \frac{1}{2}\sum_a m_{(a)}\dot{\mathbf{x}}_{(a)} \cdot \dot{\mathbf{x}}_{(a)} = \frac{1}{2}\sum_a m_{(a)}\left(\boldsymbol{\omega} \times \mathbf{x}_{(a)}\right) \cdot \left(\boldsymbol{\omega} \times \mathbf{x}_{(a)}\right) \\ &= \frac{1}{2}\boldsymbol{\omega} \cdot \sum_a m_{(a)}\mathbf{x}_{(a)} \times \left(\boldsymbol{\omega} \times \mathbf{x}_{(a)}\right) = \frac{1}{2}\boldsymbol{\omega} \cdot \mathbf{l} = \frac{1}{2}\boldsymbol{\omega} \cdot k\boldsymbol{\omega} \qquad (4.9) \\ &= \frac{1}{2}\boldsymbol{\Omega} \cdot \mathbf{L} = \frac{1}{2}\boldsymbol{\Omega} \cdot K\boldsymbol{\Omega}. \qquad (4.10) \end{aligned}$$

Equation (4.10) follows immediately from (4.9) because R preserves inner products.

From (4.8) we see that K is real and symmetric. Since $T \geqslant 0$ we learn from (4.10) that K is also positive definite.* Because of these properties, K can always be diagonalized with its eigenvalues appearing down the diagonal. The eigenvalues are real, positive, and we denote them as I_1, I_2, I_3. They are called the *principal moments of inertia.*

If the eigenvalues of K are all different, then the eigenvectors are uniquely determined to within a normalization constant. We normalize these eigenvectors to have unit length, using the usual Euclidean metric in B. For unique eigenvalues these eigenvectors are automatically orthogonal. Should an eigenvalue be degenerate, then the corresponding eigenvector is not uniquely determined. For a doubly degenerate eigenvalue the space of eigenvectors is two-dimensional, for triple degeneracy, this space is three-dimensional. Corresponding to a degenerate eigenvalue, we choose a number of linearly independent eigenvectors equal to the degree of degeneracy. These linearly independent eigenvectors are made orthogonal, using perhaps the familiar Schmidt orthogonalization procedure, and then normalized. Consequently, regardless of possible degeneracies, an orthonormal set of eigenvectors $\{\mathbf{e}_i\}$ may always be chosen. The directions along the $\{\mathbf{e}_i\}$ are called *principal axes* of the rigid body. If the principal moments of inertia are not all different, then, like the eigenvectors, the principal axes are not uniquely determined. Choosing $\{\mathbf{e}_i\}$ as a basis for B results in both K and T having a particularly simple form. K is diagonal with the principal moments of inertia down the diagonal. The angular momentum components from (4.7) are given by

$$(L^1, L^2, L^3) = \left(I_1\Omega^1, I_2\Omega^2, I_3\Omega^3\right), \tag{4.11}$$

and the kinetic energy has the form

$$T = \tfrac{1}{2}\left[I_1(\Omega^1)^2 + I_2(\Omega^2)^2 + I_3(\Omega^3)^2\right]. \tag{4.12}$$

If principal axes have been chosen for B, then from (4.8) we obtain

$$I_1 = \int_{\text{body}} \rho(\mathbf{X})\left[(X^2)^2 + (X^3)^2\right] d^3X. \tag{4.13}$$

The perpendicular distance of the mass element $\rho(\mathbf{X})\, d^3X$ from the $\mathbf{e}_1$ axis is

*A positive definite, symmetric matrix K_{ij} is one such that $\Omega^i\Omega^j K_{ij} \geqslant 0$ for arbitrary $\mathbf{\Omega}$. We have equality iff $\mathbf{\Omega} \equiv 0$.

given by $[(X^2)^2 + (X^3)^2]^{1/2}$. If we call this distance R_1, then

$$I_1 = \int_{\text{body}} \rho(\mathbf{X}) R_1^2(\mathbf{X})\, d^3X, \tag{4.14}$$

and similarly for the other principal moments of inertia. Summarizing for all three principal axes:

$$I_i = \int_{\text{body}} \rho(\mathbf{X}) R_i^2(\mathbf{X})\, d^3X. \tag{4.15}$$

Equation (4.15) is a common elementary expression for the moments of inertia.

Since the inertia tensor and the moments of inertia along the various axes are determined solely by the mass distribution of a rigid body, it is useful to define the inertia ellipsoid for a rigid body. The *inertia ellipsoid* is the ellipsoid determined in the B system by setting $T = 1/2$ in (4.12). This gives

$$\mathbf{\Omega} \cdot \mathbf{L} = I_1(\Omega^1)^2 + I_2(\Omega^2)^2 + I_3(\Omega^3)^2 = 1. \tag{4.16}$$

Equation (4.16) determines a set of vectors $\{\mathbf{\Omega}\}$, which trace out the inertia ellipsoid in the body frame B. The semiaxes of this ellipsoid are given by $a_i = \sqrt{(1/I_i)}$. Note that if a body is elongated along one axis, then the moment of inertia with respect to that axis will be small in comparison to moments of inertia for other axes. Thus a_i for that axis will be large and the inertia ellipsoid will also be stretched out along that axis. As an example see Fig. 4.2.

If two rigid bodies have the same inertia ellipsoid given by (4.16), then they behave the same under rotations. Given the same initial conditions, the bodies move identically and generate the same trajectory in $SO(3)$.

The symmetries of a rigid body have an effect on the shape of the corresponding inertia ellipsoid. If a rigid body is symmetric under a rotation of $2\pi/n$, then we refer to the axis of rotation as an *n-fold axis of symmetry*. The symmetries of the rigid body and its inertia ellipsoid are connected by

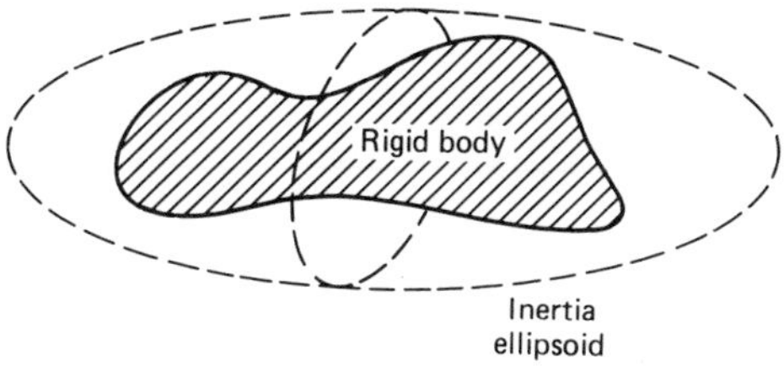

FIGURE 4.2. The inertia ellipsoid.

the fact that an n-fold axis of symmetry for the rigid body must be an n-fold axis of symmetry for the corresponding inertia ellipsoid. However, an ellipsoid with three different semiaxes does not have any symmetries for $n > 2$. Hence if $n > 2$, then an n-fold axis of symmetry for the body must be an axis of rotation for the inertia ellipsoid. An axis of rotation for the inertia ellipsoid is necessarily a principal axis. In effect, the inertia ellipsoid has at least as much symmetry as the rigid body it represents, and in a case for $n > 2$ it has more symmetry. For example, a rigid body in the form of a square has a four-fold axis of symmetry perpendicular to the plane of the square and passing through its center. This axis is a principal axis for the body and this axis is an axis of rotation for the inertia ellipsoid. See Fig. 4.3.

4.3. EULER'S EQUATIONS OF MOTION

In the absence of any outside forces, the angular momentum of a rigid body in the S frame is constant [cf. (2.11)]. Also recall that $\mathbf{l} = R\mathbf{L}$ and hence

$$\frac{d\mathbf{l}}{dt} = R\dot{\mathbf{L}} + \dot{R}\mathbf{L} = R\dot{\mathbf{L}} + \dot{R}R'\mathbf{l} = R\dot{\mathbf{L}} + \boldsymbol{\omega} \times \mathbf{l} = R[\dot{\mathbf{L}} + \boldsymbol{\Omega} \times \mathbf{L}] = 0, \tag{4.17}$$

where we have used the result of Problem 1.8. Since R is arbitrary

$$\dot{\mathbf{L}} + \boldsymbol{\Omega} \times \mathbf{L} = 0 \tag{4.18}$$

for torque-free motion in B.

The equations for the components of $\boldsymbol{\Omega}$ that follow from (4.18) are referred to as *Euler's equations*. These equations are most conveniently written in the principal axis coordinate system in B. Since $\mathbf{L} = K\boldsymbol{\Omega}$, then (4.18) gives in component form

$$\begin{aligned} I_1\dot{\Omega}^1 + \Omega^2\Omega^3(I_3 - I_2) &= 0, \\ I_2\dot{\Omega}^2 + \Omega^1\Omega^3(I_1 - I_3) &= 0, \\ I_3\dot{\Omega}^3 + \Omega^1\Omega^2(I_2 - I_1) &= 0. \end{aligned} \tag{4.19}$$

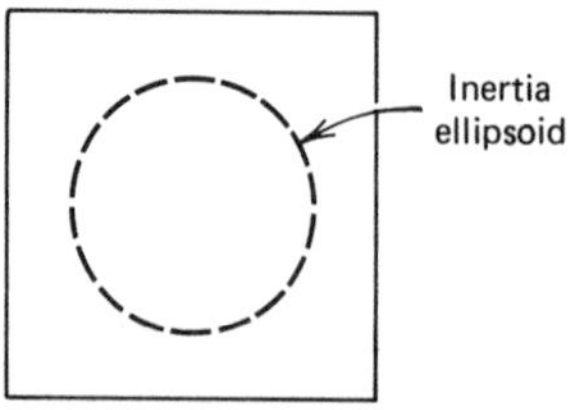

FIGURE 4.3. End-on view of inertia ellipsoid for a square.

The principal moments of inertia for a rigid body are fixed, and thus (4.19) is a set of differential equations for the components of the angular velocity vector in system B.

The most convenient way to analyze these equations is to use the Euler angles for the rotation R between B and S. These angles were defined in Chapter 3 and are sketched in Fig. 4.4. In Fig. 4.4 the individual angular velocity vectors $\dot{\boldsymbol{\phi}}$, $\dot{\boldsymbol{\theta}}$, and $\dot{\boldsymbol{\psi}}$, which constitute $\boldsymbol{\omega}$ and $\boldsymbol{\Omega}$, are also sketched.

By reference to Fig. 4.4, we take projections of $\dot{\boldsymbol{\phi}}$, $\dot{\boldsymbol{\theta}}$, and $\dot{\boldsymbol{\psi}}$ on the axes of the S system to obtain

$$\begin{aligned}\omega^1 &= \dot{\theta}\cos\phi + \dot{\psi}\sin\theta\sin\phi,\\ \omega^2 &= \dot{\theta}\sin\phi - \dot{\psi}\sin\theta\cos\phi,\\ \omega^3 &= \dot{\phi} + \dot{\psi}\cos\theta.\end{aligned} \tag{4.20}$$

A similar projection onto the axis of the B system gives the components of

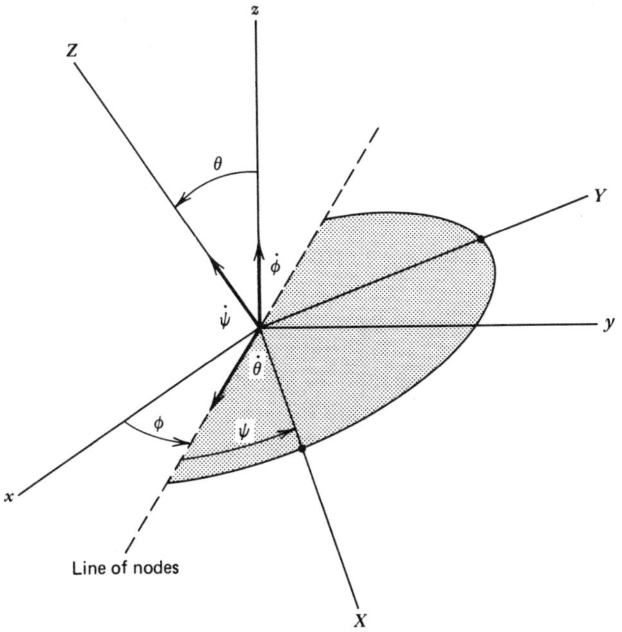

FIGURE 4.4. Passive view of the rotation R in terms of Euler angles relating B and S. The individual angular velocity vectors for the Euler angle rotations are also given.

the vector $\boldsymbol{\Omega}$ in terms of the Euler angles.

$$\begin{aligned}
\Omega^1 &= \dot{\theta}\cos\psi + \dot{\phi}\sin\theta\sin\psi,\\
\Omega^2 &= -\dot{\theta}\sin\psi + \dot{\phi}\sin\theta\cos\psi,\\
\Omega^3 &= \dot{\psi} + \dot{\phi}\cos\theta.
\end{aligned} \tag{4.21}$$

For torque-free motion described by (4.18) and (4.19), the constants of the motion allow us to describe the possible rigid-body motions qualitatively. The energy is entirely kinetic energy given by (4.12). The energy constant gives one invariant, quadratic in the components of the angular momentum. Since $\mathbf{l}$ is constant and $\mathbf{l}\cdot\mathbf{l} = \mathbf{L}\cdot\mathbf{L} = L^2$, there is a second quadratic constant. In terms of the components of $\mathbf{L}$ in a principal-axes coordinate system for B, these two constants give

$$\frac{(L^1)^2}{2TI_1} + \frac{(L^2)^2}{2TI_2} + \frac{(L^3)^2}{2TI_3} = 1, \tag{4.22}$$

$$(L^1)^2 + (L^2)^2 + (L^3)^2 = (L)^2. \tag{4.23}$$

For the vector $\mathbf{L}$ in B, (4.22) is the equation of an ellipsoid and (4.23) is the equation of a sphere. The semiaxes of the ellipsoid are given by $a_i = \sqrt{(2TI_i)}$ and in order for (4.22) and (4.23) to be simultaneously satisfied $a_1 \geqslant L \geqslant a_3$, where the axes are labeled such that $I_1 > I_2 > I_3$. Note that the relative dependence of the semiaxes of (4.22) on the principal moments of inertia is opposite from the semiaxes of the inertia ellipsoid of (4.16). The ellipsoid (4.22) is elongated along the direction with maximum principal moment of inertia, which is opposite from the inertia ellipsoid. The inertia ellipsoid is generated in B by the angular velocity, whereas (4.22) is an ellipsoid in B for angular momentum.

If L is not bounded above by a_1 and below by a_3, then no intersections of the energy ellipsoid (4.22) and the angular momentum sphere (4.23) are possible. The intersections of these two surfaces show the possible paths in the B system of the angular momentum vector. For $|\mathbf{L}|$ slightly smaller than a_1, the intersection curves are little ellipses around the end of the ellipsoid along the $\mathbf{e}_1$ axis. Likewise for $|\mathbf{L}|$ slightly larger than the smallest semiaxis a_3, the paths of $\mathbf{L}$ in B consist of small ellipses around the $\mathbf{e}_3$ axis. For $|\mathbf{L}| = a_2$ we obtain two possible paths that intersect at the $\mathbf{e}_2$ axis. Sample paths are sketched in Fig. 4.5.

It is evident from the paths of $\mathbf{L}$ sketched in Fig. 4.5 that the three points labeled 1, 2, 3, and their opposite points on the ellipsoid are special. If $\mathbf{L}$ is along one of the principal axes, then the only nonzero component of **L**, denoted L^i, is given by $L^i = I_i \Omega^i$ (no sum). The vectors **L** and **Ω** are colinear. An examination of Euler's equations (4.19) shows that for the nonzero component, $\dot{L}^i$ is zero. There will be no change of this component. Likewise, the other components that start at zero remain zero. Points 1, 2, and 3 in Fig. 4.5 are examples of fixed points of **L** for force-free, rigid-body motion. These points are similar in character to the fixed points in phase space discussed in Example (2) of Section 2.4. The points on the $\mathbf{e}_1$ and $\mathbf{e}_3$ axes are "0" points (elliptic fixed points) and the points on the $\mathbf{e}_2$ axis are "X" points (hyperbolic fixed points). If $\dot{\boldsymbol{\Omega}} = 0$, then the only solutions to (4.19) correspond to having two components of **Ω** equal to zero, since $I_1 > I_2 > I_3$. Thus there are no other fixed points. In other words, a rigid body can have a stationary rotation around any one of the three principal axes, and only around these axes.

However, the nature of these fixed points for **L** shows the stationary rotations around the principal axes are not all stable. The curves for the angular momentum that start out near point 1 in Fig. 4.5 will follow a curve that keeps it close to 1. Similar remarks hold for angular momenta that follow curves in the neighborhood of the fixed point denoted 3 in Fig. 4.5. The angular momenta $\mathbf{L} = L\mathbf{e}_1$ and $\mathbf{L} = L\mathbf{e}_3$ are stationary rotations which are stable. In contrast, point 2, the intersection point of the $\mathbf{e}_2$ principal axis and the ellipsoid of (4.22), is a hyperbolic point. Angular momenta that start initially near 2 will move far away from this point as time increases. The stationary rotations around $\mathbf{e}_2$ are unstable.

If two principal moments are equal, say $I_2 = I_3$, then the angular momentum paths on the ellipsoid are circles around the $\mathbf{e}_1$ principal axis.

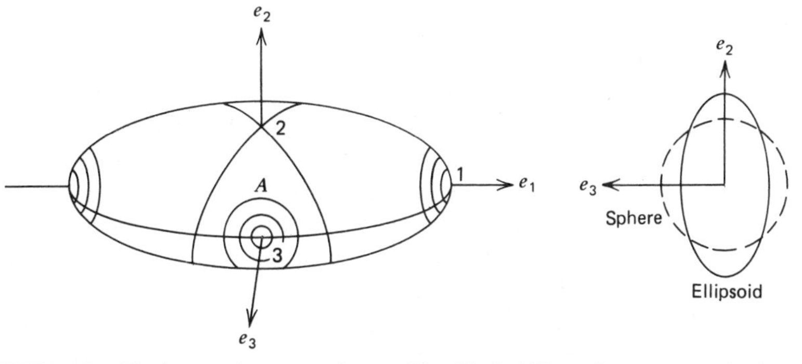

FIGURE 4.5. The intersection curves for an ellipsoid of (4.22) and a sequence of spheres of (4.23) where $\sqrt{2TI_1} \geqslant L \geqslant \sqrt{2TI_3}$. The sketch to the right shows the end-on view of the intersection that gives the curve A.

The middle circle, which intersects both the $\mathbf{e}_2$ and the $\mathbf{e}_3$ axes, consists entirely of fixed points. However, all points on this circle are unstable, as can be seen by noting that any small amount of Ω^1, no matter how small, will carry the angular momentum vector $\mathbf{L}$ in a circular path about the $\mathbf{e}_1$ axis.

The difference in the stability of rotations around these axes is easily demonstrated by tossing a rotating book or eraser into the air and comparing the rotations around each of the axes in turn. A square book has only one axis for stable rotations, the one perpendicular to the plane of the book.

There is another qualitative description of the free rotations of a rigid body, called Poinsot's description, which is occasionally helpful for visualizing the motion. This description proceeds by again considering the constants of the motion $\mathbf{l}$ and T. The energy constant T in (4.12) determines an ellipsoid in the B system for the angular velocity components. The angular velocity components are constrained to satisfy the equation

$$f(\Omega^1, \Omega^2, \Omega^3) = \frac{I_1}{2T}(\Omega^1)^2 + \frac{I_2}{2T}(\Omega^2)^2 + \frac{I_3}{2T}(\Omega^3)^2 - 1 = 0. \quad (4.24)$$

This is an ellipsoid fixed in the B system with the components $(\Omega^1, \Omega^2, \Omega^3)$ as independent variables. The ellipsoid of (4.24) is just the inertia ellipsoid with arbitrary energy T rather than $T = 1/2$. The normal vector to the surface $f(\Omega^1, \Omega^2, \Omega^3) = 0$, as given in (4.24), is found by computing the gradient of f with respect to the independent variables Ω^i. We find $\nabla f = \mathbf{L}/T$ at the point $\boldsymbol{\Omega}$ on the surface of the inertia ellipsoid (4.24). At the point $\boldsymbol{\Omega}$ on the inertia ellipsoid, there is a tangent plane and $\nabla f = \mathbf{L}/T$ is normal to this plane.

Now view these relationships in the stationary system S. In S the normal to the inertia ellipsoid (4.24) is $R(\nabla F) = \mathbf{l}/T$ and is constant in time, since both $\mathbf{l}$ and T are constant. Hence the plane tangent to the inertia ellipsoid is fixed in space and $\mathbf{l}$ is normal to this plane. This plane is called the *invariable plane*. Furthermore, the angular velocity vector $\boldsymbol{\omega}$ terminates on this plane at the point of tangency. The ellipsoid of (4.24) rotates around $\boldsymbol{\omega}$, as seen in the S system. The vector $\boldsymbol{\omega}$ may be viewed as an instantaneous axis of rotation. The point of tangency P for the inertia ellipsoid (4.24) on the invariable plane is on this instantaneous axis of rotation. Therefore P is momentarily at rest and we may view the inertia ellipsoid (4.24) as "rolling" without slipping on the invariable plane. The path of the point of contact on the invariable plane is called the *herpolhode* and the path of the point P on the inertia ellipsoid is called the *polhode*. Generally these curves will not be closed curves. A special case of this type of motion will be studied in

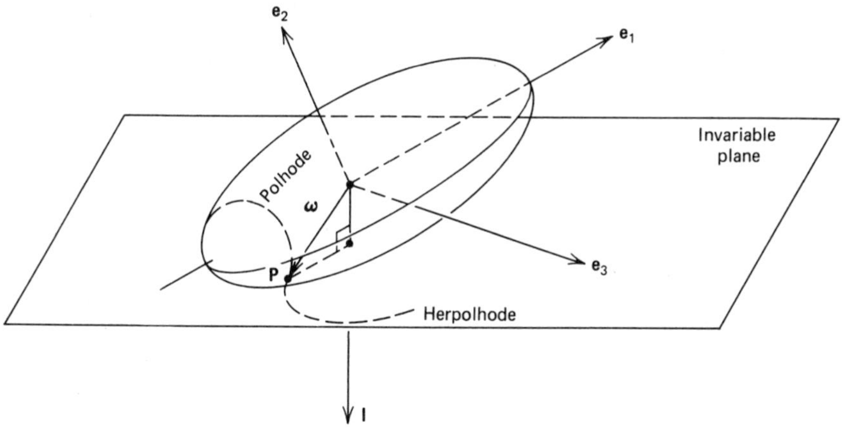

FIGURE 4.6. Rolling inertia ellipsoid on the invariable plane.

Problem 4.3. A sketch of this "rolling" of the inertia ellipsoid on the invariable plane is given in Fig. 4.6.

4.4. EULER'S EQUATIONS WITH TORQUE

If there are outside forces acting on a rigid body, then no longer do Euler's equations (4.19) for torque-free motion apply. Let $\mathbf{t}$ denote the total torque due to these forces in S; the corresponding torque vector in system B is $\mathbf{T}$. Then (4.18) becomes

$$\dot{\mathbf{L}} + \boldsymbol{\Omega} \times \mathbf{L} = \mathbf{T}. \tag{4.25}$$

As in other cases previously encountered, a Lagrangian approach frequently facilitates the solution of a dynamical problem for rigid bodies (cf. Section 3.3). From the expression for kinetic energy (4.10) we have $L = (1/2)\boldsymbol{\Omega} \cdot K\boldsymbol{\Omega} - U$. If principal axes are chosen in the B system, then

$$L = \tfrac{1}{2}\left[I_1(\Omega^1)^2 + I_2(\Omega^2)^2 + I_3(\Omega^3)^2\right] - U.$$

This expression becomes even more specific if we use (4.21) to express the angular velocity components in terms of Euler angles.

$$L = \tfrac{1}{2}\Big[I_1(\dot{\theta}\cos\psi + \dot{\phi}\sin\theta\sin\psi)^2 + I_2(-\dot{\theta}\sin\psi + \dot{\phi}\sin\theta\cos\psi)^2 + I_3(\dot{\psi} + \dot{\phi}\cos\theta)^2\Big] - U(\phi, \theta, \psi). \tag{4.26}$$

As an interesting example for an application of Euler's equations with torque (4.25), and the Lagrangian (4.26), we consider the motion of a heavy symmetrical top in a uniform gravitational field. See Fig. 4.7. To use (4.25) involving torque directly, we must have **T**, the torque in the B system. This torque **T** may be obtained by using the rotation matrix R of (3.43) and the equation $\mathbf{T} = R^t\mathbf{t}$. For **t** we use Fig. 4.7 to get the correct geometrical relations, and then **t** in S is given by $\mathbf{t} = \mathbf{r} \times \mathbf{f}$. We find

$$\mathbf{r} = a[\hat{\mathbf{x}}\sin\theta\sin\phi - \hat{\mathbf{y}}\sin\theta\cos\phi + \hat{\mathbf{z}}\cos\theta],$$

$$\mathbf{f} = -mg\hat{\mathbf{z}}.$$

Alternatively, the vectors **r** and **f** may be expressed in the B system directly, giving $\mathbf{T} = \mathbf{R} \times \mathbf{F}$. The vectors **R** and **F** are seen to be

$$\mathbf{R} = a\hat{\mathbf{Z}},$$

$$\mathbf{F} = -mg\big[\hat{\mathbf{X}}\sin\theta\sin\psi + \hat{\mathbf{Y}}\sin\theta\cos\psi + \hat{\mathbf{Z}}\cos\theta\big].$$

In the body system B, **R** is easily given, and in S the force **f** is readily found.

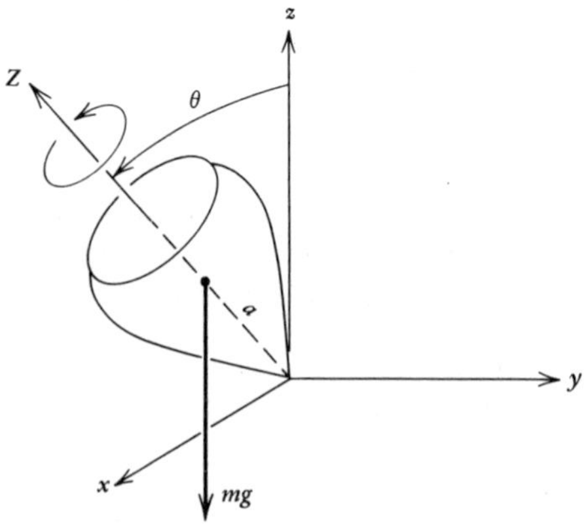

FIGURE 4.7. A symmetrical top in a uniform gravitational field.

With either method we obtain

$$\mathbf{T} = mga\left[\hat{\mathbf{X}} \sin\theta \cos\psi - \hat{\mathbf{Y}} \sin\theta \sin\psi\right]. \tag{4.27}$$

Since the top is symmetric, we have that $I_1 = I_2 \equiv I > I_3$. Using these expressions for the principal moments of inertia, the result for the torque (4.27), and substituting into (4.25), we obtain

$$\begin{aligned} I\dot{\Omega}^1 + \Omega^2\Omega^3(I_3 - I) &= mga \sin\theta \cos\psi, \\ I\dot{\Omega}^2 - \Omega^1\Omega^2(I_3 - I) &= -mga \sin\theta \sin\psi, \\ I_3\dot{\Omega}^3 &= 0. \end{aligned} \tag{4.28}$$

The last equation of (4.28) gives immediately one constant of the motion $I_3\Omega^3 = L^3$.

Rather than continuing with the analysis of (4.28), we consider the Lagrangian approach. The potential for the heavy symmetrical top is $U = mg\cos\theta$. Using (4.26) the Lagrangian becomes

$$L = \frac{I}{2}\left[\dot{\theta}^2 + \dot{\phi}^2 \sin^2\theta\right] + \frac{I_3}{2}\left[\dot{\psi} + \dot{\phi}\cos\theta\right]^2 - mga\cos\theta. \tag{4.29}$$

The energy is conserved because U does not depend explicitly on time. The total energy is constant and given by

$$E = \frac{I}{2}\left[\dot{\theta}^2 + \dot{\phi}^2 \sin^2\theta\right] + \frac{I_3}{2}\left[\dot{\psi} + \dot{\phi}\cos\theta\right]^2 + mga\cos\theta. \tag{4.30}$$

Examination of the Lagrangian shows that the coordinates ϕ and ψ are cyclic. Thus we immediately obtain two constants of the motion.

$$p_\phi = \frac{\partial L}{\partial \dot{\phi}} = I\dot{\phi}\sin^2\theta + I_3(\dot{\psi} + \dot{\phi}\cos\theta)\cos\theta, \tag{4.31}$$

$$p_\psi = \frac{\partial L}{\partial \dot{\psi}} = I_3(\dot{\psi} + \dot{\phi}\cos\theta). \tag{4.32}$$

The constant p_ψ is the constant already noted from the last of Eqs. (4.28). The quantity $I_3\Omega^3 = p_\psi$ is the component of the angular momentum along the symmetry axis of the top. The other constant of the motion p_ϕ may also be obtained from (4.28) by manipulation of the first two equations in (4.28). The contrast in labor to obtain p_ϕ is significant, however, when compared with the Lagrangian approach.

From Fig. 4.7 we see that the torque $\mathbf{t}$ is perpendicular to the plane defined by the unit vectors $(\hat{\mathbf{z}}, \hat{\mathbf{Z}})$. Thus the component of the angular momentum along both of these axes is constant. The angular momentum component along $\hat{\mathbf{Z}}$ is given in (4.32). The component of angular momentum in the vertical direction must also be constant and we suspect that this constant is just p_ϕ in (4.31). We calculate l^3 by using $\mathbf{l} = R\mathbf{L}$, where the components of $\mathbf{L}$ are

$$L^1 = I(\dot{\theta}\cos\psi + \dot{\phi}\sin\theta\sin\psi),$$

$$L^2 = I(-\dot{\theta}\sin\psi + \dot{\phi}\sin\theta\cos\psi),$$

$$L^3 = I_3(\dot{\psi} + \dot{\phi}\cos\theta),$$

and the rotation R is given in terms of Euler angles in (3.43). A straightforward calculation shows that indeed $l^3 = p_\phi$.

The constants of the motion (4.30), (4.31), and (4.32) allow a reduction of this problem to quadratures. Write $\Omega^3 = \dot{\psi} + \dot{\phi}\cos\theta = p_\psi/I_3$ and find $p_\phi = I\dot{\phi}\sin^2\theta + p_\psi\cos\theta$ so that $\dot{\phi}\sin\theta = (p_\phi - p_\psi\cos\theta)/I\sin\theta$. Substitute these results into (4.30) for the energy, solve for $\dot{\theta}$ and find

$$\dot{\theta} = \left\{\frac{2}{I}(E - mga\cos\theta) - \frac{p_\psi^2}{II_3} - \frac{(p_\phi - p_\psi\cos\theta)^2}{I^2\sin^2\theta}\right\}^{1/2}. \tag{4.33}$$

The motions implied by (4.33) may be discussed in qualitative terms. For this purpose it is convenient to use the variable $\mu = \cos\theta$. For brevity we also collect the constants of (4.33) in the form $A = 2E/I - p_\psi^2/II_3$, $B = 2mga/I > 0$, $C = p_\phi/I$, and $D = p_\psi/I$. Then (4.33) is equivalent to

$$\dot{\mu}^2 = f(\mu) \equiv (1 - \mu^2)(A - B\mu) - (C - D\mu)^2. \tag{4.34}$$

In terms of the new variable μ, the angular velocity $\dot{\phi}$ is related to the constants of the motion by

$$\dot{\phi} = \frac{C - D\mu}{1 - \mu^2}. \tag{4.35}$$

The function $f(\mu)$ is a cubic function of μ. To analyze the behavior of this function and its roots we first observe from (4.34) that as $\mu \to \pm\infty$, then $f(\mu) \to \pm\infty$. Also for arbitrary values of C and D, we see that $f(\mu) < 0$ when $\mu = \pm 1$. Furthermore, for physically acceptable values of μ, $f(\mu) = \dot{\mu}^2 \geqslant 0$ and the allowed range for μ is $-1 \leqslant \mu \leqslant 1$. Thus for a physically possible motion the function must look qualitatively like the sketch of Fig. 4.8.

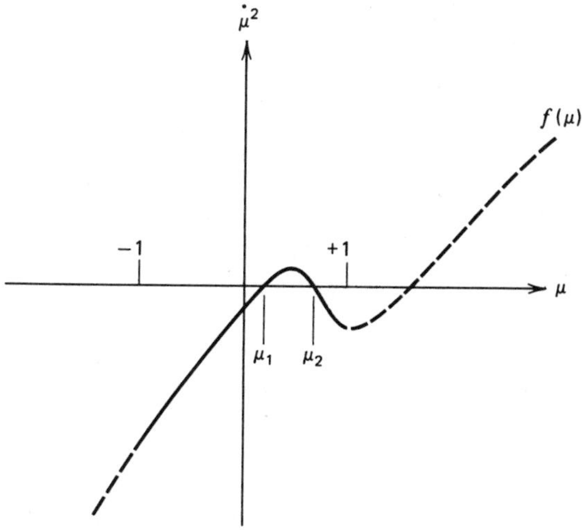

FIGURE 4.8. Qualitative sketch of the function $f(\mu)$ defined in (4.34).

For a physical solution of (4.34) there must be two real roots μ_1, μ_2 where $\dot{\mu} = 0$; or else there are none. If the value of the constants $C/D = p_\phi/p_\psi$ is such that $\mu_1 < p_\phi/p_\psi < \mu_2$, then from (4.35) it is evident that $\dot{\phi}$ will change sign as μ goes back and forth between μ_1 and μ_2. If p_ϕ/p_ψ lies outside this range, then $\dot{\phi}$ has just one sign as θ goes between θ_1 and θ_2 in time. Finally it is possible that $p_\phi/p_\psi = \mu_2$, in which case $\dot{\phi}$ vanishes at $\theta = \theta_2$. These three

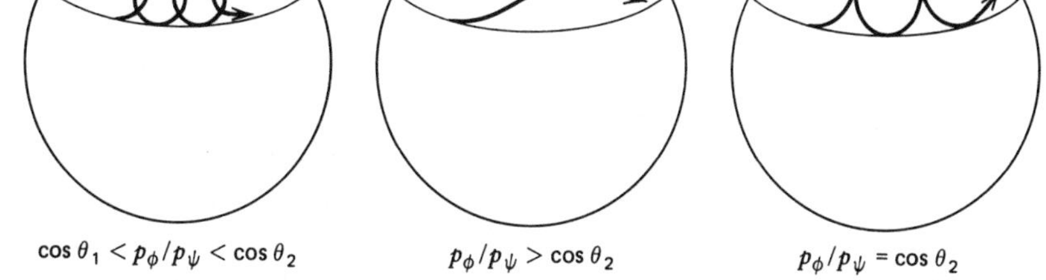

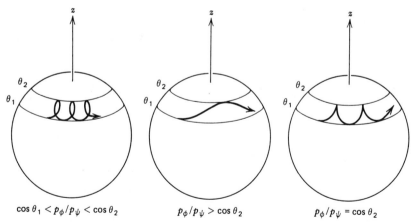

FIGURE 4.9. Possible motions for the symmetry axis of a heavy symmetrical top with lower end fixed.

cases are sketched in Fig. 4.9, where the motion of the body symmetry axis on the unit sphere is drawn. The periodic motion of the top axis between the values θ_1 and θ_2 is called *nutation*. The motion of the top axis around the z-axis of the stationary system S is called *precession*.

The complete motion of the top is given by the time evolution of the coordinates ϕ, θ, ψ. These motions consist of precession, nutation, and rotation of the top around its symmetry axis. If the frequencies of these three periodic motions are commensurable, then the top will return to its original position; otherwise it will not.

PROBLEMS

4.1. Prove the parallel axis theorem: Let I denote the moment of inertia relative to some axis passing through the center of mass. Let r denote the distance to a second axis that is parallel to the first. Then the moment of inertia of the body relative to the second axis is just $I + mr^2$, where m is the total mass of the body. Obtain the tensor form of this theorem.

4.2. Calculate the principal moments of inertia for a cone with base radius r, height h, and mass m. Take the origin at the vertex of the cone. Now what are they if the origin is taken to be the center of mass of the cone?

4.3. Analyze the Poinsot construction in the case where $I_2 = I_3 < I_1$. Find the curves called the polhode and the herpolhode. What are the periods of these two types of motion?

4.4. A light rod of length b has two weights of mass m at each end and is constrained to rotate at a constant angular velocity ω about a vertical axis, maintaining a constant angle θ with the vertical. Calculate $\mathbf{I}, k, \boldsymbol{\omega}, \mathbf{L}, K, \boldsymbol{\Omega}$ and show that $\mathbf{L} = K\boldsymbol{\Omega}$. Calculate the torque of constraint on the system and discuss physically how this torque is provided. Describe the motion that would result if the constraint were suddenly removed.

4.5. Find the angular velocity $\boldsymbol{\omega}$, the kinetic energy T, and the angular momentum $\mathbf{L}$ of a uniform solid right circular cone of mass M, height h, and radius R at the base, rolling without slipping on its side on a horizontal plane. Find the total torque $\boldsymbol{\tau}$ on the cone and analyze how the gravitational and constraint forces provide this torque. (Assume the cone returns to its initial position with period P.) Find the minimum value of the coefficient of friction for which the cone will not slip. Find the minimum value of P for which the cone will not tilt. Construct a trajectory in $SO(3)$ similar to Fig. 4.1 for the case that $R\sqrt{3} = h$. What is the limiting trajectory as the apex angle of the cone $\rightarrow \pi$?

4.6. A small ball of radius r and mass m starts at the top of large ball of radius R and mass M resting on a table. At what angle from the vertical will the small ball leave the large ball if the small ball starts at rest. Assume that all balls move without slipping.

4.7. A homogeneous ellipsoid of semiaxes a, b, c is to be mounted on a horizontal axis a distance L from the center-of-mass ($a > b > c \gg L$) in such a way that it may oscillate as a compound pendulum with the smallest possible periodic time. What position of the axis, relative to the axes of the ellipsoid, should be selected?

CHAPTER FIVE

THE DYNAMICS OF SMALL OSCILLATIONS

There can not be a language (geometry) more universal and more simple, more free from errors and from obscurities, that is to say more worthy to express the invariable relations of natural things.

FOURIER

A class of frequently studied problems consists of small departures of a system from equilibrium. Some systems can only be studied in the neighborhood of equilibrium because their dynamics is too complicated to be successfully analyzed in general. In many cases, the linear analysis forms a starting point for understanding what may be an intrinsically nonlinear problem. In this chapter we deal with the problem of small oscillations around equilibrium.

5.1. EQUILIBRIUM

In Section 2.5 we showed that by defining new variables $\mathbf{y} = \dot{\mathbf{x}}$ and $\boldsymbol{\xi} = (\mathbf{y}, \mathbf{x})$, Newton's second law can be written in the general form [cf. (2.51)]

$$\frac{d\boldsymbol{\xi}}{dt} = \mathbf{v}(\boldsymbol{\xi}, t), \tag{5.1}$$

where the phase space velocity vector $\mathbf{v}$ is given by $\mathbf{v}(\boldsymbol{\xi}, t) = (\mathbf{f}(t, \mathbf{x}, \mathbf{y}), \mathbf{y})$ with $\mathbf{f}$ the force per unit mass. All mechanical systems may have their dynamical equations put into this general form.

For the Lagrangian formulation we let the variables be $\boldsymbol{\xi} = (\dot{\mathbf{q}}, \mathbf{q})$. With n degrees of freedom, the first n components of $\mathbf{v}(\boldsymbol{\xi}, t)$ are obtained from the Euler–Lagrange equations by writing them in the form $\ddot{q}^i = f^i(\mathbf{q}, \dot{\mathbf{q}}, t)$. The remaining n components of $\mathbf{v}$ would then be equal to $\dot{q}^i$ for $i = 1, \dots, n$.*

We begin our study of (5.1) by considering equilibrium points for dynamical systems. An *equilibrium point* $\boldsymbol{\xi}_0$ for the system (5.1) is one for which $\mathbf{v}(\boldsymbol{\xi}_0, t) = 0$. Using the Euler–Lagrange equations for a system in which $T = (1/2)m_{ij}(\mathbf{q})\dot{q}^i\dot{q}^j \geqslant 0$ and $U = U(\mathbf{q})$, we then obtain for a Lagrangian system the "velocity" $v^i = (m^{-1})^{il}[\dot{q}^k\dot{q}^j((1/2)m_{jk,l} - m_{lj,k}) - (\partial U/\partial q^l)]$, $i = 1, \dots, n$, and $v^{i+n} = \dot{q}^i$, $i = 1, \dots, n$. We have made the reasonable assumption for $m_{ij}(\mathbf{q})$ that it is nonsingular. In order that $\mathbf{v}$ equal zero at $\boldsymbol{\xi}_0$, we must have that $\dot{\mathbf{q}}_0 = 0$ and $(\partial U/\partial \mathbf{q}_0) = 0$. This agrees with our former notions of an equilibrium point as discussed in Section 2.4.

Equilibrium points corresponding to local minima in the potential are Liapunov stable, as we now show. Suppose that an equilibrium point $\mathbf{q}_0$ corresponds to a minimum in the potential $U(\mathbf{q})$ and let us choose the zero of energy so that $U(\mathbf{q}_0) = 0$. Then to ensure that the phase trajectories always stay close to the point $(0, \mathbf{q}_0)$ in the $(\dot{\mathbf{q}}, \mathbf{q})$ phase space, we need only choose initial conditions such that the total energy is small enough (i.e., $E = T + U = \varepsilon$, where ε is small). Since energy is conserved the subsequent phase points $(\dot{\mathbf{q}}(t), \mathbf{q}(t))$ that evolve from such initial conditions can never get very far from $(0, \mathbf{q}_0)$. Thus the phase point $(0, \mathbf{q}_0)$ is a Liapunov stable.

To linearize the system of (5.1) we expand the vector function $\mathbf{v}$ in a Taylor's series expansion about an equilibrium position $\boldsymbol{\xi}_0$, remembering that $\mathbf{v}(\boldsymbol{\xi}_0, t) = 0$. Then

$$\mathbf{v}(\boldsymbol{\xi}, t) = \left.\frac{\partial \mathbf{v}}{\partial \xi^i}\right|_{\boldsymbol{\xi}_0} (\xi^i - \xi_0^i) + O(|\boldsymbol{\xi} - \boldsymbol{\xi}_0|^2). \tag{5.2}$$

In index form this becomes

$$v^i(\boldsymbol{\xi}, t) = F^i_{\ j}(t) X^j + O(|\mathbf{X}|^2), \tag{5.3}$$

where we have defined $F^i_{\ j}(t) \equiv (\partial v^i/\partial \xi^j)|_{\boldsymbol{\xi}_0}$ and $\mathbf{X} \equiv (\boldsymbol{\xi} - \boldsymbol{\xi}_0)$. Then the

*The Hamiltonian formulation of mechanics, to be introduced in Chapter 7, has its differential equations of motion already in the form of (5.1).

linearized version of (5.1) becomes

$$\frac{dX^i}{dt} = F^i{}_j(t)X^j. \tag{5.4}$$

For the moment we distinguish between a solution, $\boldsymbol{\xi}(t)$, to (5.1) and $\mathbf{X}(t)$, a solution to (5.4). For any arbitrary interval of time $[0, T]$, one can always choose the initial value $\boldsymbol{\xi}(0)$ close enough to $\boldsymbol{\xi}_0$ so that the difference between $(\boldsymbol{\xi}(t) - \boldsymbol{\xi}_0)$ and $\mathbf{X}(t)$ will be less than some preset amount for all t satisfying $0 \leqslant t \leqslant T$. Thus we can always choose initial conditions sufficiently close to the equilibrium position that the linearized solution will be an arbitrarily good approximation to the actual motion in the given interval. However, for arbitrarily given initial conditions and time interval, the solution $\mathbf{X}(t)$ of the linearized problem may, or may not, give a good approximation to the actual motion $\boldsymbol{\xi}(t)$.

5.2. SMALL OSCILLATIONS

In the case of mechanical systems where $F^i{}_j$ is determined by the kinetic and potential energies, we need only keep T and U up to quadratic terms, since it is the derivative of these terms that give the linear terms in the equations of motion. Let $\mathbf{q}_0$ denote an equilibrium point. Since $T = (1/2)m_{ij}(\mathbf{q})\dot{q}^i\dot{q}^j$ is already quadratic in the generalized velocities, it suffices to take m_{ij} evaluated at $\mathbf{q} = \mathbf{q}_0$, that is, at the equilibrium point. Since $\partial U/\partial \mathbf{q}_0 = 0$ we have $U(\mathbf{q}) \simeq (1/2)k_{ij}q^iq^j$, where $k_{ij} = (\partial^2 U/\partial q^i \partial q^j)|_{\mathbf{q}_0}$. Thus for the linearized problem we have the Lagrangian

$$L = \tfrac{1}{2}m_{ij}\dot{q}^i\dot{q}^j - \tfrac{1}{2}k_{ij}q^iq^j. \tag{5.5}$$

Both the matrices m_{ij} and k_{ij} are symmetric. Furthermore, we know that m_{ij} is positive definite since $T \geqslant 0$. The Euler–Lagrange equations obtained from (5.5) are

$$m_{ij}\ddot{q}^j + k_{ij}q^j = 0. \tag{5.6}$$

Motivated by the one-dimensional form of (5.6), we seek an oscillatory solution in the form $\mathbf{q} = \operatorname{Re} \mathbf{a} e^{i\omega t}$, or equivalently, $\mathbf{q} = \mathbf{a}e^{i\omega t} + \mathbf{a}^* e^{-i\omega t}$. Substitution into (5.6) gives

$$\left(k_{ij} - \omega^2 m_{ij}\right)a^j = 0. \tag{5.7}$$

For a nontrivial **a**, the matrix operator must be singular, that is,

$$\det\left(k_{ij} - \omega^2 m_{ij}\right) = 0. \tag{5.8}$$

The characteristic roots of this equation $\omega_1^2, \ldots, \omega_n^2$ are called the *eigenfrequencies* of the system. For each choice of $\omega^2 = \omega_\alpha^2$ we then obtain an eigenvector $\mathbf{a}_\alpha$ from (5.7). For a root of multiplicity l we choose l linearly independent, orthogonal eigenvectors (with inner product defined below in (5.10)) following the same procedure outlined in Section 4.2 for degenerate eigenvalues of the inertia operator. A general solution of (5.6) is then a linear combination of these eigensolutions:

$$\mathbf{q}(t) = \mathbf{a}_\alpha\left(A^\alpha e^{i\omega_\alpha t} + A^{*\alpha} e^{-i\omega_\alpha t}\right) \qquad (\text{sum on } \alpha), \tag{5.9}$$

where A^α denotes the arbitrary (complex) amplitude for the αth eigensolution. The A^α are determined by the initial conditions.

Often it is desirable to change to a set of coordinates, called *normal coordinates*, each of which oscillates at a single characteristic frequency rather than a combination as given by (5.9). In (5.9) we see that $\mathbf{q}(t)$ is indeed a sum of terms, each of which oscillates at a single characteristic frequency. These individual terms are almost the desired normal coordinates. But to be able to invert (5.9) and solve for the normal coordinates in terms of $\mathbf{q}(t)$ we normalize the eigenvectors $\mathbf{a}_\alpha$ according to the equation

$$a_\alpha^i m_{ij} a_\beta^j = \delta_{\alpha\beta}, \qquad \alpha, \beta = 1, \ldots, n. \tag{5.10}$$

Since m_{ij} is a positive definite matrix, it can be used in this way as a metric in the space of vectors $\mathbf{a}_\alpha$. Then in terms of the new coordinates $Q^i(t)$, we write (5.9) in the form

$$q^i(t) = a_\alpha^i Q^\alpha(t). \tag{5.11}$$

To solve for the $Q^\alpha(t)$ multiply both sides of (5.11) by $a_\beta^j m_{ji}$ and use (5.10) to find

$$Q^\beta(t) = a_\beta^j m_{ji} q^i(t). \tag{5.12}$$

It is straightforward to check, using (5.6), (5.7), and the fact that k_{ij} and m_{ij} are symmetric, that

$$\ddot{Q}^\alpha + \omega_\alpha^2 Q^\alpha = 0 \qquad (\text{no sum on } \alpha). \tag{5.13}$$

Transforming the Lagrangian (5.5) by transforming to normal coordinates results in

$$L = \tfrac{1}{2}\delta_{\alpha\beta}\dot{Q}^\alpha\dot{Q}^\beta - \tfrac{1}{2}\omega_{\alpha\beta}Q^\alpha Q^\beta, \tag{5.14}$$

where $\omega_{\alpha\beta} = \omega_\alpha^2\,\delta_{\alpha\beta}$ (no sum on α). The matrix $\omega_{\alpha\beta}$ is a diagonal matrix with the eigenvalues ω_α^2 along the diagonal.

Solving (5.13) we see that there are three possibilites for the motion of the Q^α coordinate: (1) $\omega_\alpha^2 > 0$, then $Q^\alpha(t)$ is a stable oscillation of the form $Q^\alpha(t) = C_1^\alpha \cos\omega_\alpha t + C_2^\alpha \sin\omega_\alpha t$. (2) $\omega_\alpha^2 = 0$, then $Q^\alpha(t)$ is linearly unstable of the form $Q^\alpha(t) = C_1^\alpha + C_2^\alpha t$. (3) $\omega_\alpha^2 < 0$, then $Q^\alpha(t)$ is exponentially unstable of the form $Q^\alpha(t) = C_1^\alpha \cosh\omega'_\alpha t + C_2^\alpha \sinh\omega'_\alpha t$, where $(\omega'_\alpha)^2 = -\omega_\alpha^2$.

We have achieved the decomposition of an arbitrary small oscillation into a sum of characteristic oscillations. The sum of characteristic oscillations is generally not periodic. Only in those special cases where the ratio of each pair of characteristic frequencies is a rational number will the system be periodic.

To illustrate the foregoing analysis we consider the system depicted in Fig. 5.1. The first task is to find the potential and kinetic energies, respectively.

$$U = -m_1 g l_1 \cos\theta_1 - m_2 g l_2 \cos\theta_2$$

$$+ \frac{k}{2}\left\{\left[(l_2\sin\theta_1 - d - l_2\sin\theta_2)^2 + (l_2\cos\theta_1 - l_2\cos\theta_2)^2\right]^{1/2} - d\right\}^2$$

$$T = \frac{1}{2}m_1 l_1^2\dot{\theta}_1^2 + \frac{1}{2}m_2 l_2^2\dot{\theta}_2^2.$$

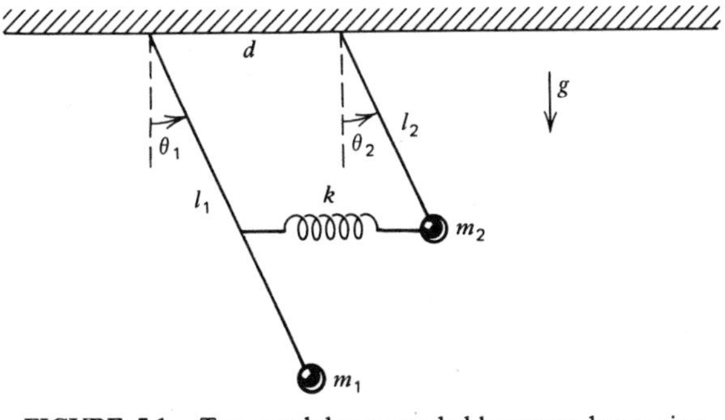

FIGURE 5.1. Two pendulums coupled by a massless spring.

Next compute the appropriate derivatives of U to find the matrix

$$k_{ij} = \begin{bmatrix} m_1 g l_1 + k l_2^2 & -k l_2^2 \\ -k l_2^2 & m_2 g l_2 + k l_2^2 \end{bmatrix}.$$

For the mass matrix in the kinetic energy we obtain readily

$$m_{ij} = \begin{bmatrix} m_1 l_1^2 & 0 \\ 0 & m_2 l_2^2 \end{bmatrix}.$$

To calculate the characteristic frequencies we form the determinant.

$$\det\left(k_{ij} - \omega^2 m_{ij}\right)$$
$$= \begin{bmatrix} m_1 g l_1 + k l_2^2 - \omega^2 m_1 l_1^2 & -k l_2^2 \\ -k l_2^2 & m_2 g l_2 + k l_2^2 - \omega^2 m_2 l_2^2 \end{bmatrix}.$$

This gives the characteristic equation

$$\left(m_1 g l_1 + k l_2^2 - \omega^2 m_1 l_1^2\right)\left(m_2 g l_2 + k l_2^2 - \omega^2 m_2 l_2^2\right) - k^2 l_2^4 = 0.$$

There are several natural frequencies associated with individual pieces of this system, and it is helpful to express this characteristic equation in terms of these equations: $\alpha_1^2 = g/l_1$; $\alpha_2^2 = g/l_2$; $\beta_1^2 = k/m_1$; $\beta_2^2 = k/m_2$ and we let $r = l_2/l_1$. The characteristic equation becomes

$$\left(\alpha_1^2 + r^2\beta_1^2 - \omega^2\right)\left(\alpha_2^2 + \beta_2^2 - \omega^2\right) - r^2\beta_1^2\beta_2^2 = 0.$$

This is a quadratic equation for ω^2. In the limit that $k \to 0$, β_1 and $\beta_2 \to 0$, and we obtain the frequencies for two uncoupled pendulums. As the spring connecting the two pendulums becomes stiffer, that is as $k \to \infty$, then one of the two characteristic frequencies goes to infinity and the other goes to the limiting result:

$$\omega_\infty^2 = \frac{g(m_1 l_1 + m_2 l_2)}{m_1 l_1^2 + m_2 l_2^2}.$$

See problem 5.2. The general behavior for the two roots is plotted in Fig. 5.2.

As an illustration of the normal modes let us consider the special case where $l_1 = l_2 = l$ and $m_1 = m_2 = m$. The characteristic equation reduces to

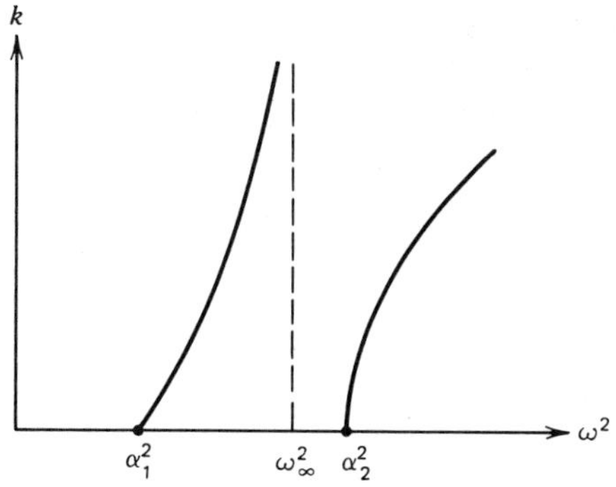

FIGURE 5.2. Plot of frequency ω for pendulums of unequal length coupled by spring with constant k.

$(\alpha^2 + \beta^2 - \omega^2)^2 - \beta^4 = 0$, or $\alpha^2 + \beta^2 - \omega^2 = \pm\beta^2$. We obtain for the two characteristic frequencies $\omega_1^2 = \alpha^2 = g/l$ and $\omega_2^2 = \alpha^2 + 2\beta^2 = (g/l) + 2(k/m)$. The next task is to find the eigenvectors of (5.7).

$$\begin{bmatrix} \alpha^2 + \beta^2 - \omega^2 & -\beta^2 \\ -\beta^2 & \alpha^2 + \beta^2 - \omega^2 \end{bmatrix} \begin{bmatrix} a_1 \\ a_2 \end{bmatrix} = 0$$

In the case that $\omega^2 = \omega_1^2$ we can take as an eigenvector solution (unnormalized) $\mathbf{a}_1 = (1, 1)$. For the other choice of frequency, $\omega^2 = \omega_2^2$, we have as eigenvector $\mathbf{a}_2 = (1, -1)$. The first eigenvector $\mathbf{a}_1 = (1, 1)$ leads through (5.12) to the normal mode $Q^1(t) \propto \theta_1(t) + \theta_2(t)$. The second eigenvector $\mathbf{a}_2 = (1, -1)$ similarly leads to $Q^2(t) \propto \theta_1(t) - \theta_2(t)$. Thus we see physically that the first normal mode has the pendulums oscillating in phase, with equal amplitudes, and at the frequency for a free, unconnected pendulum. The second normal mode has the pendulums oscillating 180° out of phase, with equal amplitudes, and at a frequency determined by the spring constant and the length of the pendulums. See Fig. 5.3.

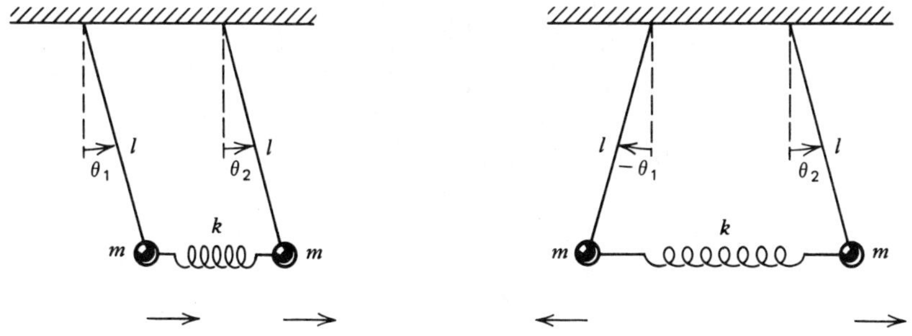

FIGURE 5.3. Normal modes for two coupled pendulums.

Continuing with this example, we normalize the eigenvectors with respect to the mass matrix as in (5.10). As shown in Problem 5.1, distinct eigenfrequencies imply that the eigenvectors are orthogonal. In the present example $\omega_1 \neq \omega_2$ and hence $\mathbf{a}_1$ and $\mathbf{a}_2$ are orthogonal. It remains only to multiply by an appropriate constant to normalize them: $\mathbf{a}_1 = (ml^2)^{-1/2}(1, 1)$ and $\mathbf{a}_2 = (ml^2)^{-1/2}(1, -1)$. From (5.12) we obtain $Q^1 = (ml^2)^{1/2}(\theta_1 + \theta_2)$ and $Q^2 = (ml^2)^{1/2}(\theta_1 - \theta_2)$.

It is also instructive in this system to consider the phenomenon of beats. The general solutions for $\theta_1(t)$ and $\theta_2(t)$ are from (5.9)

$$\theta_1 = A^1 e^{i\omega_1 t} + A^{1*} e^{-i\omega_1 t} + A^2 e^{i\omega_2 t} + A^{2*} e^{-i\omega_2 t},$$

$$\theta_2 = A^1 e^{i\omega_1 t} + A^{1*} e^{-i\omega_1 t} - A^2 e^{i\omega_2 t} - A^{2*} e^{-i\omega_2 t}.$$

Make the following choice of initial conditions: $\theta_1(0) = \theta_2(0) = \dot{\theta}_1(0) = 0$ and $\dot{\theta}_2(0) = \Theta$. Using the foregoing equations for $\theta_1(t)$ and $\theta_2(t)$, we solve for the complex amplitudes. One finds $A^1 = -i\Theta/4\omega_1$ and $A^2 = -\omega_1 A^1/\omega_2$. The previous expressions for $\theta_1(t)$ and $\theta_2(t)$ simplify to

$$\theta_1 = \frac{\Theta}{2}\left[\frac{1}{\omega_1}\sin\omega_1 t - \frac{1}{\omega_2}\sin\omega_2 t\right],$$

$$\theta_2 = \frac{\Theta}{2}\left[\frac{1}{\omega_1}\sin\omega_1 t + \frac{1}{\omega_2}\sin\omega_2 t\right].$$

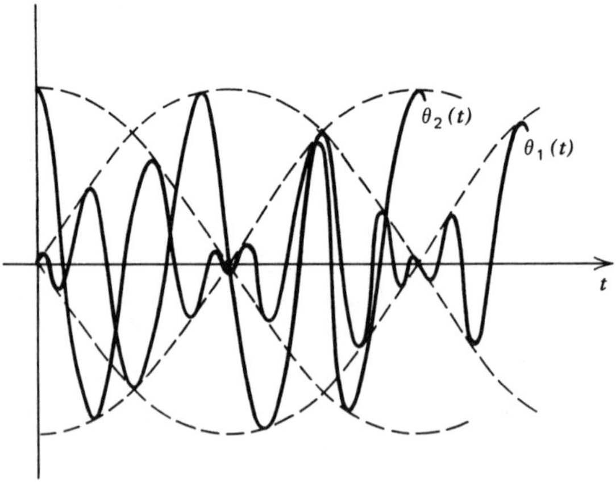

FIGURE 5.4. The oscillations for two weakly coupled pendulums.

For brevity we let $\varepsilon = (k/2m\omega_1^2)$ and assume $\varepsilon \ll 1$. Then $\omega_2 \simeq \omega_1(1 + 2\varepsilon)$ and the angles $\theta_1(t)$ and $\theta_2(t)$ are approximately

$$\theta_1 \simeq -\frac{\Theta}{\omega_2}\sin \varepsilon\omega_1 t \cos[(1+\varepsilon)\omega_1 t]$$

$$\theta_2 \simeq \frac{\Theta}{\omega_2}\cos \varepsilon\omega_1 t \sin[(1+\varepsilon)\omega_1 t].$$

In these results θ_1 and θ_2 oscillate with frequency $\omega_1(1+\varepsilon) \simeq \omega_1$ with a slowly varying amplitude. After a time $T \simeq \pi/2\varepsilon\omega_1$ only the first pendulum will be oscillating. The two pendulums alternate in this fashion with a period on the order of $\pi/\varepsilon\omega_1$. See Fig. 5.4.

5.3. PARAMETRIC DEPENDENCE

It is interesting to consider how the characteristic frequencies for a system depend on the parameters such as the masses, spring constants, and so on. Figure 5.2 is a sketch of just this sort of change in frequency as the system parameter k changes. The relative behavior of the frequencies depicted in Fig. 5.2 is an example of a very general relationship among the frequencies for systems with changing "spring constants." This general result, obtained later, can be useful for understanding the behavior of mechanical systems with changing parameters. In this section we consider such questions and always assume the use of normal coordinates.

In the normal coordinates the kinetic and potential energies have the forms $T = \frac{1}{2}\delta_{ij}\dot{Q}^i\dot{Q}^j > 0$ and $U = \frac{1}{2}\omega_{ij}Q^iQ^j > 0$. In a spirit similar to our considerations of the inertia ellipsoid in the previous chapter, we consider ellipsoids determined by the potential U. A $U =$ constant surface is an ellipsoid in the n-dimensional configuration space of the coordinates Q^i. If the constant equals $\frac{1}{2}$, then we have for the $U(Q) = \frac{1}{2}$ surface the ellipsoid given by the equation

$$\frac{(Q^1)^2}{1/\omega_1^2} + \cdots + \frac{(Q^n)^2}{1/\omega_n^2} = 1. \tag{5.15}$$

This ellipsoid is called the *frequency ellipsoid* and has its semiaxes given by $a_1 = 1/\omega_1, \ldots, a_n = 1/\omega_n$. (A constant other than $\frac{1}{2}$ for the definition of the frequency ellipsoid would give semiaxes related to these by a constant scale factor.) We assume that the characteristic frequencies have been ordered as $\omega_1 \leqslant \omega_2 \leqslant \cdots \leqslant \omega_n$; consequently, we have that the semiaxes are ordered according to $a_1 \geqslant a_2 \geqslant \cdots \geqslant a_n$.

If $U(Q) = (U_0/2) > 1/2$, then

$$\frac{(Q^1)^2}{U_0 a_1'^2} + \frac{(Q^2)^2}{U_0 a_2'^2} + \cdots + \frac{(Q^n)^2}{U_0 a_n'^2} = 1. \tag{5.16}$$

Thus $U_0 a_i'^2 = a_i^2$ and since $U_0 > 1$ we have that $a_i' < a_i$ or $\omega_i' > \omega_i$ for all i. We will call a second system more *rigid* than the first when, with the kinetic energy held constant, the second system has a larger potential energy for the same value of the coordinates. As we have seen earlier, the frequencies are larger for the more rigid system. If we denote the ordered sequence of characteristic frequencies for the more rigid system as $\omega_1' \leqslant \cdots \leqslant \omega_n'$, then $\omega_i \leqslant \omega_i'$ and $a_i \geqslant a_i'$ for all $i = 1, \ldots, n$.

For the example depicted in Fig. 5.1, we studied the behavior of the frequencies as $k \to \infty$ with this behavior sketched in Fig. 5.2. This sketch shows that both frequencies have $(d\omega_i/dk) > 0$. That is to say, as the spring stiffens, or the system becomes more rigid, the frequencies increase. In the limit that $k \to \infty$, the system is two pendulums connected by a rigid rod, and in effect we have removed one degree of freedom. We have placed a constraint on the system. To see the effect of constraints on a system we consider the following theorem, which relates the frequencies of the constrained system to the former set.

Let $\omega_1 \leqslant \cdots \leqslant \omega_n$ denote the characteristic frequencies for an unconstrained system. Let $\omega_1' \leqslant \cdots \leqslant \omega_{n-1}'$ denote the set of frequencies for a constrained system, where the constraint has removed one degree of freedom. We will show that these two sets of characteristic frequencies are related in the following way: $\omega_1 \leqslant \omega_1' \leqslant \omega_2 \leqslant \cdots \leqslant \omega_{n-1}' \leqslant \omega_n$. That is to say, the frequencies of the constrained system lie in between the frequencies of the unconstrained system. Our proof follows Arnold (1978).

Let the frequency ellipsoid for the unconstrained system be denoted by E, and the frequency ellipsoid for the constrained system be denoted by E'. E is contained in $\mathbb{R}^n$ and E' in $\mathbb{R}^{n-1}$. Let $|\mathbf{A}|$ denote the length of a vector $\mathbf{A}$ in a Euclidean space with the usual Euclidean metric. Let k be an integer such that $k \leqslant (n-1)$ and consider the intersection of $\mathbb{R}^k$ with E'. The space $\mathbb{R}^k$ is constrained to contain the origin of E'. Then there is a set of vectors $\mathbf{Q}$ from the origin, terminating on the surface E', which lie in this intersection. For all possible choices of $\mathbf{Q}$ in the intersection $\mathbb{R}^k \cap E'$, we let b_k' denote the minimum length of such vectors. We write $b_k' = \min_{\mathbf{Q} \in \mathbb{R}^k \cap E'} \{|\mathbf{Q}|\}$. This minimum is found by looking at all possible vectors in a given intersection $\mathbb{R}^k \cap E'$. With k held fixed, let the possible choices of $\mathbb{R}^k$ change, subject to the previously mentioned restrictions, and find the maximum value of these b'_k. Consideration of the examples of Fig. 5.5 is helpful in seeing that this maximum is a'_k. For $k = 1$ this process will select the largest semiaxis of E', a'_1; for $k = 2$ we obtain the next largest, a'_2, etc. We write

$$a'_k = \max_{\{\mathbb{R}^k\}} b'_k = \max_{\{\mathbb{R}^k\}} \left[\min_{\mathbf{Q} \in \mathbb{R}^k \cap E'} \{|\mathbf{Q}|\} \right]. \tag{5.17}$$

The k-dimensional subspace $\mathbb{R}^k$ is some subspace of $\mathbb{R}^{n-1}$ containing E'. If we add another dimension to a given $\mathbb{R}^k$, giving a particular $\mathbb{R}^{k+1}$, then we know that the minimum of $|\mathbf{Q}|$ over the new intersection $\mathbb{R}^{k+1} \cap E'$ must be less than or equal to b'_k because the set of possibilities from which the minimum is selected is larger. In other words, $\mathbb{R}^{k+1} \cap \mathbb{R}^{n-1}$ contains $\mathbb{R}^k$, and possibly much more. Thus

$$b'_k \geqslant c'_k = \min_{\mathbf{Q} \in \mathbb{R}^{k+1} \cap E'} \{|\mathbf{Q}|\}$$

The space $\mathbb{R}^{k+1}$ is a subspace of $\mathbb{R}^n$ and if we now maximize over the various choices of $(k + 1)$-dimensional spaces we find

$$a'_k \geqslant \max_{\{\mathbb{R}^{k+1}\}} c'_k = \max_{\{\mathbb{R}^{k+1}\}} \left[\min_{\mathbf{Q} \in \mathbb{R}^{k+1} \cap E'} \{|\mathbf{Q}|\} \right]$$

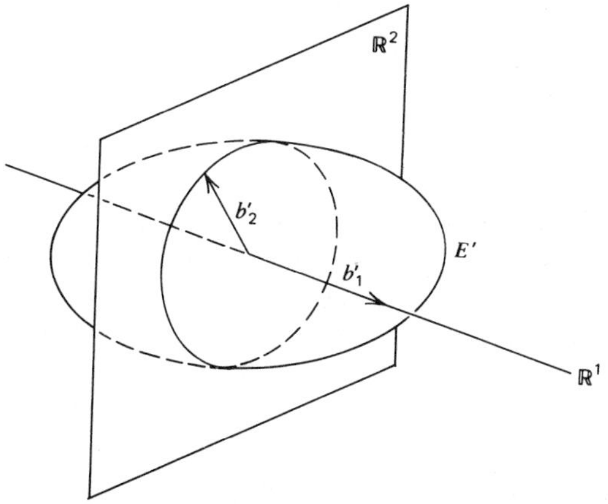

FIGURE 5.5. The intersection of the frequency ellipsoid for the constrained problem E' with two examples of $\mathbb{R}^k$. The minimum lengths for these two examples are also indicated.

If in the expression for c'_k we enlarge E' to E, we will get a smaller minimum. that is,

$$c'_k = \min_{\mathbf{Q} \in \mathbb{R}^{k+1} \cap E'} \{|\mathbf{Q}|\} \geqslant \min_{\mathbf{Q} \in \mathbb{R}^{k+1} \cap E} \{|\mathbf{Q}|\}.$$

Finding the maximum of the foregoing inequality over the possible choices of $\mathbb{R}^{k+1}$ makes the rightmost expression equal to a_{k+1}, according to (5.17). Thus we have the result

$$a'_k \geqslant \max_{\{\mathbb{R}^{k+1}\}} c'_k \geqslant a_{k+1}$$

and hence $a'_k \geqslant a_{k+1}$. As we stiffen the system we already know from the discussion immediately following (5.16) that $a_k \geqslant a'_k$ and thus combining these results:

$$a_k \geqslant a'_k \geqslant a_{k+1} \quad \text{or} \quad \omega_k \leqslant \omega'_k \leqslant \omega_{k+1}, \tag{5.18}$$

which is the desired result.

As the spring stiffens to a rigid rod in the simple system of Fig. 5.1, we see that $\omega_1 \leqslant \omega_\infty \leqslant \omega_2$ as shown in Fig. 5.2.

5.4. PARAMETRIC RESONANCE

Not only is it possible for the parameters of a system to change monotonically, as we have considered previously, but the parameters of a system may also change with time in a periodic fashion. Many interesting resonance effects are associated with a periodic change in parameters. Perhaps the most familiar example of this effect is the swing. Most people have experienced the increased amplitude of a swing made possible by raising and lowering their center of gravity. The key to doing this successfully is that the "raising and lowering" must take place at the right time, that is, in resonance with the natural frequency of the swing. The addition of small amounts of energy to an oscillatory system, at times resonant with the natural period, can make the oscillations grow in amplitude and the system become unstable. We consider these small amounts of energy to enter the oscillatory systems via small changes in system parameters and refer to this process as *parametric resonance*. Thus we seek to determine the values for small changes in system parameters that make the system unstable. Pumping up a swing may be modeled as changing the effective length of a simple pendulum. See Fig. 5.6. Other systems for which parametric resonance is

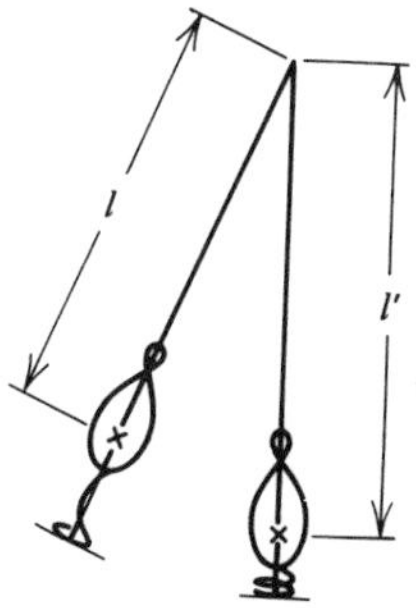

FIGURE 5.6. Change in the length of the effective pendulum for a swing, $l' > l$.

important are particle motion in an Alternating Gradient Synchrotron and the motion of a particle in the gravitational field of a body that is itself undergoing periodic motion.

We formulate the mechanical problems with parametric resonance in the form of (5.1), where the time dependence of the phase space velocity enters through the periodic change of the system parameters. We write

$$\frac{d\boldsymbol{\xi}}{dt} = \mathbf{v}(\boldsymbol{\xi}, t); \qquad \mathbf{v}(\boldsymbol{\xi}, t + T) = \mathbf{v}(\boldsymbol{\xi}, t), \tag{5.19}$$

where the general phase flow vector $\mathbf{v}(\boldsymbol{\xi}, t)$ is periodic with period T. For our example of the simple pendulum with changing length, we have $\ddot{\theta} + g\theta/l(t) = 0$. Or in the form corresponding to (5.19), $\boldsymbol{\xi} = (\psi, \theta)$ and $\dot{\boldsymbol{\xi}} = \mathbf{v} = (\dot{\psi}, \dot{\theta}) = (-g\theta/l(t), \psi)$, with an appropriate periodicity for $l(t)$.

In the case where the flow vector $\mathbf{v}$ of (5.19) depends on time, we have that the phase flow no longer forms a one-parameter group. Thus no longer is it true that $g^{t+s} = g^t \circ g^s$. This difference is sketched in Fig. 5.7. However, as long as $\mathbf{v}$ is periodic with period T we do find that $q^{T+s} = g^s \circ g^T$ and so we do have the mappings at a period T forming a group. Thus $g^{nT} = g^T \circ \cdots \circ g^T = (g^T)^n$, where n is an integer. For brevity we let the mapping $g^T = A$ and call A the *period map*.

$$\boldsymbol{\xi}(T) = A(\boldsymbol{\xi}_0). \tag{5.20}$$

The period map A plays the central role in our analysis of parametric resonance.

We caution the reader not to confuse the period of $\mathbf{v}(\boldsymbol{\xi}, t)$ with possible periods of the motion. In our example with the swing, the changing length $l(t)$ has a period associated with this change in length. This period we have denoted as T. Also a pendulum has a period of the motion $\tau = 2\pi(l/g)^{1/2}$. In principle the periods T and τ are not, and need not be, related. The

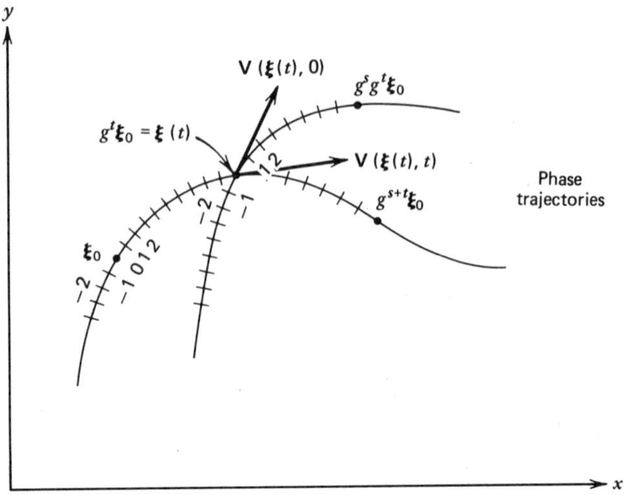

FIGURE 5.7. Sketch showing the difference in the paths $g^{s+t}\xi_0$ and $g^s g^t \xi_0$ that results because $\mathbf{v}$ of Eq. (5.19) is time dependent, in particular $\mathbf{v}(\xi(t), t) \neq \mathbf{v}(\xi(t), 0)$ as sketched. Some sample time marks are indicated on the trajectories.

period of the motion and the period T are related when A has a fixed point, as we now show. If ξ_0 is a fixed point of A, then $\xi(T+s) = g^s A \xi_0 = g^s \xi_0 = \xi(s)$. We see that the initial point ξ_0 is a fixed point of the mapping A iff the solution $\xi(t)$, with initial point ξ_0, is periodic with period T.

We note further without proof: (1) If (5.19) is a Hamiltonian system then A preserves area (cf. Section 7.4 and Liouville's theorem). (2) If $\mathbf{v}(\xi, t)$ is linear, that is, $\mathbf{v}(\xi, t) = D(t)\xi$, then A is also linear. (3) If the periodic solution $\mathbf{x}(t)$ is Liapunov stable, then the fixed point $\mathbf{x}_0$ of the mapping A is also Liapunov stable and vice versa. We will often refer to the mapping as being stable and will mean thereby that the fixed point of the mapping is stable.

For a first example we look at a period map A with arbitrary period, that is, one for which $\mathbf{v}(\xi, t) = \mathbf{v}(\xi)$. Our example is the simple harmonic oscillator and we let $\xi = (y, x)$ and $\dot{\xi} = (\dot{y}, \dot{x}) = (-x, y)$. The period of the motion is 2π and the period for A is arbitrary. In this case $\mathbf{v}(\xi, t)$ is linear and we may write

$$\dot{\xi} = D\xi, \tag{5.21}$$

where

$$D = \begin{bmatrix} 0 & -1 \\ 1 & 0 \end{bmatrix}, \quad D^2 = \begin{bmatrix} -1 & 0 \\ 0 & -1 \end{bmatrix} = -I. \tag{5.22}$$

Equations (5.21) have the solution $\boldsymbol{\xi}(t) = e^{Dt}\boldsymbol{\xi}_0$, where

$$e^{Dt} = I + tD + \frac{t^2}{2!}D^2 + \frac{t^3}{3!}D^3 + \cdots . \tag{5.23}$$

This exponential is the period map, $A = e^{Dt}$. Using (5.22) we can sum (5.23) to find

$$A = e^{Dt} = I + Dt - \frac{t^2}{2!}I - \frac{t^3}{3!}D + \cdots$$

$$= I\cos t + D\sin t = \begin{bmatrix} \cos t & \sin t \\ -\sin t & \cos t \end{bmatrix}. \tag{5.24}$$

The period map A is a rotation operator. The points at time t follow from $\boldsymbol{\xi}(t) = A\boldsymbol{\xi}_0$, or specifically $(y(t), x(t)) = (y_0\cos t + x_0\sin t, -y_0\sin t + x_0\cos t)$. For arbitrary t the only fixed point is $(0,0)$. For $t = 2\pi$, all points are fixed points.

In general, an arbitrary infinitesimal area in the space of vectors ξ is given by $d\xi^1\,d\xi^2$. (For higher dimensionality the infinitesimal area is $d\xi^1\,d\xi^2\,d\xi^3\,d\xi^4\ldots$) As an active transformation a period map A may be viewed as a coordinate transformation $(\xi^1, \xi^2) \to (\xi^{\hat{1}}, \xi^{\hat{2}})$. For A linear, then $\boldsymbol{\xi}(T) = A\xi_0$, and

$$\frac{\partial \xi^{\hat{i}}}{\partial \xi^j} = A^{\hat{i}}_{\ j}. \tag{5.25}$$

The infinitesimal areas are related by $d\xi^{\hat{1}}\,d\xi^{\hat{2}} = (\text{Jacobian})\,d\xi^1\,d\xi^2$. The Jacobian of this transformation is the determinant of A, as one sees from (5.25). Thus for a linear period map A to preserve area

$$\det A = 1. \tag{5.26}$$

In the foregoing example, it is clear the A of (5.24) satisfies (5.26).

Since all points $\boldsymbol{\xi}$ undergo the same rotation by the period map of (5.24), all points are Liapunov stable for this map. Any two points initially nearby remain nearby. The period map (5.24) is a stable mapping.

This stability is intimately related to the eigenvalues of A. For the period map in (5.24) these eigenvalues are $e^{\pm it}$, as is appropriate for a rotation [cf. discussion preceding (3.40)]. A simple case corresponding to a hyperbolic rotation (unstable) is the subject of Problem 5.6.

A large number of interesting cases for linear systems have a single degree of freedom and thus the results for two-dimensional phase space find frequent application. Let λ_1 and λ_2 denote the characteristic values (eigenvalues) of the matrix A. Using (5.26) the characteristic equation $\det(A - \lambda I) = 0$ can be written in the form $\lambda^2 - \lambda \operatorname{tr} A + 1 = 0$, where $\operatorname{tr} A = \lambda_1 + \lambda_2$ and $\lambda_1\lambda_2 = \det A = 1$. Applying the quadratic formula to find the roots of this equation, real roots are obtained for $|\operatorname{tr} A| > 2$ and complex roots for $|\operatorname{tr} A| < 2$. In the case of real roots, one root is less than 1 and one root is greater than 1. In the complex case, there are two complex roots that are conjugates of each other. For the complex case we take $\lambda_{1,2} = e^{\pm i\alpha}$ and see that A is equivalent to a rotation through an angle α. Again, A is stable. When $|\operatorname{tr} A| = 2$ we say that A is *marginally stable*.

Hence for linear systems with a phase space of two dimensions, every stability question near an equilibrium point reduces to the question of computing the trace of a matrix A. Only in special cases can this trace be found analytically. Often a numerical integration of (5.19) is necessary to compute A.

We now examine one special case amenable to analytic treatment, in order to illustrate the type of analysis possible. Furthermore, the case we treat is an approximation to the swing. The length of the equivalent pendulum is assumed to change in a time that is short when compared to the period of the swing.

Let the system be described by a Lagrangian $L = (\dot{x}^2/2) - \omega^2(t)(x^2/2)$. This gives the equation of motion

$$\ddot{x} + \omega^2(t)x = 0, \tag{5.27}$$

or written in the form of (5.19)

$$\begin{bmatrix} \dot{y} \\ \dot{x} \end{bmatrix} = \begin{bmatrix} -\omega^2(t)x \\ y \end{bmatrix}. \tag{5.28}$$

The function $\omega(t)$ is periodic and the time scale for (5.27) and (5.28) has been chosen to be the period of $\omega(t)$. We assume $\omega(t)$ is a step function given by

$$\omega(t) = \begin{cases} \omega_0 + \varepsilon, & m < t < m + \frac{1}{2} \\ \omega_0 - \varepsilon, & m + \frac{1}{2} < t < m + 1 \end{cases} \tag{5.29}$$

where $(\varepsilon/\omega_0) \ll 1$ and $m = 0, 1, 2, \ldots$. This variable frequency is sketched in Fig. 5.8.

Let us calculate the period map A and examine its stability properties. We represent the systems for different values of ε in the (ε, ω) parameter

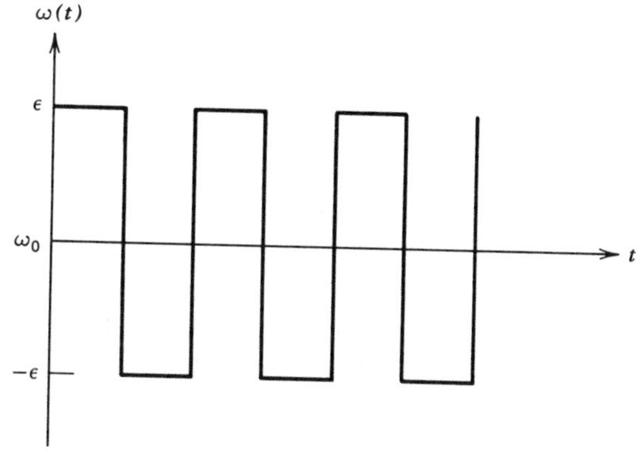

FIGURE 5.8. A frequency function that undergoes discontinuous jumps of amplitude 2ε.

space. The stable systems are represented in the (ε, ω) plane by open sets where $|\mathrm{tr}\, A| < 2$. The unstable regions correspond to open sets where $|\mathrm{tr}\, A| > 2$. The transition region or points of marginal stability correspond to $|\mathrm{tr}\, A| = 2$. Denote $\omega_{\pm} \equiv \omega_0 \pm \varepsilon$ and let $\hat{\omega}$ denote either of these two frequencies. A general solution to (5.28) is

$$y(t) = -a\hat{\omega}\sin(\hat{\omega}t + b)$$
$$x(t) = a\cos(\hat{\omega}t + b). \tag{5.30}$$

Define the mappings A_+ and A_- as follows: Let m be any integer and then A_+ is the mapping where $(y, x)|_{t=m} \to (y, x)|_{t=m+1/2}$ and A_- is the mapping such that $(y, x)|_{t=m+1/2} \to (y, x)|_{t=m+1}$. Then the map A is composed of the mappings A_+ and A_-, where $A = A_-A_+$. To compute the first and second columns of A_+ we select for the initial values $(y(0), x(0))$ the choices $(1, 0)$ and $(0, 1)$, respectively. We make identical choices for $(y(1/2), x(1/2))$ to compute the respective columns of A_-. From (5.30) we find

$$\begin{bmatrix} y(1/2) \\ x(1/2) \end{bmatrix} = A_+ \begin{bmatrix} y(0) \\ x(0) \end{bmatrix}$$
$$\begin{bmatrix} y(1) \\ x(1) \end{bmatrix} = A_- \begin{bmatrix} y(1/2) \\ x(1/2) \end{bmatrix}$$

where

$$A_{\pm} = \begin{bmatrix} \cos(\omega_{\pm}/2) & -\dfrac{1}{\omega_{\pm}}\sin(\omega_{\pm}/2) \\ \omega_{\pm}\sin(\omega_{\pm}/2) & \cos(\omega_{\pm}/2) \end{bmatrix}. \tag{5.31}$$

The mapping $A = A_- A_+$ has its trace given by

$$\operatorname{tr} A = 2\cos\left(\frac{\omega_+}{2}\right)\cos\left(\frac{\omega_-}{2}\right) - \left(\frac{\omega_+}{\omega_-} + \frac{\omega_-}{\omega_+}\right)\sin\left(\frac{\omega_+}{2}\right)\sin\left(\frac{\omega_-}{2}\right). \tag{5.32}$$

In the (ε, ω) plane we find the regions of stability by examining the values for marginal stability, that is, those curves satisfying $\operatorname{tr} A = \pm 2$. These curves are the boundary separating the regions of stability and instability. As a beginning we compute the intersection of these curves with the ω-axis, where $\varepsilon = 0$. Setting $\varepsilon = 0$ in (5.32), we find for the trace

$$\left|2\cos^2\left(\frac{\omega_0}{2}\right) - 2\sin^2\left(\frac{\omega_0}{2}\right)\right| = |2\cos\omega_0| = 2. \tag{5.33}$$

Thus $\cos\omega_0 = \pm 1$ and we have

$$\omega_{0n} = \begin{cases} 2n\pi, & \operatorname{tr} A = 2 \\ (2n+1)\pi, & \operatorname{tr} A = -2 \end{cases} \tag{5.34}$$

where $n = 0, 1, 2, \ldots$. Any choice of ω_0 other than those given in (5.34) results in a value of $|\cos\omega_0| < 1$ and hence all other values on the ω-axis are points of stability for the mapping A. There are no points of instability on the ω-axis. This is reasonable since it is not possible to increase the amplitude, or to have A be unstable, without having some energy input via a changing parameter. Thus we wish to consider how the curves of marginal stability proceed outward into the (ε, ω) plane from those intersection points with the ω-axis, as given in (5.34).

We must further analyze (5.32) with $\varepsilon \neq 0$. We use

$$\cos\left(\frac{\omega_+}{2} + \frac{\omega_-}{2}\right) = \cos\omega_0 = \cos\left(\frac{\omega_+}{2}\right)\cos\left(\frac{\omega_-}{2}\right) - \sin\left(\frac{\omega_+}{2}\right)\sin\left(\frac{\omega_-}{2}\right)$$

and substitute into (5.32) to find

$$\operatorname{tr} A = 2\cos\omega_0 - \frac{4\varepsilon^2}{\omega_0^2 - \varepsilon^2}\sin\left(\frac{\omega_+}{2}\right)\sin\left(\frac{\omega_-}{2}\right). \tag{5.35}$$

To simplify (5.35) we exploit the assumption $(\varepsilon/\omega_0) \ll 1$ and expand the product of sine terms.

$$\sin\left(\frac{\omega_\pm}{2}\right) = \sin\left(\frac{\omega_0}{2}\right)\left(1 - \frac{\varepsilon^2}{8}\right) \pm \frac{\varepsilon}{2}\cos\left(\frac{\omega_0}{2}\right) + O(\varepsilon^3). \tag{5.36}$$

We thus find

$$\sin\left(\frac{\omega_+}{2}\right)\sin\left(\frac{\omega_-}{2}\right) = \sin^2\left(\frac{\omega_0}{2}\right) - \frac{\varepsilon^2}{4} + O(\varepsilon^3). \tag{5.37}$$

Substituting into (5.35)

$$\operatorname{tr} A = 2\cos\omega_0 - \frac{4\varepsilon^2}{\omega_0^2 - \varepsilon^2}\sin^2\left(\frac{\omega_0}{2}\right) + \frac{\varepsilon^4}{\omega_0^2 - \varepsilon^2} + O(\varepsilon^5). \tag{5.38}$$

To proceed further we must deal separately with the opposite sign choices for $\operatorname{tr} A$. If $\operatorname{tr} A = +2$ then (5.38), with a minor amount of algebra, becomes

$$\varepsilon^4 = 4\omega_0^2\sin^2\left(\frac{\omega_0}{2}\right). \tag{5.39}$$

With $\operatorname{tr} A = -2$ we find in a similar way that

$$\varepsilon^4 - 4\varepsilon^2 + 4\omega_0^2\cos^2\left(\frac{\omega_0}{2}\right) = 0. \tag{5.40}$$

We assume

$$\omega_0 = \omega_{0n} \pm |\beta|, \tag{5.41}$$

where ω_{0n} is given by (5.34) and $|\beta|$ is assumed small. Then to lowest order we find

$$\varepsilon^2 = \begin{cases} |\beta|^2, & \omega_0 = \pm|\beta| \\ 2n\pi|\beta|, & \omega_0 = 2n\pi \pm |\beta|, \quad n = 1, 2, \ldots \\ \left[\dfrac{(2n+1)\pi|\beta|}{2}\right]^2, & \omega_0 = (2n+1)\pi \pm |\beta|, \quad n = 0, 1, 2, \ldots \end{cases} \tag{5.42}$$

These results enable us to extend the points on the axis for marginal stability into the (ε, ω) plane for nonzero values of ε. Figure 5.9 is a sketch of these stability regions for the period map A of (5.28). The changing period $\omega(t)$ is given in (5.29) and sketched in Fig. 5.8. The shaded regions in Fig. 5.9 correspond to the values of $\omega = \omega_0 \pm \varepsilon$ where the amplitude of the swing may be increased. It is evident from this sketch that the region of instability is much broader for the fundamental frequency and its first harmonic than it is for the higher harmonics. It is much easier to pump a swing up at its fundamental frequency than it is at the higher harmonics.

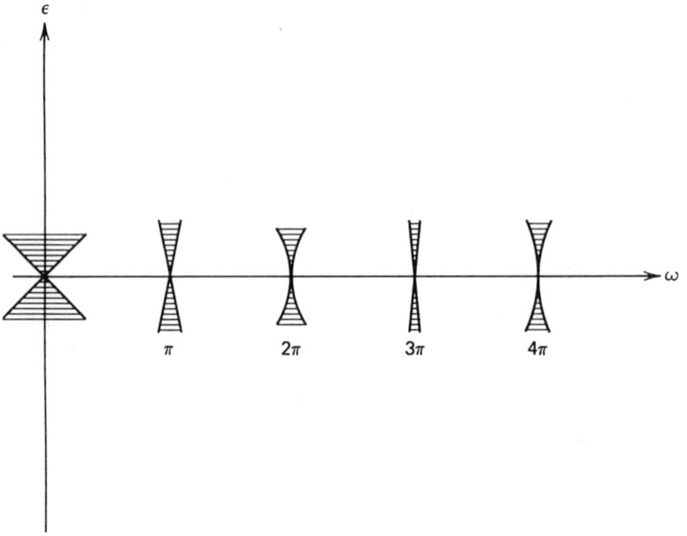

FIGURE 5.9. The regions of stability and instability in the (ε, ω) plane for small values of ε. Shaded regions are regions of instability.

The regions of instability become very narrow as they approach the $\varepsilon = 0$ axis for large values of n. The resonance frequencies must satisfy very rigid bounds as n gets large. Thus in practice one typically observes parametric resonance when the period T is close to the natural period of the system.

If some small amount of dissipation in the system exists in the form of friction, then no longer does the period map A preserve area. Furthermore, such dissipation can only increase the stability of the system. The points in the (ε, ω) plane on the ω-axis, where $\varepsilon = 0$, will be stable, since the oscillations will be damped and die away. Thus the instability regions depicted in Fig. 5.9 must move up off the axis. If one does not pump hard enough in the swing, then friction damps the oscillations and they eventually cease. The parameter ε must be large enough to overcome the dissipation.

PROBLEMS

5.1. Show for distinct eigenfrequencies ω_α and ω_β that $a^i_\alpha m_{ij} a^j_\beta = 0$.

5.2. Obtain the limiting result ω^2_∞ as given in the worked example of Fig. 5.1.

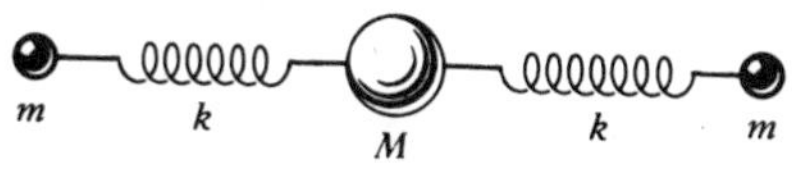

FIGURE 5.10. Sketch of system for Problem 5.3.

5.3. Consider three masses connected by springs and constrained to move on a line as sketched in Fig. 5.10. Find the normal modes and frequencies and investigate whether this system will exhibit beat phenomena.

5.4. Find the eigenfrequencies and normal modes of the physical system sketched in Fig. 5.11. The system is depicted in its equilibrium configuration. Check the limit as $k \to \infty$ to see whether results agree with those obtained for the example worked in the text.

5.5. Consider a pendulum constrained to oscillate in a plane. Ordinarily the equilibrium point where the pendulum is vertical *above* the point of support is unstable. Show that by an appropriate oscillation of the point of support this unstable equilibrium point can be stabilized. This problem can be analyzed analytically by assuming that the acceleration of the point of support is constant, changing sign in each half cycle (Arnold, 1978).

5.6. Analyze the period map for $\xi = (y, x)$, $\dot{\xi} = (\dot{y}, \dot{x}) = (x, y)$ similar to the way the example of (5.21) was analyzed in the text.

5.7. Consider a system of three balls connected by four identical springs as shown in Fig. 5.12. The balls have masses m, $2m$, and m, respectively, and the springs have spring constant k. The balls are confined to horizontal motion. Find the natural frequencies of this system, describe the normal modes, and find normal coordinates.

5.8. Obtain the frequencies and normal modes of a linear triatomic molecule where the central atom is bound to the origin and to the atoms on each side with a spring with constant k. All motions take place along a line.

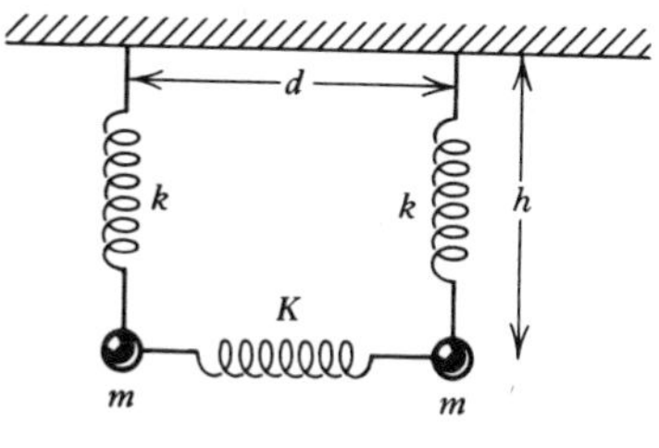

FIGURE 5.11. Physical system to be analyzed in Problem 5.4.

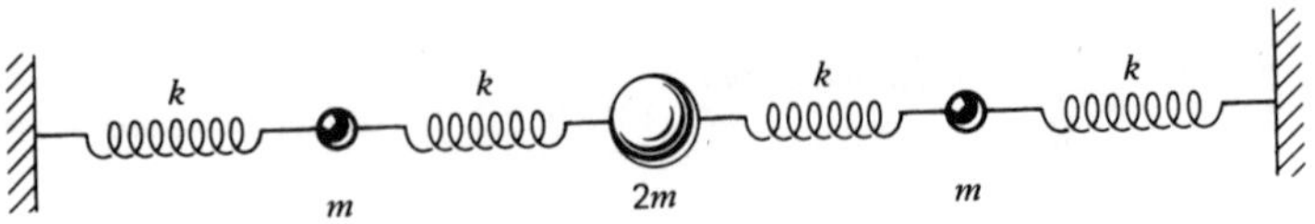

FIGURE 5.12. Masses and springs for Problem 5.7.

5.9. A pendulum is constrained by a spring whose free length is l, where $l < 2a$. See Fig. 5.13. Find the period and the criterion of stability for small oscillations about the vertical. The motion is constrained to the vertical plane of spring and pendulum. Now consider point A, where the pendulum is attached, to undergo rapid oscillations in the vertical direction of very small amplitude. Consider the acceleration of the point A to be $\pm\alpha$, where α is a constant. How does the stability of the system now depend on the system parameters?

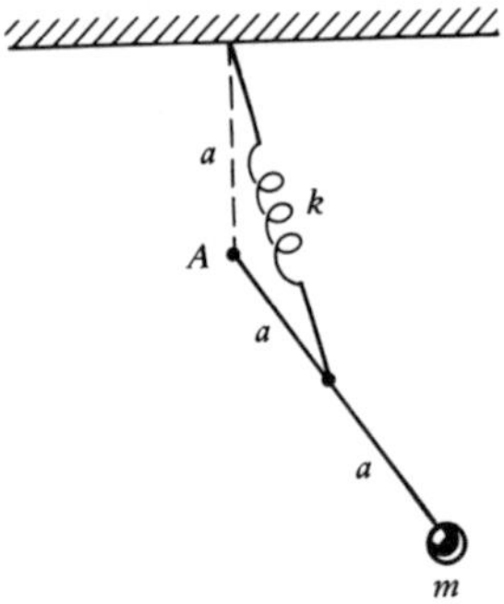

FIGURE 5.13. Spring and pendulum for Problem 5.9.

CHAPTER SIX

INVARIANTS OF THE MOTION

The book of nature is written in mathematical language.

GALILEO

The primary objective of this chapter is to consider the symmetries of mechanical systems and the manner in which symmetries, and their associated invariants, lead to a partial solution of the equations of motion. This objective is pursued within the context of the Lagrangian formulation. Since Lagrangians depend on the generalized velocities, we choose to discuss thoroughly tangent vectors, tangent bundles, and induced mappings of these vectors. These topics are considered as a prelude to a discussion of symmetry. Noether's theorem is derived, and then applied to obtain momentum and angular momentum as the constants of the motion associated with translation invariance and rotation invariance, respectively. Constants of the motion generated by mappings that are not symmetries are also discussed.

6.1. TANGENT VECTORS

Consider once again a pendulum with its point of support attached to a mass that can move along a horizontal line. This is the mechanical system sketched in Fig. 3.2 with configuration manifold depicted in Fig. 3.8. At each point along a trajectory of this system, there is a tangent vector that corresponds to the velocity of the pendulum bob. Because the configuration manifold is a cylinder, arrows drawn to represent vectors cannot lie in the manifold itself but rather in a plane that is tangent to the cylinder at the point of interest on the trajectory. Figure 6.1 shows a few of these "tangent vectors" along a typical trajectory.

Tangent vectors are important not only because they are of intrinsic interest but because they are one of the arguments of the Lagrangian function. Moreover, through the kinetic energy, the velocity vectors form an important piece of the energy of the system. Motivated by these observations concerning the tangent vectors to a trajectory in configuration space, we proceed to define and study tangent vectors.

We first give a definition for tangent vectors that is easily visualized. Let M be an arbitrary manifold that can be embedded in $\mathbb{R}^{n+k}$ (cf. Section 3.6 on embedded manifolds). We assume that the dimension of M is n, U is a subset of $\mathbb{R}^{n+k}$, and the functions f_α: $U \to \mathbb{R}$ ($\alpha = 1, \ldots, k$) are such as to satisfy the requirements for M to be an embedded submanifold of $\mathbb{R}^{n+k}$. The vectors ∇f_α are then linearly independent at each point $x_0 \in M$. These vectors form the basis for a linear vector space at x_0 in $\mathbb{R}^{n+k}$. The *tangent vector space to M at the point* x_0, which is denoted TM_{x_0}, is the orthogonal complement to the vector space generated by the ∇f_α. This definition implies that the tangent space TM_{x_0} is just a copy of $\mathbb{R}^n$ and therefore has the same dimension as the manifold M.

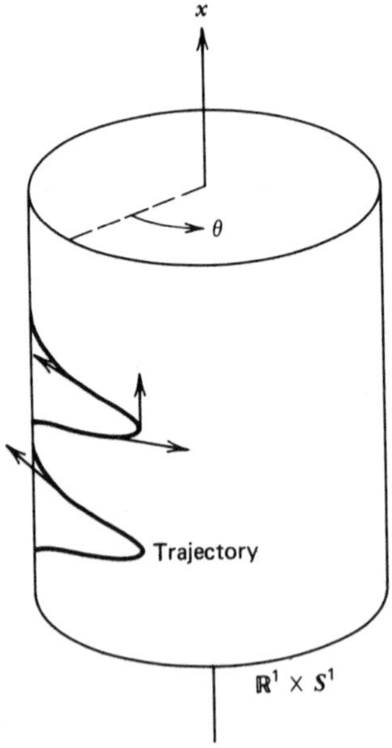

FIGURE 6.1. The configuration manifold of Fig. 3.8. The viewpoint has been rotated so that now only limited sections of a possible trajectory are visible. At four points on this trajectory tangent vectors to the curve are shown.

In our example of Fig. 6.1, the embedded manifold in $\mathbb{R}^3$ is defined by a single constraint function written in cylindrical coordinates: $f(r, \theta, z) = r - L = 0$. Clearly, $\nabla f = \hat{\mathbf{r}}$ and the tangent planes on the cylinder in Fig. 6.1 are orthogonal to $\hat{\mathbf{r}}$.

With M considered as an embedded submanifold we could define tangent vectors without reference to the functions f_α or their gradients. Let $\mathbb{R}^{n+k}$ have a fixed cartesian basis and consider curves in $M \cap \mathbb{R}^{n+k}$, that is, functions $x\colon I \to M$, where I is some interval of $\mathbb{R}$. The function is denoted $x(t)$. Such curves represent a sequence of vectors in $\mathbb{R}^{n+k}$ that can be added. We require this function $x(t)$ to be differentiable, that is, the components $x^i(t)$ are differentiable functions. Then the *tangent vector* $\mathbf{v}(t_0)$ to the curve $x(t)$ at the point $x_0 = x(t_0)$ is defined by

$$\mathbf{v}(t_0) = \lim_{\Delta t \to 0} \frac{1}{\Delta t}\left[x(t_0 + \Delta t) - x(t_0)\right] = \left.\frac{d}{dt}x(t)\right|_{t_0} = \dot{x}(t_0) \quad (6.1)$$

The notation has been deliberately chosen so as to remind the reader that if $x(t)$ is a motion, then the tangent vector to the curve is nothing more than the familiar velocity vector. Since the vectors $x(t_0 + \Delta t)$ and $x(t_0)$ are in M, the vector $\dot{x}$ is orthogonal to the vectors ∇f_α. Thus $\dot{x}$ lies in TM_x.

These tangent vectors at a point can be added together and multiplied by scalars since the vector functions $x(t)$ can. In short, they define a vector space that is identical with TM_{x_0} defined previously. Note that many curves through a given point $x_0 \in M$ may have the same tangent vector.

One of the most appealing and useful characterizations of tangent vectors is in terms of a directional derivative, that is, the derivative of a function in a specified direction. We explore this connection of tangent vectors to directional derivatives by using an arbitrary, differentiable, real-valued function on M, $f\colon M \to \mathbb{R}$. We examine the behavior of this function in the neighborhood of $x_0 \in M$ by using a chart (U, ϕ). Refer to Fig. 6.2 for a sketch depicting the relationships among the quantities of interest.

The *local representative of the function* f we denote by F and define by

$$F(\phi(x)) = f(x) \qquad \text{for all } x \in U. \quad (6.2)$$

The local representative is familiar and easy to work with, since it is a real-valued mapping from $\mathbb{R}^n$ to $\mathbb{R}$. We define f to be differentiable if F is differentiable. Let $x(t)$ be a curve in M such that $x(0) = x_0$. This curve has an image in $\mathbb{R}^n$ under the bijection ϕ. Let us call this curve $X(t)$. The *local representative of the tangent vector* is then $\mathbf{V} = \dot{X}(0)$.

With these ideas in place we examine the way in which the function f changes along the curve $x(t)$ in the neighborhood of the point x_0. This is

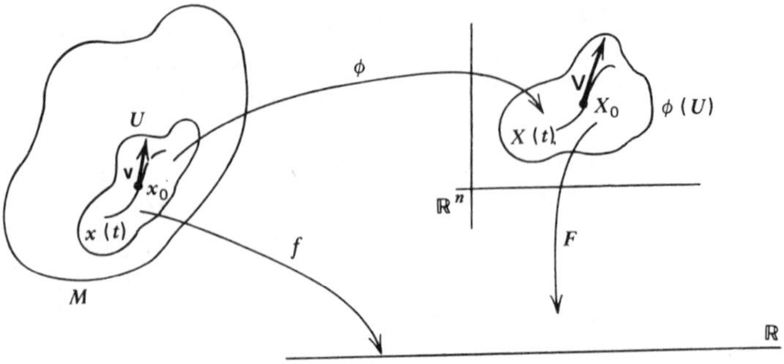

FIGURE 6.2. Schematic showing the relation between a real-valued map f, a chart bijection ϕ, and the local representative F.

done by using the local representatives F and $\mathbf{V}$.

$$f(x(t)) = F(X(t)) = F(X(0)) + t\mathbf{V} \cdot \nabla F + \cdots \tag{6.3}$$

Recall that $\mathbf{V}$ and F refer to a standard Euclidean vector space $\mathbb{R}^n$ and so quantities on the right of (6.3) may be interpreted in the familiar way. The quantity $\mathbf{V} \cdot \nabla F$ is a scalar and under any change of coordinates (change of chart) this scalar is invariant. We write

$$\mathbf{V} \cdot \nabla F = \partial_{\mathbf{v}}[f]. \tag{6.4}$$

The notation $\partial_{\mathbf{v}}[f]$ is to convey the meaning that the tangent vector $\mathbf{v}$ is operating on the function f. From (6.3) we identify $\partial_{\mathbf{v}}[f]$ as the first-order change in the function f away from the point x_0 in the direction $\mathbf{v}$. Thus we interpret $\partial_{\mathbf{v}}$ as a directional derivative operator on real-valued functions on M. This derivative property of tangent vectors is often made the basis of a definition for tangent vectors (Choquet-Bruhat, DeWitt-Morette, 1982; Hermann, 1968). In an abuse of language, we occasionally refer to $\mathbf{v}$ as an "operator," when in fact we mean the directional derivative operator $\partial_{\mathbf{v}}$.

One useful application of the relationship between a function and its local representative is provided by the coordinate functions. Let the function F be the coordinate function X^i, $X^i\colon \mathbb{R}^n \to \mathbb{R}$. To get a function f on the manifold M we compose this with the chart bijection ϕ. We denote this composite function as the real-valued mapping $x^i = X^i \circ \phi$. As a concrete example of these relationships consider a point P in the manifold S^2, that is, a point on a sphere. If (U, ϕ) denotes a chart on this sphere, then $\phi(P)$ is a point in the plane $\mathbb{R}^2$. Once we assign coordinates to this plane (i.e., place labels on the axes), we have a pair of numbers associated with the point P.

The X^i correspond to the labels on the axes of $\mathbb{R}^2$. The bijection ϕ assigns the correspondence between the point P in S^2 and the point in the plane $\mathbb{R}^2$. The composition of these two gives the coordinates of the point P, $x^i(P)$.

Using X^i and x^i, we calculate according to (6.3) and obtain the local components of the tangent vector $\mathbf{v}$.

$$x^i(t) = X^i(t) = X^i(0) + t\mathbf{V} \cdot \nabla X^i + \cdots = X^i(0) + tV^i + \cdots \quad (6.5)$$

From (6.4) we conclude $\partial_{\mathbf{v}}[x^i] = V^i$. For the components of the tangent vector in a local coordinate system we drop the distinction between large and small case letters and simply write $v^i = \partial_{\mathbf{v}}[x^i]$. Similarly, we will refer to x^i as the coordinates of the point x in the manifold M, without any specific reference to a local chart. As is the common practice in physics, we very often speak of vectors and coordinates in this way that, to have precise meaning, would require the specification of a chart. When no confusion can result we adopt this practice (as, for example, in Chapter 1) and speak of a vector by reference to its components without ever really giving a specification of what the local coordinate system is. Using a component notation without specifying a local coordinate system is consistent with the fact that tangent vectors and tangent spaces exist independent of any particular coordinate system.

These foregoing considerations also lead to a useful operator representation for basis vectors $\{\mathbf{e}_i\}$. Consider a tangent vector $\mathbf{v}$ operating on a function f. Since $(\partial f/\partial x^i)|_0 \equiv (\partial F/\partial X^i)|_0$ we obtain

$$\partial_{\mathbf{v}}[f] = \mathbf{V} \cdot \nabla F = V^i \left.\frac{\partial F}{\partial X^i}\right|_0 = v^i \left.\frac{\partial f}{\partial x^i}\right|_0. \quad (6.6)$$

In terms of local representatives the operator $\partial_{\mathbf{v}}$ is $\mathbf{V} \cdot \nabla$. It is evident from (6.6) that in local coordinates the operator $\partial_{\mathbf{v}}$ may be written as

$$\partial_{\mathbf{v}} = v^i \frac{\partial}{\partial x^i} \equiv v^i \partial_i, \quad (6.7)$$

where by definition $(\partial/\partial x^i) \equiv \partial_i$. The set of partial differential operators $\{\partial_i\}$ forms a basis for the tangent space TM_{x_0} insofar as vectors in TM_{x_0} are interpreted as operators. They will similarly constitute a basis for the tangent space at all points of the manifold where the local coordinates (chart) work. The basis $\{\partial_i\}$ is called the *natural basis*. There are, of course, other basis sets for the tangent space at a point and they may be more useful for some purposes, but the natural basis is the one presented to us by the coordinate system.

Thinking of the partial derivative operators ∂_i as vectors is very much like thinking of the gradient operator ∇ of vector calculus as a vector operator. Since $\mathbf{e}_i = \delta_i^j \mathbf{e}_j$ the components of a basis vector $\mathbf{e}_i$ are δ_i^j, and from (6.6) we obtain

$$\partial_{\mathbf{e}_i}[f] = \delta_i^j F_{,j} = F_{,i} = \frac{\partial F}{\partial x^i} = \frac{\partial f}{\partial x^i} = \partial_i[f]$$

The end results in this chain of equalities give

$$\partial_{\mathbf{e}_i} = \partial_i, \tag{6.8}$$

since f is arbitrary.

We use for the arbitrary vector $\mathbf{v}$ the two notations $\mathbf{v} = v^i \mathbf{e}_i$ and $\partial_{\mathbf{v}} = v^i \partial_i$ interchangeably, depending on which property of a vector we may wish to emphasize or use. Vectors as "operators" and vectors as "tangent vectors to curves" are connected in the following way. Let $\mathbf{B}$ be a vector in the tangent space TM_x. Here we choose to call the arbitrarily chosen vector "$\mathbf{B}$" rather than "$\mathbf{v}$" to emphasize that all vectors at the point x (e.g., the magnetic field) are elements of the tangent space TM_x, not just the vectors of velocity or acceleration. Let $\phi(t)$ be a curve in the manifold that passes through the point x and has tangent $\mathbf{B}$ at this point. For $\phi(t)$ one might choose the curve identified physically as a field line. As before we let $f\colon M \to \mathbb{R}$ be an arbitrary, real-valued function on the manifold. Then

$$\partial_{\mathbf{B}}[f] = \left.\frac{d}{dt}\right|_{t=0} f(\phi(t)) = \left.\frac{d}{dt}\right|_{t=0} f \circ \phi(t), \tag{6.9}$$

where $\phi(0) = x$ and $\mathbf{B} = \dot{\phi}(0)$. Equation (6.9) makes explicit the relationship between the "tangent vector" $\mathbf{B}$ and the "directional-derivative operator" $\partial_{\mathbf{B}}$. In particular, when $f \to x^i$ in (6.9), then $\partial_{\mathbf{B}}[x^i] = B^i$, which are the contravariant components of $\mathbf{B}$.

As a nontrival example of a tangent space we consider the manifold $SO(3)$, which was discussed in Section 3.6. The results obtained are important in their own right because rotations play such a crucial role in the physics of atomic and nuclear systems. The vectors tangent to the curves in $SO(3)$, corresponding to rotations around the x, y, z axes, are intimately connected to the angular momentum operators of quantum mechanics.

Let the rotation R denote an arbitrary "point" in $SO(3)$. If we use the Euler angles, then a basis for the tangent space at R is $\{\partial_\phi, \partial_\theta, \partial_\psi\}$. However, as an illustration of computational techniques let us construct a basis $(\partial_x, \partial_y, \partial_z)$ in which the basis vectors are tangent to curves in $SO(3)$ corresponding to rotations around the x, y, z axes, respectively.

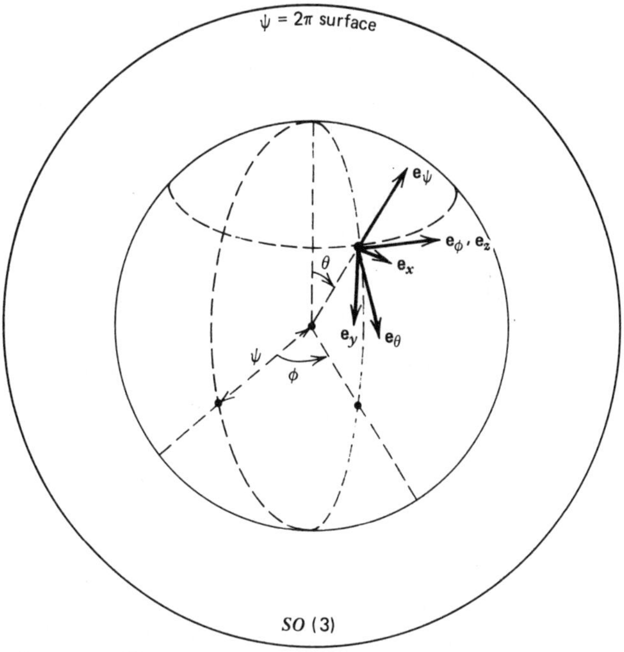

FIGURE 6.3. The manifold $SO(3)$ pictured as a sphere in $\mathbb{R}^3$ where points with radius $\psi = 2\pi$ and $\psi = 0$ are to be identified. At an arbitrary point (ϕ, θ, ψ) basis vectors for the tangent space are sketched. The curve $r(\lambda) = R_z(\lambda)R$ can be identified with the latitude curve to which $\mathbf{e}_z$ is tangent. The bases $(\mathbf{e}_\phi, \mathbf{e}_\theta, \mathbf{e}_\psi)$ and $(\mathbf{e}_x, \mathbf{e}_y, \mathbf{e}_z)$ are related through Eqs. (6.17)–(6.19).

As an arbitrary point in $SO(3)$, R denotes an arbitrary rotation matrix and is given in (3.43). We consider curves in $SO(3)$ passing through the arbitrary point $R \in SO(3)$. The point R is not the identity matrix because the (ϕ, θ, ψ) coordinate system for $SO(3)$ breaks down there. First consider curves generated by rotations around the z-axis, that is, $r(\lambda) = R_z(\lambda)R$. The curve parameter λ is an angle and for each choice of λ the element $r(\lambda)$ is a rotation or point in $SO(3)$. For points in a neighborhood of R we may consider λ to be infinitesimal. A sketch of coordinate curves passing through a point R [coordinates (ϕ, θ, ψ)] with associated tangent vectors is given in Fig. 6.3.

Our interest is in the tangent vector to the curve $r(\lambda)$ for $\lambda = 0$. We denote this vector as ∂_z, since it is the directional derivative for rotations around the z-axis.* We see that every point on the curve $r(\lambda)$ is related to

*In a group theory context ∂_z would be referred to as the "infinitesimal generator" for rotations around the z-axis.

the arbitrary point R by a rotation through some angle λ around the z-axis. $\mathbf{e}_z = (dr/d\lambda)|_{\lambda=0} = \dot{r}(0)$. Let $\pi^i{}_j\colon SO(3) \to \mathbb{R}$ be the projection mapping on $SO(3)$; $\pi^i{}_j(R) = R^i_j \in \mathbb{R}$. Consider the action of the tangent vector $\mathbf{e}_z$ on this function $\pi^i{}_j$. (Actually we are considering nine different functions obtained by different choices for i and j.)

$$\begin{aligned}\partial_z\left[\pi^i{}_j\right] &= \frac{d}{d\lambda}\left(\pi^i{}_j(r(\lambda))\right)\Big|_{\lambda=0} \\ &= \frac{d}{d\lambda}\left(\left(R_z(\lambda)\right)^i{}_m R^m{}_j\right)\Big|_{\lambda=0} = \frac{d}{d\lambda}\left(R_z(\lambda)\right)^i{}_m\Big|_{\lambda=0} R^m{}_j \end{aligned} \tag{6.10}$$

Referring to (3.42), approximated for small angles λ, we see that

$$\frac{d}{d\lambda}\left(R_z(\lambda)\right)^i{}_m\Big|_{\lambda=0} = \begin{bmatrix} 0 & -1 & 0 \\ 1 & 0 & 0 \\ 0 & 0 & 0 \end{bmatrix} = \delta^i_2\delta^1_m - \delta^i_1\delta^2_m \tag{6.11}$$

Thus

$$\partial_z\left[\pi^i{}_j\right] = -\delta^1_i R^2{}_j + \delta^2_i R^1{}_j. \tag{6.12}$$

We note that there is no point R in $SO(3)$ at which ∂_z is zero because to be identically zero requires that $\partial_z[f] = 0$ for all real-valued functions f and in particular for the functions $\pi^i{}_j$. But for $\partial_z[\pi^i{}_j]$ to be identically zero with $i = 1$; $j = 1, 2, 3$ would require $\det R = 0$, which is a contradiction for $R \in SO(3)$.

The vector $\mathbf{e}_z$ viewed as an operator may be written in the following form using the natural basis:

$$\partial_z = a_\phi\partial_\phi + a_\theta\partial_\theta + a_\psi\partial_\psi. \tag{6.13}$$

Using (3.43) and the abbreviated notation for sines and cosines (e.g. $\sin\alpha \equiv s_\alpha$), we find

$$\begin{aligned}\pi^1{}_2(R) &= -c_\phi s_\psi - s_\phi c_\theta c_\psi. \\ \frac{\partial\pi^1{}_2}{\partial\phi}\bigg|_R &= s_\psi s_\phi - c_\psi c_\theta c_\phi; \\ \frac{\partial\pi^1{}_2}{\partial\theta}\bigg|_R &= s_\phi s_\theta c_\psi; \\ \frac{\partial\pi^1{}_2}{\partial\psi}\bigg|_R &= -c_\psi c_\phi + s_\psi c_\theta s_\phi. \end{aligned} \tag{6.14}$$

Thus from (6.13) and (6.14),

$$\partial_z\left[\pi^1{}_2\right]\big|_R = a_\phi(s_\psi s_\phi - c_\psi c_\theta c_\phi) + a_\theta(s_\phi s_\theta c_\psi) + a_\psi(-c_\psi c_\phi + s_\psi c_\theta s_\phi). \tag{6.15}$$

From (6.12) we know that

$$\partial_z\left[\pi^1{}_2\right] = -R^2{}_2 = s_\psi s_\phi - c_\psi c_\theta c_\phi. \tag{6.16}$$

Comparing (6.15) and (6.16), we have that $a_\phi = 1$ and the other coefficients in (6.13) must be zero. Thus

$$\partial_z = \partial_\phi. \tag{6.17}$$

In a similar way, one can show that (Problem 6.1)

$$\partial_x = \cos\phi\,\partial_\theta - \sin\phi\left(\cot\theta\,\partial_\phi - \frac{1}{\sin\theta}\partial_\psi\right), \tag{6.18}$$

$$\partial_y = \sin\phi\,\partial_\theta + \cos\phi\left(\cot\theta\,\partial_\phi - \frac{1}{\sin\theta}\partial_\psi\right). \tag{6.19}$$

The last two vectors generate the rotations around the x and y axes, respectively; that is, they are tangent at $R \in SO(3)$ to curves generated by rotations around the x- and y-axes, respectively. The vectors $(\mathbf{e}_x, \mathbf{e}_y, \mathbf{e}_z)$ form a basis for expressing the angular velocity vector of a rigid body in the (S)pace system. These connections are explored in Section 10.6.

6.2. TANGENT BUNDLE

Previously we have pointed out the dependence of Lagrangians on the "$\dot{q}^i$'s" as well as the "q^i's". In obtaining the Euler–Lagrange equations, the generalized velocities are treated as independent coordinates. Thus the configuration manifold for a mechanical system is not the domain space for the Lagrangian function. We must have the velocity coordinates as well. Having now some familiarity with tangent spaces, we can introduce the notion of a tangent bundle, which is the proper domain manifold for a Lagrangian.

Let M be an n-dimensional manifold and let $\mathbf{x}$ be an arbitrary point in M. The tangent space at this point is $TM_{\mathbf{x}}$ and each point has such a tangent

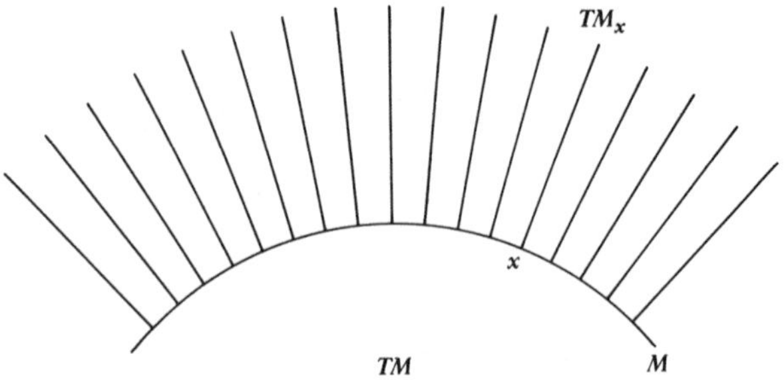

FIGURE 6.4. Qualitative representation of a tangent bundle TM. The curved arc represents the base manifold M and the tangent spaces TM_x are represented by the straight lines above each point of M.

space. We denote the *tangent bundle* as TM and define it as the union of all these tangent spaces; $TM = \cup_{\mathbf{x}\in M} TM_{\mathbf{x}}$. Points in TM consist of points $\mathbf{x}$ and tangent vectors $\mathbf{v} \in TM_{\mathbf{x}}$. Thus if $\{x^i\}$ is a set of local coordinates in a neighborhood of the point $\mathbf{x} \in M$, then the functions $\{v^i, x^i\}$ form a set of local coordinates in a neighborhood about a selected point in TM. The tangent bundle TM is itself a manifold because of the local coordinate system that it inherits from M. Often we will denote points in TM in the form $(\mathbf{v}, \mathbf{x})$, where it is to be understood that $\mathbf{v} \in TM_{\mathbf{x}}$. If π: $TM \to M$ is the natural projection on TM which projects points $(\mathbf{v}, \mathbf{x})$ to $\mathbf{x}$, then the inverse image of $\mathbf{x}$ under π is just the tangent space $TM_{\mathbf{x}}$.* A qualitative sketch of a tangent bundle TM is given in Fig. 6.4. The tangent bundle, which uses velocity and points as coordinates, is a generalization of the phase space discussed in Section 2.6. The Lagrangian $L(\dot{\mathbf{q}}, \mathbf{q})$ is a real-valued mapping on TM, L: $TM \to \mathbb{R}$. Often we must consider systems in which the Lagrangian is an explicit function of time $L(\dot{\mathbf{q}}, \mathbf{q}, t)$. Then in this case the Lagrangian is the mapping L: $(TM) \times \mathbb{R} \to \mathbb{R}$, where $L(\dot{\mathbf{q}}, \mathbf{q}, t) \in \mathbb{R}$. In such a case one or both of the kinetic and potential energies will depend on time.

As an example we consider again the system with configuration manifold shown in Fig. 6.1. A Lagrangian for this system we recall from (3.15).

$$L = \tfrac{1}{2}(M + m)\dot{x}^2 + \tfrac{1}{2}m(L^2\dot{\theta}^2 + 2\dot{x}\dot{\theta}L\cos\theta) + mgL\cos\theta. \quad (6.20)$$

The velocity at each point is given by specifying $\dot{x}$ and $\dot{\theta}$. The tangent bundle manifold is $T(\mathbb{R}^1 \times S^1)$ and an arbitrary point in this manifold is

*This tangent space in the context of a vector bundle is referred to by mathematicians as the *fiber* of the tangent bundle over the point $\mathbf{x}$.

denoted by $(\dot{x}, \dot{\theta}, x, \theta)$. Since the tangent bundle is the generalization of phase space considered in Section 2.4, all the benefits derived from qualitative considerations of the motion are present. In this example a constant energy surface in $T(\mathbb{R}^1 \times S^1)$ is a three-dimensional volume in the four-dimensional tangent-bundle manifold. The energy is constant since the potential energy is time independent. For later convenience we take the potential energy to be $U = mgL(1 - \cos\theta)$ and write for the energy constant.

$$E = \tfrac{1}{2}(M + m)\dot{x}^2 + \tfrac{1}{2}m(L^2\dot{\theta}^2 + 2\dot{x}\dot{\theta}L\cos\theta) + mgL(1 - \cos\theta). \tag{6.21}$$

From (6.20) it is clear that x is cyclic and hence

$$\frac{\partial L}{\partial \dot{x}} = \text{constant} = p_x = (M + m)\dot{x} + mL\dot{\theta}\cos\theta. \tag{6.22}$$

Since we have two constants p_x and E, the motion is confined to a two-dimensional surface in the tangent bundle $T(\mathbb{R}^1 \times S^1)$. To obtain the

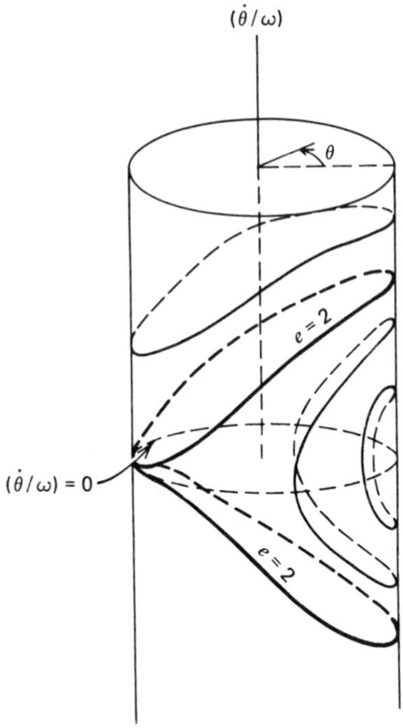

FIGURE 6.5. Trajectories for the system with Lagrangian (6.20) in the tangent bundle $T(\mathbb{R}^1 \times S^1)$. The curves are confined to this two-dimensional surface because the energy and the momentum in the x-direction are constants. The heavy line marks the trajectory that separates these motions where $-\pi < \theta < \pi$ from those where θ takes on all values.

phase curves on this two-dimensional surface we eliminate $\dot{x}$ from (6.21) using (6.22). Also, scale all quantities in (6.21) by $mL^2\omega^2$, where $\omega^2 = g/L$, the natural frequency of the pendulum. Let $e = [E - (p_x^{\,2})/(2(M + m))]/(mL^2\omega^2)$ and then (6.21) can be written as

$$\frac{(e - 1 + \cos\theta)}{(1 - [m/(M+m)]\cos^2\theta)} = \frac{1}{2}\left(\frac{\dot{\theta}}{\omega}\right)^2. \tag{6.23}$$

The qualitative behavior of curves for different values of e are sketched in Fig. 6.5.

6.3. THE DERIVATIVE MAP

The primary objective of this chapter is to consider symmetries of systems and the corresponding invariants. A symmetry in qualitative terms means some mapping or transformation under which the mechanical system is invariant. Here *mechanical system* means all objects of interest plus their interactions with each other, and externally applied fields. Mechanical systems are represented by their Lagrangian functions, which depend on the velocities. If we are to study how Lagrangians remain invariant, we must first clarify how symmetry mappings transform tangent vectors. To this end we define the derivative of mappings between manifolds. This derivative map may be viewed as a linearization of the manifold map—in other words, the linear piece.

Let $f\colon M \to N$ be a mapping between two manifolds M and N. As with real-valued maps on manifolds, we say that f is *differentiable* at a point if its local representative is differentiable. Let $\mathbf{v} \in TM_x$ and let $\phi(t)$ be a curve in M with $\phi(0) = x$, such that the tangent to this curve at x is $\mathbf{v}$. That is to say, $\phi\colon \mathbb{R} \to M$ and $\phi(0) = x$, where $\dot{\phi}(0) = \mathbf{v}$. The *derivative map* of a differentiable mapping $f\colon M \to N$ at the point $x \in M$ is a linear map of the tangent spaces: $f_{*x}\colon TM_x \to TN_{f(x)}$. This map, f_{*x} is defined as follows: the vector $f_{*x}(\mathbf{v}) \in TN_{f(x)}$ is the tangent vector to the curve $(f \circ \phi)\colon \mathbb{R} \to N$. See Fig. 6.6. We write

$$f_{*x}(\mathbf{v}) = \frac{d}{dt}(f(\phi(t)))\bigg|_{t=0} = \frac{d}{dt}(f \circ \phi(t))\bigg|_{t=0}.$$

This foregoing equation defining $f_{*x}(\mathbf{v})$ and (6.9) look similar. The difference lies in the meaning of f. In (6.9) f is a real-valued mapping on the manifold, whereas the f giving the derivative map f_{*x} is a mapping from one manifold to another.

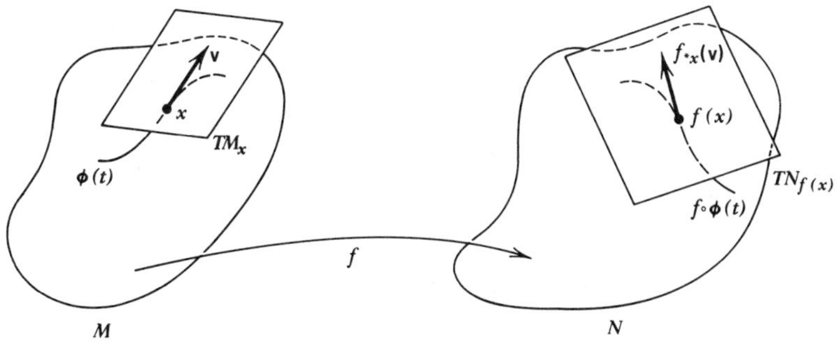

FIGURE 6.6. The relationship of the derivative map f_{*x}: $TM_x \to TN_{f(x)}$ and the manifold map f: $M \to N$.

Consider the meaning of the derivative map in local coordinates. In local coordinates, $\phi(t)$ is given by $x^i(t)$ and $(f \circ \phi(t))$ by $y^j(x^i(t))$, where $\{x^i\}$ is a set of local coordinates on M and $\{y^j\}$ is a set of local coordinates on N.

$$\frac{d}{dt}\left[y^j(x^i(t))\right] = \frac{\partial y^j}{\partial x^i}\frac{dx^i}{dt} = v^i \frac{\partial y^j}{\partial x^i}.$$

If the local coordinates of $\mathbf{v}$ are v^i, then the coordinates of $f_{*x}(\mathbf{v})$ are given by $v^i(\partial y^j/\partial x^i)|_{t=0}$. From this form in local coordinates it is obvious that f_{*x} is linear. The local representative of the derivative of the map f: $M \to N$ is the $n \times m$ matrix $\partial y^j/\partial x^i$ where $j = 1, \ldots, n$ and $i = 1, \ldots, m$. The dimension of M is m and the dimension of N is n.

We now consider two examples.

Example 1. Consider $M = \mathbb{R}^3$ and $N = \mathbb{R}^2$ with f: $\mathbb{R}^3 \to \mathbb{R}^2$ given by $f(x, y, z) = (x^2 - z, yz)$. If (v^1, v^2, v^3) are the contravariant components of $\mathbf{v}$ belonging to the tangent space at $(x, y, z) = \mathbf{x}$, then the contravariant components of $f_{*x}(\mathbf{v})$ are $(2xv^1 - v^3, zv^2 + yv^3)$ at the point $(x^2 - z, yz)$ in $\mathbb{R}^2$.

Example 2. Let $M = N = \mathbb{R}^3$ and let f be a rotation R. Then $y^j(t) = R^j{}_i x^i(t)$ and we compute $dy^j(t)/dt = R^j{}_i(dx^i/dt)$. In this case f_* at each x is given by the rotation matrix itself and so the tangent vectors in TN_{Rx} are obtained by a rotation applied to those in TM_x.

Taking the union of all these maps f_{*x} on the tangent spaces, plus the mapping on the manifold itself, we obtain a mapping on the entire tangent bundle: f_*: $TM \to TN$, where $(v^i, x^i) \to (v^i(\partial y^j/\partial x^i), y^j(x^i))$ in local coordinates. This mapping f_* is then the mapping on the entire tangent bundle that transforms velocities as well as coordinates, as is necessary for a Lagrangian function. In the preceding example, where f is a rotation map R, the mapping f_* as applied to a point (v^i, x^i) in the tangent bundle $T(\mathbb{R}^3)$ gives $(R^i{}_j v^j, R^i{}_j x^j)$. In the next section we see how the bundle map f_* leads to a precise definition of symmetry.

6.4. NOETHER'S THEOREM

Among the most important global properties of any physical system are its symmetries. Any solution whatever must reflect these symmetries, independent of the initial conditions. Thus in obtaining solutions it behooves us to make maximum use of any symmetry information available. One application of a symmetry is in finding constants of the motion. By definition a *constant of the motion* is a mapping I: $TM \to \mathbb{R}$ such that $dI/dt = 0$. For example, the energy E is a mapping on the tangent bundle, which is usually a constant of the motion. The relationship of a symmetry to a corresponding constant of the motion is the subject of Noether's theorem.

In qualitative terms a symmetry of a mechanical system is generally understood to mean a transformation that may be applied without altering the system or its dynamical interactions in any way. For example, consider motion of a particle in a central force field with potential $U(|\mathbf{r}|)$. Rotations alter nothing in this system and are referred to as a symmetry. We now sharpen this qualitative notion of symmetry.

Let M denote the configuration manifold for a physical system and $L(\dot{\mathbf{q}}, \mathbf{q}, t)$ the Lagrangian. Let h: $M \to M$ denote a differentiable mapping on the manifold M and h_*: $TM \to TM$ denote the corresponding bundle map. We say that h is an *admissible map* for the system if h_*: $TM \to TM$ leaves the Lagrangian invariant. To say that the Lagrangian is invariant under h_* means that $L \circ h_* = L$. In other words, the value of the Lagrangian on a transformed point in the tangent bundle is the same as the value on the point before transformation, and this statement is true for all points $(\mathbf{v}, \mathbf{x})$ in the tangent bundle TM. The relevant relationships are sketched in Fig. 6.7.

The Lagrangian with a central potential $U(|\mathbf{r}|)$ and the rotations again provide an example. The bundle mapping corresponding to h_* takes the point (v^i, x^i) in the tangent bundle $T(\mathbb{R}^3)$ to the point $(R^i{}_j v^j, R^i{}_j x^j)$ in $T(\mathbb{R}^3)$. We call this mapping R_*. Since rotations preserve inner products the kinetic energy as well as the potential energy is preserved under a rotation. Thus it is clear that $L \circ R_* = L$ and a rotation is an admissible map for this Lagrangian.

By a *symmetry* of the mechanical system we mean a one-parameter group of diffeomorphisms that are admissible mappings for the Lagrangian. If we

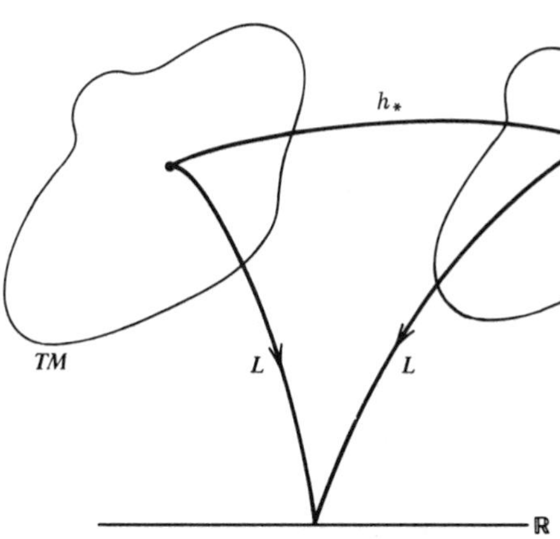

FIGURE 6.7. Bundle mapping h_* corresponding to an admissible map n for the Lagrangian L.

denote such a one-parameter group as $\{h^s\}$, then $\{h^s\}$ is a symmetry of the mechanical system with Lagrangian L if

$$L \circ h^s{}_* = L \qquad \text{for all } s. \tag{6.24}$$

We encountered one-parameter groups of diffeomorphisms for the first time when we considered phase flows in Section 2.6. In that case the diffeomorphism g^t: $M \to M$ was generated by the system itself as it evolved in time. In the present context we contemplate one-parameter diffeomorphisms that bear no direct relationship to the phase flow. Again rotations and a Lagrangian with a central potential provide a ready example. In this case the parameter s would be the rotation angle and the one-parameter group is a group of rotations.

With symmetries defined we proceed to obtain the associated invariants. Let $\{h^s\}$ denote a one-parameter group of diffeomorphisms on the configuration manifold M of a physical system. First consider such diffeomorphisms in general and then specialize to symmetries of the system. Let $\mathbf{Q}(t)$, $\mathbf{Q}$: $\mathbb{R} \to M$, denote a motion of the system, that is, a solution of the Euler–Lagrange equations, and let $\mathbf{q}(t, s) = h^s(\mathbf{Q}(t))$ denote that path that is the image of $\mathbf{Q}(t)$ under the diffeomorphism h^s. Note that $\mathbf{q}(t, 0) = h^0(\mathbf{Q}(t)) = \mathbf{Q}(t)$. In general, $L(\dot{\mathbf{q}}, \mathbf{q}, t)$ depends on the parameter s and we consider how it changes with this parameter.

$$\frac{\partial L}{\partial s} = \frac{\partial L}{\partial \dot{q}^i}\frac{\partial \dot{q}^i}{\partial s} + \frac{\partial L}{\partial q^i}\frac{\partial q^i}{\partial s}. \tag{6.25}$$

The partial derivative on s indicates that t is held fixed. Evaluate this derivative in (6.25) at $s = 0$. Since the tangent to the curve $\mathbf{q}$: $\mathbb{R} \to M$ is computed with s held fixed, $\dot{\mathbf{q}} = d\mathbf{q}/dt$. So at $s = 0$ we have

$$\left.\frac{\partial L}{\partial s}\right|_{s=0} = \frac{\partial L}{\partial \dot{Q}^i}\frac{d}{dt}\left(\left.\frac{\partial q^i}{\partial s}\right|_{s=0}\right) + \left.\frac{\partial L}{\partial Q^i}\frac{\partial q^i}{\partial s}\right|_{s=0}. \tag{6.26}$$

Since $\mathbf{Q}(t)$ is by hypothesis a motion of the system, $\partial L/\partial Q^i = d(\partial L/\partial \dot{Q}^i)/dt$. Substituting into (6.26) we find

$$\left.\frac{\partial L}{\partial s}\right|_{s=0} = \frac{d}{dt}\left(\left.\frac{\partial L}{\partial \dot{Q}^i}\frac{\partial q^i}{\partial s}\right|_{s=0}\right). \tag{6.27}$$

In the case that $\{h^s\}$ is a symmetry of the system, $\partial L/\partial s = 0$ because L is independent of s. Then from (6.27)

$$I = \left.\frac{\partial L}{\partial \dot{Q}^i}\frac{\partial}{\partial s}\left(h^s(Q^i)\right)\right|_{s=0} = \frac{\partial L}{\partial \dot{q}^i}\frac{\partial q^i}{\partial s} \tag{6.28}$$

is a constant of the motion. Note that we need not even evaluate at $s = 0$ as is done in (6.27) because if $\{h^s\}$ is a symmetry, then $\mathbf{q}(t, s)$ is also a motion. This follows because L is invariant and hence the action (3.5) is unchanged along the path $\mathbf{q}(t, s)$ and thus also an extremum.

To consolidate our results we restate *Noether's theorem*: If $\{h^s\}$ is a one-parameter group of diffeomorphisms such that (6.24) is satisfied, then I given by (6.28) is constant.

Not all constants of the motion can be associated with a symmetry of the system; however, it is generally true that the additive constants of the motion are associated with some general symmetry. Mappings that are symmetries (for appropriate Lagrangians) include translations, rotations, and combinations of rotations plus translations.

Let us consider some examples:

Example 1. Let L be a free particle Lagrangian, $L = m\dot{\mathbf{X}} \cdot \dot{\mathbf{X}}/2$. Let h^s: $\mathbb{R}^3 \to \mathbb{R}^3$ be given by $\mathbf{x} = h^s(\mathbf{X}) = \mathbf{X} + s\mathbf{a}$, where $\mathbf{a}$ is a fixed vector. Clearly, L is invariant and $\partial x^i/\partial s = a^i$. From (6.28) $m\dot{\mathbf{x}} \cdot \mathbf{a} =$ constant. In other words, the component of momentum in the direction of $\mathbf{a}$ is a constant. But since $\mathbf{a}$ is any arbitrary vector we have that all components of the momentum are constant. Even if the particle is not free and there exists a potential function such that the potential is invariant under translations along a fixed vector $\mathbf{a}$, then the component of the momentum along $\mathbf{a}$ is conserved.

Example 2. Suppose we consider the Lagrangian for a particle in a central force field: $L = (m/2)\dot{\mathbf{X}} \cdot \dot{\mathbf{X}} - U(|\mathbf{X}|)$. Because rotation matrices satisfy (1.32) and because $|\mathbf{X}|$ is invariant under a rotation, L is invariant under any rotation. Thus $\mathbf{x}(t, s) = R(s)\mathbf{X}(t)$ and $(\partial \mathbf{x}/\partial s) = R'\mathbf{X}$, where prime denotes differentiation with respect to s. We may write this equation as $(\partial \mathbf{x}/\partial s) = R'R^{-1}\mathbf{x} = R'R^t\mathbf{x} = \boldsymbol{\omega} \times \mathbf{x}$, where we have used (1.40) and (1.41) of Chapter 1, and identified the parameter t there, with the parameter s in the present context. Since $(\partial L/\partial \dot{\mathbf{x}}) = m\dot{\mathbf{x}}$, the invariant associated with an arbitrary rotation around the axis defined by $\boldsymbol{\omega}$ is $I = m\dot{\mathbf{x}} \cdot (\boldsymbol{\omega} \times \mathbf{x}) = m\boldsymbol{\omega} \cdot (\mathbf{x} \times \dot{\mathbf{x}})$. We recognize this as the component of the angular momentum $\mathbf{L}$ along the direction of $\boldsymbol{\omega}$. In the context of a central field, R may be any arbitrary rotation and hence the angular momentum component along any direction whatsoever is a constant and so $\mathbf{L}$ itself is a constant. If the potential is no longer central, but such that a rotation around a particular direction leaves it invariant, then only the component of the angular momentum along that direction will be constant. Further examples are given as problems.

We note that momentum, angular momentum, and energy are additive constants since for a system of many particles the value of these quantities for the entire system is just the sum of the constants for the individual particles. This additive property was shown in Chapter 2. The foregoing application of Noether's theorem shows that linear momentum is a constant of the motion when configuation space is translation invariant, that is, homogeneous (the same everywhere). Similarly, angular momentum is a constant when configuration space is rotationally invariant, that is, isotropic (has no preferred direction). The energy constant results from invariance under translation in time, as is shown in the following discussion.

Consider a somewhat different view of this situation and in particular concentrate on the notion of "constant of the motion" and not so much on symmetry. There are constants of the motion in general that are not given by a symmetry. In (6.27) $(\partial L/\partial s)|_{s=0}$ is no longer a function of s. If indeed there exists a function $G(\dot{\mathbf{q}}, \mathbf{q}, t)$ such that $(\partial L/\partial s)|_{s=0} = dG/dt$, then we have the constant

$$I = \left(\frac{\partial L}{\partial \dot{q}^i} \frac{\partial q^i}{\partial s} \right)\Bigg|_{s=0} - G. \tag{6.29}$$

Hence a symmetry is not required in order for a one-parameter group of diffeomorphisms to lead us to a constant of the motion. Sometimes for a given one-parameter group $\{h^s\}$, G exists and is easy to find. More often there will be no such function. The trick in applying (6.29) is to be able to recognize the $\{h^s\}$ that will lead to a constant. Generally, only a few are recognizable, but their discovery and use lead to a reduction in the complexity of the problem.

As an example to compare with the ones giving the momentum and angular momentum constants, let $\{h^s\}$ be the one-parameter group generated by the phase flow, that is, $h^s(\mathbf{Q}(t)) = \mathbf{q}(t, s) = \mathbf{Q}(t + s)$. We consider a system for which the Lagrangian does not depend on time explicitly. Let $\tau = t + s$ and then

$$\frac{\partial q^i}{\partial s} = \frac{dQ^i}{d\tau}\frac{\partial \tau}{\partial s} = \frac{dQ^i}{d\tau}\frac{\partial \tau}{\partial t} = \frac{dQ^i}{dt} = \frac{dq^i}{dt} = \dot{q}^i.$$

Since L depends on s only through $\mathbf{q}$ and $\dot{\mathbf{q}}$ in similar fashion we obtain $\partial L/\partial s = dL/dt$ and thus $G = L$ will work. Thus (6.29) implies that $H = (\partial L/\partial \dot{q}^i)\dot{q}^i - L$ is constant. The function H is called the *Hamiltonian* and is the constant of the motion associated with invariance under time translation. In Section 7.1 we show that the Hamiltonian is often equal to the total energy.

We subsequently refer to time-independent constants of the motion as *integrals of the motion* or *first integrals*. First integrals involve only the coordinates and velocities and not the time explicitly.

As a final example of obtaining constants of the motion from mappings let us consider a system with 2 degrees of freedom and with a Lagrangian given by

$$L = \tfrac{1}{2}m\dot{\mathbf{Q}} \cdot \dot{\mathbf{Q}} - \mathbf{A} \cdot \mathbf{Q}, \tag{6.30}$$

where $\mathbf{Q} = (Q^1, Q^2)$. The vector $\mathbf{A} = (a^1, a^2)$ is constant and we use the usual cartesian metric $g_{ij} = \delta_{ij}$. Because L does not depend explicitly on time, we know immediately that $E = \frac{1}{2}m\dot{\mathbf{Q}} \cdot \dot{\mathbf{Q}} + \mathbf{A} \cdot \mathbf{Q}$ is a constant of the motion. Choosing a vector $\mathbf{A}' = (a^2, -a^1)$ orthogonal to $\mathbf{A}$, there is one transformation of the manifold that leaves the Lagrangian invariant: $\mathbf{q} = \mathbf{Q} + s\mathbf{A}'$. Note that $\dot{\mathbf{q}} = \dot{\mathbf{Q}}$ and that $\mathbf{A} \cdot \mathbf{q} = \mathbf{A} \cdot \mathbf{Q} + s\mathbf{A} \cdot \mathbf{A}' = \mathbf{A} \cdot \mathbf{Q}$. Thus the one-parameter group of diffeomorphisms $\{h^s\}$ given by $h^s(\mathbf{Q}(t)) = \mathbf{q}(t, s) = \mathbf{Q}(t) + s\mathbf{A}'$ is a symmetry and gives a constant of the motion. We compute as before: $\partial q^i/\partial s = A'^i$ and $\partial L/\partial \dot{q}^i = m\dot{q}^i$, and we obtain from (6.28) the constant $\mathbf{I} = m\dot{\mathbf{q}} \cdot \mathbf{A}'$. This constant is the momentum in the direction of the vector $\mathbf{A}'$. From the derivative of the potential in the Lagrangian (6.30), one obtains the force $\mathbf{F} = -\nabla U = -\mathbf{A}$. Since $\mathbf{A}'$ is orthogonal to $\mathbf{A}$ (or $\mathbf{F}$), the momentum in the direction perpendicular to the force is conserved, as we would expect.

A direct integration of the Euler–Lagrange equations is straightforward and we find that $Q^i(t) = Q^i(0) + \dot{Q}^i(0)t - A^i t^2/m$ for $i = 1, 2$. Here we have four constants determined by the initial conditions. The energy and momentum constants plus these four are not all independent. They are related by $E = m\dot{Q}_i(0)\dot{Q}^i(0)/2 + A_i Q^i(0)$ and $I = m(\dot{Q}_1(0)a^2 - \dot{Q}_2(0)a^1)$. As independent constants we might choose E, I, $Q_1(0)$, $\dot{Q}_2(0)$. Then $Q_1(0) = Q_1 - \dot{Q}_1 t - a^1 t^2/2m$ and $\dot{Q}_2(0) = \dot{Q}_2 + a^2 t/m$ are constants of the motion that have explicit time dependence. However, the combination of the explicit terms with $Q^1(t)$, $\dot{Q}^1(t)$, and $\dot{Q}^2(t)$ is just so as to yield constants in time. We can find one-parameter diffeomorphisms that will lead through (6.29) to the given constants of the motion, but we leave this to an exercise (Problem 6.3). Only E and I are first integrals for this system.

Before leaving this example we wish to note that the form of the potential in the Lagrangian suggests a coordinate transformation: $q_1' = \mathbf{A} \cdot \mathbf{Q} = a^1 Q^1 + a^2 Q^2$. Then choosing $q_2' = a^2 Q^1 - a^1 Q^2$ so that the kinetic energy depends only on velocities, the Lagrangian in terms of these new coordinates is

$$L = \frac{m}{2|\mathbf{A}|^2}|\dot{\mathbf{q}}'|^2 - q_1'. \tag{6.31}$$

The coordinate q_2' is cyclic. Thus $\partial L/\partial \dot{q}_2' = (m/|\mathbf{A}|^2)\dot{q}_2'$ is constant, and this is essentially the same constant as I.

At some point in the solution of most problems it becomes simpler to continue with a straightforward integration of the reduced Euler–Lagrange equations than it is to hunt for further one-parameter diffeomorphisms that lead to constants of the motion. Indeed, since (6.24) is really the equation stating the invariance of a Lagrangian, we should perhaps really talk about a diffeomorphism h^s: $TM \to TM$ on the tangent bundle itself rather than focus on the induced map $h^s{}_*$. We will indeed exploit invariances under bundle mappings at a later time in a Hamiltonian context.

In this chapter we have seen how to visualize motion as trajectories on the tangent bundle. Such motions may be viewed as mappings on this manifold. Symmetries are interpreted as mappings on the tangent bundle that preserve the Lagrangian. Such symmetries lead to constants of the motion as given in Noether's theorem. If the Lagrangian is not an explicit function of time, then the mappings generated by the system motion leads to the energy invariant.

After a discussion of Hamilton's equations we return in Section 7.4 to the question of finding a sufficient set of integrals of the motion for a complete solution to the dynamical problem.

PROBLEMS

6.1. Derive (6.18) and (6.19) as given in the text.

6.2. Let $f\colon M \to N$ and $g\colon N \to K$ be differentiable mappings between manifolds. Show that $(g \circ f)_* = g_* \circ f_*$.

6.3. Consider the example in the text with the Lagrangian $L = \frac{1}{2}m\dot{\mathbf{Q}} \cdot \dot{\mathbf{Q}} - \mathbf{A} \cdot \mathbf{Q}$ where $\mathbf{A} = (a^1, a^2)$ and $\mathbf{Q} = (Q^1, Q^2)$. This problem is to find one-parameter mappings corresponding to the constants

$$I_1 = Q_1 - \dot{Q}_1 t - \frac{a^1 t^2}{2m}; \qquad I_2 = \dot{Q}_2 + \frac{a^2 t}{m}.$$

(a) Show that if G can be found to be independent of $\dot{\mathbf{Q}}$, then a mapping of the form $q^i = Q^i(t) + sz^i(t)$ where $z^i(t)$ is given by the equation $z^i = W^{ij}(\partial I/\partial \dot{Q}^j)$, may be used.

(b) Construct the mappings $\{h^s\}$ for the two constants I_1 and I_2.

6.4. Construct a one-parameter diffeomorphism and from symmetry considerations find a conserved quantity (in addition to the energy) for the Lagrangian

$$L = \tfrac{1}{2}m|\dot{\mathbf{x}}|^2 - V_0 x \sin\left(\frac{2\pi z}{R}\right) - V_0 y \cos\left(\frac{2\pi z}{R}\right),$$

where V_0 and R are constants. What is the symmetry of L. Find generalized coordinates one of which is cyclic.

6.5. The Lagrangian for the three-dimensional isotropic harmonic oscillator is

$$L = \tfrac{1}{2}m|\dot{\mathbf{X}}|^2 - \tfrac{1}{2}k|\mathbf{X}|^2.$$

Consider the matrix

$$F_{ij} = \frac{m}{2}\dot{X}_i\dot{X}_j + \frac{k}{2}X_i X_j.$$

(a) How many independent components of F_{ij} are there?

(b) Show that the F_{ij} are constants of the motion.

(c) Using the result of Problem 6.3(a), find one-parameter diffeomorphisms on $\mathbb{R}^3$ that yield the F_{ij} as constants of the motion through (6.29).

(d) Show that these constants are not all independent.

(e) Give an independent set of constants based on symmetries of L insofar as possible.

6.7. Consider motion that takes place on a surface of revolution in a uniform gravitational field. Show that angular momentum in the direction of the axis of revolution is conserved. Reduce the problem to quadratures. By setting the mass equal to 1 and $g = 0$, study the geodesics on this surface of revolution.

6.8. Show that the Runge–Lenz vector $\mathbf{A} = \dot{\mathbf{x}} \times \mathbf{L} - \alpha\mathbf{x}/x$ is a first integral for Kepler's problem. Find a one-parameter diffeomorphism that generates this constant as in Problem 6.3.

6.9. Investigate the conditions under which the small-oscillations Lagrangian (5.14) has a symmetry. Are there other constants of the motion obtainable for this Lagrangian through (6.29)?

CHAPTER SEVEN

HAMILTONIAN DYNAMICS

The language of truth is simple.

SENECA

Following a discussion of canonical momentum, Hamilton's equations are obtained and shown to be equivalent to the Euler–Lagrange equations of dynamics. The possibility of dynamic calculations using Hamiltonians depending on "coordinates" (p_i, q^i) motivates a consideration of covectors, cotangent spaces, and cotangent bundles. Poisson brackets are introduced and form the basis for a discussion of canonical transformation and symmetries of the Hamiltonian. Completely integrable systems are subsequently discussed, as are Hamiltonian phase flow and Liouville's theorem.

7.1. HAMILTON'S CANONICAL EQUATIONS

In obtaining the Euler–Lagrange equations from the Lagrangian, the coordinates and velocities are treated as independent variables. Nevertheless, the Euler–Lagrange equations are second-order differential equations for the functions $q^i(t)$, $i = 1, \ldots, n$. The function $\dot{q}^i(t)$ is not considered as independent from $q^i(t)$ once the differentiations of the Lagrangian have been performed. In contrast, the Hamiltonian view fully embraces the notion of dealing with $2n$ independent variables and first-order differential equations.

The usual path to a Hamiltonian formulation is through the Lagrangian, and we follow that procedure here.

Let $L: TM \to \mathbb{R}$ be the Lagrangian for a mechanical system where M denotes the configuration manifold. We let $\{q^i\}$ denote a local system of coordinates on M. Then the momentum *conjugate* to the coordinate q^i is defined as

$$p_i \equiv \frac{\partial L}{\partial \dot{q}^i}. \tag{7.1}$$

The quantity p_i is referred to as the *canonical momentum* conjugate to the coordinate q^i. We write simply $\mathbf{p}$ to denote the vector with components $(p_1, p_2, \dots, p_n)$.

We have written the canonical momentum with subscripts because, as we now show, $\partial L/\partial \dot{q}^i$ transforms like a covariant vector. Consider a coordinate transformation $q^{\hat{i}} = q^{\hat{i}}(q^1, q^2, \dots, q^n)$. Then $\dot{q}^{\hat{i}} = (\partial q^{\hat{i}}/\partial q^j)\dot{q}^j$ and $(\partial \dot{q}^{\hat{i}}/\partial \dot{q}^j) = (\partial q^{\hat{i}}/\partial q^j)$. Thus

$$p_i = \frac{\partial L}{\partial \dot{q}^i} = \frac{\partial L}{\partial \dot{q}^{\hat{i}}} \frac{\partial \dot{q}^{\hat{i}}}{\partial \dot{q}^i} = p_{\hat{i}} \frac{\partial q^{\hat{i}}}{\partial q^i}. \tag{7.2}$$

We emphasize that canonical momentum and mechanical momentum ($m\mathbf{v}$) are not necessarily the same. For many important physical systems they are different. In fact, canonical momenta may not even carry the same units as mechanical momentum.

With the notion of canonical momenta in hand, we shift our view of mechanics to the Hamiltonian perspective. Let us consider the differential of a Lagrangian, that is, a real-valued function $L: TM \times \mathbb{R} \to \mathbb{R}$.

$$dL = \dot{p}_i\, dq^i + p_i\, d\dot{q}^i + \frac{\partial L}{\partial t} dt, \tag{7.3}$$

where we have made use of the definition for canonical momenta and the Euler–Lagrange equations. Now rather than thinking of $(\dot{q}^i, q^i)$ as the proper coordinates for describing a mechanical system, let us consider as coordinates (p_i, q^i). In order to transform (7.3) to the differential of a function with these coordinates, we write $p_i\, d\dot{q}^i = d(p_i \dot{q}^i) - \dot{q}^i\, dp_i$, where we think of the $\dot{q}^i$ as real-valued functions of (p_i, q^i). Then with some rearrangement we can write (7.3) in the form

$$d\left(p_i \dot{q}^i - L\right) = \dot{q}^i\, dp_i - \dot{p}_i\, dq^i + \frac{\partial}{\partial t}\left(p_i \dot{q}^i - L\right) dt. \tag{7.4}$$

Identifying the Hamiltonian $H = p_i \dot{q}^i - L$, we then have *Hamilton's canonical equations*

$$\dot{q}^i = \frac{\partial H}{\partial p_i} \tag{7.5}$$

$$\dot{p}_i = -\frac{\partial H}{\partial q^i}, \tag{7.6}$$

with the additional relation for explicit time dependence

$$\frac{\partial H}{\partial t} = -\frac{\partial L}{\partial t}. \tag{7.7}$$

It is important to realize from these results that if $H = p_i \dot{q}^i - L$, then Hamilton's canonical equations completely determine the motion of the system and are entirely equivalent to the Euler–Lagrange equations. When using (7.5)–(7.7) it is important to keep in mind which variables are held constant when taking partial derivatives. Indeed, (7.7) results because $(\partial/\partial t)$ means hold (p_i, q^i) fixed (or $(\dot{q}^i, q^i)$ in the Lagrangian case). To gain reassurance as to the equivalence of (7.5), (7.6) with the Euler–Lagrange equations, let us obtain the Euler–Lagrange equations directly from these equations. To do this successfully, one must keep track of the variables that are being held constant. We do this with a subscript following a right-hand parenthesis. For example, $\partial H/\partial q^i = \partial H/\partial q^i)_p$. This notation is to remind us that all the momenta and all other coordinates other than q^i are to be held constant. With this in mind we have for (7.6), where we now write $H = p_j \dot{q}^j - L$ and think of H as a function of $\mathbf{q}$ and $\dot{\mathbf{q}}$:

$$\frac{d}{dt}\left(\frac{\partial L}{\partial \dot{q}^i}\right) = \dot{p}_i = -\left.\frac{\partial H}{\partial q^i}\right)_p = -p_j \left.\frac{\partial \dot{q}^j}{\partial q^i}\right)_p + \left.\frac{\partial L}{\partial q^i}\right)_p. \tag{7.8}$$

Using the chain rule for the last term in (7.8)

$$\left.\frac{\partial L}{\partial q^i}\right)_p = \left.\frac{\partial L}{\partial q^i}\right)_{\dot{q}} + \left.\frac{\partial \dot{q}^j}{\partial q^i}\right)_p \left.\frac{\partial L}{\partial \dot{q}^j}\right)_q. \tag{7.9}$$

We thus obtain from (7.8) the Euler–Lagrange equations. A similar treatment applied to (7.5) shows it to yield an identity.

If the Lagrangian is of the usual form, $L = T - V$, where $T = T(\dot{\mathbf{q}}, t)$ and $V = V(\mathbf{q}, t)$, then $\partial L/\partial \dot{q}^i = \partial T/\partial \dot{q}^i = p_i$. Furthermore, if the kinetic energy is a homogeneous, quadratic function of the velocities, then (see Problem 7.1) $\dot{q}^i(\partial T/\partial \dot{q}^i) = 2T$ and we have that $H = p_i\dot{q}^i - L = T + V$. In such a case the Hamiltonian is the total energy. Even though in many cases we can construct the Hamiltonian from the kinetic and potential energies in a fashion similar to the Lagrangian, we must always remember that it must be expressed in terms of the canonical momenta and the coordinates. This is in contrast to the Lagrangian, which must always be expressed in terms of the q^i and $\dot{q}^i$. Let us now look at an example.

Once again we consider the example of a moving pendulum given by the Lagrangian of (6.20).

$$L = (M + m)\dot{x}^2/2 + m(L^2\dot{\theta}^2 + 2\dot{x}\dot{\theta}L\cos\theta)/2 + mgL\cos\theta.$$

We find $p_x = \partial L/\partial \dot{x} = (M + m)\dot{x} + m\dot{\theta}L\cos\theta$ and $p_\theta = \partial L/\partial \dot{\theta} = mL^2\dot{\theta} + m\dot{x}L\cos\theta$. Solve for $\dot{x}$ and $\dot{\theta}$ in terms of the canonical momenta to find $\dot{x} = p_x mL^2/\Delta - p_\theta mL\cos\theta/\Delta$ and $\dot{\theta} = p_\theta(M + m)/\Delta - p_x mL\cos\theta/\Delta$, where $\Delta = (M + m)mL^2 - m^2L^2\cos^2\theta$. The Hamiltonian is

$$H = \frac{mL^2}{2\Delta}p_x^2 + \frac{M + m}{2\Delta}p_\theta^2 - \frac{mL\cos\theta}{\Delta}p_x p_\theta - mgL\cos\theta. \qquad (7.10)$$

Other examples will be considered in the exercises.

7.2. COVECTORS AND THE COTANGENT BUNDLE

In the foregoing section the Hamiltonian formulation of mechanics was introduced. The Hamiltonian formulation uses as independent variables the generalized coordinates and the corresponding conjugate momenta. Rather than second-order differential equations, first-order differential equations of motion result. A solution of this system gives in principle the functions $\mathbf{p}(t)$ and $\mathbf{q}(t)$. But in what kind of manifold should we view this motion? In answer to this question, we discuss the space for the vectors of canonical momentum and the proper domain manifold for the Hamiltonian $H(\mathbf{p}, \mathbf{q})$.

Corresponding to the finite dimensional vector space $TM_\mathbf{x}$ for each point $\mathbf{x}$ is a second vector space. This "dual," which we call the *cotangent space* and denote $T^*M_\mathbf{x}$, consists of all linear, real-valued mappings on $TM_\mathbf{x}$. To be explicit, if $\omega \in T^*M_\mathbf{x}$, then $\omega[\mathbf{v}] \in \mathbb{R}$, $\omega[\mathbf{v} + \mathbf{u}] = \omega[\mathbf{v}] + \omega[\mathbf{u}]$, and $\omega[a\mathbf{v}] = a\omega[\mathbf{v}]$, for all $a \in \mathbb{R}$ and for all $\mathbf{v}, \mathbf{u} \in TM_\mathbf{x}$. If ω and σ belong to $T^*M_\mathbf{x}$, then addition and scalar multiplication are defined, respectively, according to

$$(\omega + \sigma)[\mathbf{v}] = \omega[\mathbf{v}] + \sigma[\mathbf{v}] \in \mathbb{R} \qquad \text{and} \qquad a\omega[\mathbf{v}] = a(\omega[\mathbf{v}]) \in \mathbb{R},$$

for all $\mathbf{v} \in TM_{\mathbf{x}}$ and for all $a \in \mathbb{R}$. With these definitions the elements of $T^*M_{\mathbf{x}}$ do indeed form a vector space. We show later that $T^*M_{\mathbf{x}}$ is the natural home of momentum vectors with components p_i.

Tangent vectors to curves are natural residents of $TM_{\mathbf{x}}$. Similarly, the natural residents of $T^*M_{\mathbf{x}}$ are the differentials at $\mathbf{x} \in M$ of the real-valued functions on M. This we can see as follows. Let f: $M \to \mathbb{R}$ be a real-valued function on M with a local representative F. Each vector $\mathbf{v}$ in $TM_{\mathbf{x}}$, with local representative $\mathbf{V}$, gives a real number $\mathbf{V} \cdot \nabla F$ identified with the first-order change in f in the direction of $\mathbf{v}$. If we change $\mathbf{v} \in TM_{\mathbf{x}}$, then we change this number. Thus associated to f there is clearly a mapping from $TM_{\mathbf{x}}$ to the real numbers $\mathbb{R}$. The mapping we denote as $\mathbf{d}f$,

$$\mathbf{d}f[\mathbf{v}] = \mathbf{V} \cdot \nabla F \in \mathbb{R}. \tag{7.11}$$

Furthermore, from the properties of $\mathbf{V} \cdot \nabla F$ it is clear that $\mathbf{d}f$: $TM_{\mathbf{x}} \to \mathbb{R}$ is linear, and hence an element of $T^*M_{\mathbf{x}}$. Consequently, the differentials of real-valued mappings on the manifold M are natural residents of $T^*M_{\mathbf{x}}$.

Comparing (6.4), (7.11), and (6.6), we obtain

$$\mathbf{d}f[\mathbf{v}] = \partial_{\mathbf{v}}[f] = v^i f_{,i}, \tag{7.12}$$

where $f_{,i} = \partial f / \partial x^i$. The quantity to the left of the equal signs represents the view of $\mathbf{d}f$ (a linear, real-valued functional) operating on the vector $\mathbf{v}$. The quantity between the equal signs in (7.12) represents the view that the vector $\mathbf{v}$ is a (directional derivative) operator on the real-valued function f. Both of these quantities in (7.12) are equal to the expression on the right in terms of components. All three expressions in (7.12) represent the change in the function f along $\mathbf{v}$. Therefore $\mathbf{d}f[\mathbf{v}]$ and $\partial_{\mathbf{v}}[f]$ are two notations for exactly the same real number, where f: $M \to \mathbb{R}$, $\mathbf{d}f \in T^*M_{\mathbf{x}}$, and $\mathbf{v} \in TM_{\mathbf{x}}$. The differentials $\mathbf{d}f$ of such real-valued functions are elements of the cotangent space $T^*M_{\mathbf{x}}$. We call such elements *covectors*.

From (6.4) we found the coordinate functions to be useful in giving the components of an arbitrary tangent vector $\mathbf{v}$: $v^i = \partial_{\mathbf{v}}[x^i]$. The differentials of these coordinate functions play a central role in $T^*M_{\mathbf{x}}$. Indeed, from (7.12) we find

$$\mathbf{d}x^i[\mathbf{v}] = \partial_{\mathbf{v}}[x^i] = v^j \partial_j[x^i] = v^i. \tag{7.13}$$

Suppose $\mathbf{v} = v^i \mathbf{e}_k$ is the natural basis vector $\mathbf{e}_j$, that is, $v^k = \delta_j^k$. Then using (6.7), Eq. (7.13) takes the form

$$\mathbf{d}x^i[\mathbf{e}_j] = \delta_j^k \partial_k[x^i] = \frac{\partial x^i}{\partial x^j} = \delta_j^i. \tag{7.14}$$

Using (7.14) we show that the set $\{\mathbf{d}x^i\}$ forms a basis of $T^*M_{\mathbf{x}}$.

Let $\boldsymbol{\omega} \in T^*M_{\mathbf{x}}$ be an arbitrary covector, and let $\boldsymbol{\omega}[\mathbf{e}_i] \equiv \omega_i \in \mathbb{R}$. Then it follows that $\boldsymbol{\omega} = \omega_i \, \mathbf{d}x^i$, since for an arbitrary $\mathbf{v}$

$$\boldsymbol{\omega}[\mathbf{v}] = \boldsymbol{\omega}\left[v^i \mathbf{e}_i\right] = v^i \boldsymbol{\omega}[\mathbf{e}_i] = \omega_i v^i \tag{7.15}$$

and

$$\omega_i \, \mathbf{d}x^i[\mathbf{v}] = \omega_i \, \mathbf{d}x^i\left[v^j \mathbf{e}_j\right] = \omega_i v^j \delta^i_j = \omega_i v^i. \tag{7.16}$$

We conclude from (7.15) and (7.16) that the set $\{\mathbf{d}x^i\}$ forms a basis for *all* covectors of $T^*M_{\mathbf{x}}$. The basis $\{\mathbf{d}x^i\}$ of $T^*M_{\mathbf{x}}$ is referred to as the *dual basis* to the basis $\{\mathbf{e}_i\}$ of $TM_{\mathbf{x}}$. Other bases of $TM_{\mathbf{x}}$ will also have corresponding dual bases that are also related to each other through the Kronecker delta as in (7.14). The basis $\{\boldsymbol{\omega}^i\}$ of $T^*M_{\mathbf{x}}$ is *dual* to the basis $\{\hat{\mathbf{e}}_j\}$ of $TM_{\mathbf{x}}$ if $\boldsymbol{\omega}^i[\hat{\mathbf{e}}_j] = \delta^i_j$.

For differentials of functions we write $\mathbf{d}f = f_{,i} \, \mathbf{d}x^i$. Then using (7.14)

$$\mathbf{d}f[\mathbf{v}] = f_{,i} \, \mathbf{d}x^i\left[v^j \mathbf{e}_j\right] = f_{,i} v^j \, \mathbf{d}x^i[\mathbf{e}_j] = f_{,i} v^i. \tag{7.17}$$

This is exactly (7.12). As with any other covector, differentials can be written as linear combinations of the $\{\mathbf{d}x^i\}$.

Because of the intimate connection with the differentials of real-valued functions, the covectors of $T^*M_{\mathbf{x}}$ are often referred to as differential forms of first rank of just *differential* 1-*forms*. At times when we wish to emphasize the role that elements of $T^*M_{\mathbf{x}}$ play as operators on vectors, we will refer to them as 1-forms.

Despite the attention we have given to distinguishing between $TM_{\mathbf{x}}$ and $T^*M_{\mathbf{x}}$, when a metric tensor exists defining scalar products, there is no real difference between them, in the sense that for every vector there is a covector and vice versa. We say that $TM_{\mathbf{x}}$ and $T^*M_{\mathbf{x}}$ are *isomorphic*. Let $\mathbf{u} \in TM_{\mathbf{x}}$ be an arbitrary vector. Then $(\mathbf{u} \cdot \mathbf{v}) \in \mathbb{R}$ for all $\mathbf{v} \in TM_{\mathbf{x}}$, and further if we view the composite symbol "$\mathbf{u} \cdot$ " as an operator on $TM_{\mathbf{x}}$, this operator is real-valued, linear, and thus a covector! In other words, if we think of the metric as a real-valued mapping on $TM_{\mathbf{x}}$ with two arguments, we can write

$$g(\mathbf{u}, \mathbf{v}) = g_{ij} u^i v^j \in \mathbb{R}, \tag{7.18}$$

Then from (7.18) it follows that the mapping $g(\mathbf{u}, \ ): TM_{\mathbf{x}} \to \mathbb{R}$ is indeed a real-valued, linear mapping on $TM_{\mathbf{x}}$ and hence an element of $T^*M_{\mathbf{x}}$. The mappings $g(\mathbf{u}, \ )$ and "$\mathbf{u} \cdot$ " are two notations for the same covector. This element is the isomorphic image of **u**. If $\mathbf{u} \in TM_{\mathbf{x}}$ has components u^i in a natural basis, then the isomorphic image has components $u_i = g_{ij}u^j$. See Problem 7.6. The operations of raising and lowering indices is tantamount to switching back and forth between vectors and covectors.

Velocity components are defined in their contravariant form and canonical momenta are defined in the covariant form. For this reason we speak of velocity vectors as natural residents of $TM_{\mathbf{x}}$ and canonical momentum as natural to $T^*M_{\mathbf{x}}$. However, each of these physical quantities, along with other vectors such as acceleration, magnetic field, and so on, has a vector and a covector form. This holds as long as a metric is defined.

In exactly the same way as is done for the tangent bundle, we form the cotangent bundle as the union of all the cotangent vector spaces; $T^*M = \cup_{\mathbf{x} \in M} T^*M_{\mathbf{x}}$. If $\mathbf{p} \in T^*M_{\mathbf{x}}$ then a point in T^*M is $(\mathbf{p}, \mathbf{x})$, and we denote the local coordinates on T^*M as (p_i, x^i). The Hamiltonian function then is a mapping $H: T^*M \to \mathbb{R}$. As is standard in physics, we will henceforth refer to the cotangent bundle as *phase space* with coordinates (p_i, q^i).

For an example we look briefly at the Hamiltonian of (7.10). Hamilton's equations, (7.5) and (7.6), tell us immediately that $\dot{p}_x = 0$, and so p_x is constant. The energy function is also constant and equal to the Hamiltonian. Thus the trajectories in the four-dimensional cotangent bundle for this system are confined to a cylindrical submanifold with coordinates (p_θ, θ). The trajectories are similar to those given in Fig. 6.5, and we do not make a separate sketch. This cotangent bundle $T^*(\mathbb{R}^1 \times S^1)$ is called the phase space of the system.

The transformation from $L: TM \times \mathbb{R} \to \mathbb{R}$ to the function $H: T^*M \times \mathbb{R} \to \mathbb{R}$ where $H = p_i\dot{q}^i - L$ is called a Legendre transformation and finds frequent application in thermodynamics. Both in mechanics and thermodynamics the motivation is to change from a current set of variables to a new set, which for some reason may be more convenient. At first glance such a transformation may not seem too important, since it amounts to nothing more than a coordinate change. However, a physical system is often more easily characterized or studied in one set of coordinates than another. Indeed, one's understanding of a problem may depend critically on having the right view or perspective (coordinates). The Hamiltonian gives an alternative view to the Newtonian or Lagrangian perspectives.

7.3. POISSON BRACKETS

The study of dynamics is fundamentally a study of how physical variables change with time. In the Hamiltonian view a physical variable F is a function of the "coordinates" $(p_i(t), q^i(t))$, $i = 1, 2, \ldots, n$, where n is the dimension of the configuration manifold M. The function F may also be an explicit function of time. From a mathematical viewpoint F is a mapping $F(\mathbf{p}(t), \mathbf{q}(t), t)$, $F\colon T^*M \times \mathbb{R} \to \mathbb{R}$. Position, energy, and speed are all examples of the physical variable F.

Since we are interested in the way such functions change with time, we calculate the time rate of change of an arbitrary function F.

$$\frac{dF}{dt} = \frac{\partial F}{\partial t} + \frac{\partial F}{\partial q^i}\dot{q}^i + \frac{\partial F}{\partial p_i}\dot{p}_i. \tag{7.19}$$

Substitute in (7.19) for $\dot{q}^i$ and $\dot{p}_i$ from Hamilton's equations (7.5) and (7.6).

$$\frac{dF}{dt} = \frac{\partial F}{\partial t} + \frac{\partial F}{\partial q^i}\frac{\partial H}{\partial p_i} - \frac{\partial F}{\partial p_i}\frac{\partial H}{\partial q^i}. \tag{7.20}$$

A convenient and extremely useful notation is introduced by defining the Poisson bracket. The *Poisson bracket* of the phase-space functions f and g is defined by

$$\{f, g\} \equiv \frac{\partial f}{\partial q^i}\frac{\partial g}{\partial p_i} - \frac{\partial f}{\partial p_i}\frac{\partial g}{\partial q^i}. \quad \text{(Note sum on } i!) \tag{7.21}$$

From (7.21) we see immediately the following algebraic properties of Poisson brackets.

$$\{f, g\} = -\{g, f\} \qquad \text{(skew symmetric)}, \tag{7.22}$$

$$\{a_1 f + a_2 g, h\} = a_1\{f, h\} + a_2\{g, h\} \qquad \text{(bilinear)}, \tag{7.23}$$

where a_1, a_2 are arbitrary constants and f, g, h are arbitrary functions on phase space. In terms of the Poisson bracket (7.20) becomes

$$\frac{dF}{dt} = \frac{\partial F}{\partial t} + \{F, H\}. \tag{7.24}$$

Relation (7.24), or equivalently (7.20), is very important because it details how any function on phase space will change with time as a Hamiltonian system evolves.

We call the evolution of a mechanical system described by a Hamiltonian a *Hamiltonian flow*. Hamiltonian flow is an example of phase flow as discussed in Section 2.6. In the present context the variable $\boldsymbol{\xi}$ of (2.52) is the phase-space "position" vector $(\mathbf{p}, \mathbf{q})$. The "velocity" of (2.52) is $(\dot{\mathbf{p}}, \dot{\mathbf{q}}) =$

$(-\partial H/\partial \mathbf{q}, \partial H/\partial \mathbf{p})$. We denote the Hamiltonian flow map as g_H^t and note that if H does not depend explicitly on time, then the diffeomorphisms $\{g_H^t \mid t \in \mathbb{R}\}$ form a one-parameter group.

As a first example, let F of (7.24) be equal to the Hamiltonian itself. From (7.22) we see that $\{f, f\} = 0$ for an arbitrary function f. Thus (7.24) gives

$$\frac{dH}{dt} = \frac{\partial H}{\partial t}. \tag{7.25}$$

In words: If H is not an explicit function of time, then H is a constant of the motion. As an application of (6.29), we showed in the previous chapter that if the Lagrangian L was not an explicit function of time, then H was a constant of the motion. Equation (7.25) is a companion result. If L is not an explicit function of time, then neither is H. From (7.7) we conclude that the converse is also true.

A major advantage of Lagrangian mechanics in comparison to the Newtonian formulation was the use of generalized coordinates. These coordinates eliminate constraint equations and, in general, add theoretical insight as the motion is considered in terms of trajectories on configuration manifolds. Likewise, an advantage of Hamiltonian dynamics lies in the fact that even more general coordinates are used. Not only do we have the q^i as coordinates, but we also have the p_i. Not only can we consider transformations of coordinates on the configuration manifold M, but we can see that now the possibility exists for coordinate transformations on phase space (i.e., the cotangent bundle manifold). Such transformations may mix the "p's" and "q's" and it becomes impossible to classify certain coordinates as "momentum" and others as not. The possibilities are limitless and it is useful to restrict such coordinate transformations on phase space to those that preserve Hamilton's equations. With this in mind, let us consider a general coordinate transformation on phase space.

In some neighborhood of phase space let $\{P_i, Q^i\}$ denote the coordinates of a new coordinate chart, which does not change with time. Then $P_i = P_i(\mathbf{p}, \mathbf{q})$ and $Q^i = Q^i(\mathbf{p}, \mathbf{q})$. Similarly, we may consider $p_i = p_i(\mathbf{P}, \mathbf{Q})$ and $q^i = q^i(\mathbf{P}, \mathbf{Q})$. Apply (7.24) to $P_i(\mathbf{p}, \mathbf{q})$ and $Q^i(\mathbf{p}, \mathbf{q})$ while thinking of H as a function of $(\mathbf{p}, \mathbf{q})$ through $(\mathbf{P}, \mathbf{Q})$.

$$\frac{dQ^i}{dt} = \frac{\partial H}{\partial Q^k}\{Q^i, Q^k\}_{(\mathbf{p},\mathbf{q})} + \frac{\partial H}{\partial P_k}\{Q^i, P_k\}_{(\mathbf{p},\mathbf{q})}, \tag{7.26}$$

$$\frac{dP_i}{dt} = \frac{\partial H}{\partial Q^k}\{P_i, Q^k\}_{(\mathbf{p},\mathbf{q})} + \frac{\partial H}{\partial P_i}\{P_i, P_k\}_{(\mathbf{p},\mathbf{q})}, \tag{7.27}$$

where subscripts have been added to the Poisson brackets in order to make explicit which variables are involved in the differentiation. If the functions P_i and Q^i satisfy the Poisson bracket relations

$$\{Q^i, Q^j\}_{(\mathbf{p},\mathbf{q})} = 0; \qquad \{P_i, P_j\}_{(\mathbf{p},\mathbf{q})} = 0; \qquad \{Q^i, P_j\}_{(\mathbf{p},\mathbf{q})} = \delta^i_j, \quad (7.28)$$

then (7.26) and (7.27) show immediately that P_i and Q^i satisfy Hamilton's equations. Coordinate transformations that satisfy (7.28) play a prominent role in the theory of Hamiltonian dynamics and are called *canonical transformations*.

Since a canonical transformation implies that the new variables satisfy Hamilton's equations, (7.28) are also satisfied if we interchange the variables $(\mathbf{p}, \mathbf{q}) \leftrightarrow (\mathbf{P}, \mathbf{Q})$. Two coordinate charts related to each other by a canonical transformation are entirely equivalent insofar as the dynamical equations of motion are concerned. Canonical transformations are not like symmetry transformations in that they do not give immediately a constant of the motion. Nevertheless, the "symmetry" involved in canonical transformations, preservation of Hamilton's equations, has many consequences, which will be explored fully in the next chapter.

Before leaving this brief introduction to canonical transformations, it is important to give some examples.

Example 1. Let $Q = qp$ and $P = \ln p$. the partial derivatives of Q and P, when substituted into (7.28), show this transformation to be canonical.

Example 2. Let $Q = q$ and $P = 2p$. Following the same procedure we see that this transformation is not canonical. However, if (p, q) satisfy Hamilton's equations, then (P, Q) will also for the "new" Hamiltonian $2H$. We emphasize that our definition of a canonical transformation is independent of reference to any particular Hamiltonian; it refers only to a relationship among the coordinates on phase space. In contrast to the definition given in some textbooks, our definition requires more than that the form of Hamilton's equations be preserved. However, Hamilton's equations are always preserved as a consequence of the definition given in (7.28). The converse is not true,. as demonstrated by the foregoing example. Preservation of Hamilton's equations for some new Hamiltonian, even if one can always be found, does not insure that (7.28) will be satisfied. We remark that time-dependent canonical transformations require a slight generalization of (7.28) and are discussed in Chapter 8.

The central topic considered in the previous chapter was invariants of the motion. A new look at dynamics in the Hamiltonian formulation invites a reconsideration of symmetries and associated constants. To obtain the Hamiltonian equivalent of Noether's theorem, we consider a second function on phase space $F(\mathbf{p}, \mathbf{q})$, without explicit time dependence. Designate g_F^s as the flow map generated by the function F. The flow g_F^s is generated in exactly the same way that the Hamiltonian flow is generated. Specifically, $(d\mathbf{p}/ds, d\mathbf{q}/ds) = (-\partial F/\partial \mathbf{q}, \partial F/\partial \mathbf{p})$, which is the equivalent of (2.52). The change in H with a change in s is given by

$$\frac{dH}{ds} = \frac{\partial H}{\partial q^i}\frac{dq^i}{ds} + \frac{\partial H}{\partial p_i}\frac{dp_i}{ds} = \{H, F\}. \tag{7.29}$$

Thus H is invariant under the mapping g_F^s iff $\{H, F\} = 0$. However, for a function without explicit time dependence, $\{H, F\} = 0$, and (7.24) imply that $dF/dt = 0$. Thus we have the result that a time-independent, phase-space function $F(\mathbf{p}, \mathbf{q})$ is a first integral iff H is invariant under the phase-space mapping g_F^s generated by F.

As an example let us consider $F = p_k$. We then find that $(d\mathbf{p}/ds, d\mathbf{q}/ds)$ in component form is given by $(dp_i/ds, dq^i/ds) = (0, \delta^i_k)$. This equation defining the flow g_F^s has the immediate solution $q^k = s + q^k(0)$ and all other coordinates and momenta are constant under the flow g_F^s. Thus if $F = p_k$ is to be a constant of the motion, we must have $dH/ds = 0$. But this can be true only if H is independent of q^k. Once again we see that the momentum conjugate to a cyclic coordinate in the Hamiltonian is conserved. In analogous fashion we can show that q^k is constant if H is independent of p_k.

In comparing to Noether's theorem in Section 6.4, we see that g_F^s corresponds to h_*^s. For the Lagrangian and the Hamiltonian, it is invariance under the mappings h_*^s and g_F^s, respectively, that leads to the associated constant. Just as $L \circ h_*^s = L$, we have $H \circ g_F^s = H$. Compare Fig. 7.1 and Fig. 6.7.

Continuing a study of invariants of the motion, we show that the Poisson bracket of two constants of the motion is also a constant of the motion. We make use of the *Jacobi relation* for Poisson brackets.

$$\{F, \{G, H\}\} + \{G, \{H, F\}\} + \{H, \{F, G\}\} = 0, \tag{7.30}$$

where F, G, and H are arbitrary functions on phase space. The proof of (7.30) is straightforward and left to Problem 7.13.

Let R and S be two functions on phase space, which are constants of the motion generated by a Hamiltonian H, that is, $dR/dt = 0$ and $dS/dt = 0$. Then from (7.24)

$$\frac{d}{dt}\{R, S\} = \frac{\partial}{\partial t}\{R, S\} + \{\{R, S\}, H\}. \tag{7.31}$$

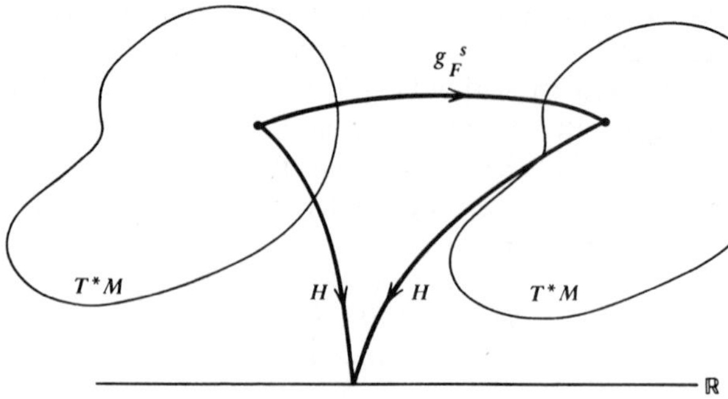

FIGURE 7.1. Invariance of the Hamiltonian under the flow map g_F^s generated by a constant of the motion F.

The commutivity of partial derivatives implies

$$\frac{\partial}{\partial t}\{R, S\} = \left\{\frac{\partial R}{\partial t}, S\right\} + \left\{R, \frac{\partial S}{\partial t}\right\}.$$

The Jacobi relation (7.30) may be written in the form

$$\{\{R, S\}, H\} = \{\{R, H\}, S\} + \{R, \{S, H\}\}.$$

Substituting these foregoing relations into (7.31) and exploiting the linearity of the Poisson brackets as in (7.23) gives

$$\frac{d}{dt}\{R, S\} = \left\{\frac{dR}{dt}, S\right\} + \left\{R, \frac{dS}{dt}\right\} = 0. \tag{7.32}$$

Hence the Poisson bracket of two constants of the motion is itself a constant of the motion (Poisson's theorem).

This theorem raises once again the question of how many constants of the motion there are. Since the Poisson bracket of two constants of the motion is a constant, it can be included in Poisson brackets to obtain more constants. However, Poisson's theorem does not guarantee that the new constant $\{R, S\}$ will be independent of R and S or that it will be nontrivial. Frequently, $\{R, S\}$ gives nothing very interesting because either it is zero or it gives nothing new. In the following section we consider what it means to find "enough" constants of the motion.

7.4. COMPLETELY INTEGRABLE SYSTEMS

As one struggles to find constants of the motion, employing whatever methods seem suitable for the problem at hand, including the exploitation of symmetries, one might well ask, "How many constants of the motion are needed?" For complex dynamical systems with more than one or two degrees of freedom, it is unlikely that there will be a sufficient number of obvious symmetries that a complete solution for the motion can be obtained. But how many constants of the motion should one seek?

At first sight one might expect that a system with n degrees of freedom would require $2n$ constants of the motion. This is suggested by the fact that the Euler–Lagrange equations, a set of n second-order differential equations, requires $2n$ constants to specify a solution completely. Or alternatively, Hamilton's equations consist of a set of $2n$ first-order equations requiring $2n$ constants for a complete solution. It is indeed true that $2n$ constants are needed to specify completely the solution to the dynamical problem. However, we show below a result attributed to Liouville: Finding n first integrals is sufficient to obtain a complete solution to the problem. Recall that a first integral is a constant of the motion that does not depend on time explicitly.

Consider a system with n degrees of freedom where the Hamiltonian is given by $H(\mathbf{p}, \mathbf{q})$. We restrict our attention to a system for which n integrals of the motion $I_i(\mathbf{p}, \mathbf{q})$, $i = 1, \ldots, n$ have been found. We further assume that these integrals are linearly independent functions of the variables $(\mathbf{p}, \mathbf{q})$ and that the I_i are single valued in some suitable neighborhood of T^*M.

We must place one further restriction on the relationship among these first integrals. They are required to satisfy a relationship among their derivatives that is termed "involutive." The first integrals I_i, I_j are said to be in *involution* when $\{I_i, I_j\} = 0$. We assume our set of n first integrals to be in involution, that is, $\{I_i, I_j\} = 0$ for all i, j. It is important to realize that the existence of a set of n first integrals satisfying these restrictions is the exception. Most mechanical systems will not have such a set.

Since the set of $\{I_i\}$ is linearly independent and satisfies $dI_i/dt = 0$ for each i, they would make a nice set of momentum coordinates if we could find a corresponding set of configuration coordinates Q^i, so that Hamilton's equations are satisfied. This we can do in the following way. First note that since the I_i are single valued and linearly independent, they can be inverted to find $p_i(\mathbf{I}, \mathbf{q})$. Now use the function

$$S(\mathbf{q}, \mathbf{I}) = \int_{\mathbf{q}_0}^{\mathbf{q}} p_i(\mathbf{I}, \mathbf{q})\, dq^i \tag{7.33}$$

to obtain the candidates for the new configuration coordinates from the relation

$$Q^i = \frac{\partial S}{\partial I_i}. \tag{7.34}$$

Functions like S are called "generating functions" and we will study them in detail in Chapter 8. For the present we simply obtain the "new coordinates" through (7.34) and note further that $\partial S/\partial q^i = p_i$. We caution the reader once again that strict attention must be paid to which variables are being held constant in such partial differentiations.

By differentiating $Q^i(\mathbf{p}, \mathbf{q})$ and using (7.33) and (7.34) we find

$$\frac{\partial Q^j}{\partial q^i} = \frac{\partial^2 S}{\partial I_k\, \partial I_j}\frac{\partial I_k}{\partial q^i} + \frac{\partial^2 S}{\partial q^i\, \partial I_j}, \tag{7.35}$$

$$\frac{\partial Q^j}{\partial p_i} = \frac{\partial^2 S}{\partial I_k\, \partial I_j}\frac{\partial I_k}{\partial p_i}. \tag{7.36}$$

These results may be used to find the Poisson brackets of Q^i with Q^j and I_j. The result we summarize along with the involutive constraint:

$$\{Q^i, Q^j\} = \{I_i, I_j\} = 0.$$

$$\{Q^i, I_j\} = \delta^i_j. \tag{7.37}$$

Thus (7.28) is satisfied and the transformation $(p_i, q^i) \to (I_i, Q^i)$ is a canonical transformation. Equations (7.26) and (7.27) ensure that Hamilton's equations are satisfied and we obtain

$$\dot{I}_i = -\frac{\partial H}{\partial Q^i}, \tag{7.38}$$

$$\dot{Q}^i = \frac{\partial H}{\partial I_i}. \tag{7.39}$$

Recall that the I_i are constants and $\dot{I}_i = 0$ for all i. Hence, from (7.38) $\partial H/\partial Q^i = 0$ for all i, that is, $H(\mathbf{I}, \mathbf{q}) = H(\mathbf{I})$. The Hamiltonian must depend *only* on the n constants of the motion I_i, $i = 1, \ldots, n$. So $\dot{Q}^i = \partial H/\partial I_i$ is constant, that is, $dQ^i/dt = \alpha^i$, and we find $Q^i(t) = \alpha^i t + \beta^i$, where α^i and β^i are both constant in time.

Such a system, for which n constants of the motion may be found satisfying the stated conditions, has been completely solved! Note that the Hamiltonian, which is usually taken to be one of the constants of the motion, is in involution with all other constants of the motion.

We consider again the example with Lagrangian (6.30) but make a change in notation to lower-case letters for the original coordinates.

$L = m\dot{\mathbf{q}} \cdot \dot{\mathbf{q}}/2 - \mathbf{A} \cdot \mathbf{q}$ and $H = \mathbf{p} \cdot \mathbf{p}/2m + \mathbf{A} \cdot \mathbf{q}$, where $\mathbf{q} = (q^1, q^2)$, $\mathbf{A} = (a^1, a^2)$, and $p_i = m\dot{q}^i$. H is one constant of the motion and we showed that $I = \mathbf{p} \cdot \mathbf{A}'$ is another constant, where $\mathbf{A}' = (a^2, -a^1)$. Since $dI/dt = \{H, I\} = 0$, the constants $I_1 \equiv H$ and $I_2 \equiv I$ are sufficient to find a complete solution. Inverting the expressions for these constants in terms of momenta, we solve for p_1 and p_2 to find

$$p_1 = \frac{\left\{a^2 I_2 \pm a^1\left[2m|\mathbf{A}|^2(I_1 - \mathbf{A} \cdot \mathbf{q}) - I_2^2\right]^{1/2}\right\}}{|\mathbf{A}|^2}$$

$$p_2 = \frac{\left\{-a^1 I_2 \pm a^2\left[2m|\mathbf{A}|^2(I_1 - \mathbf{A} \cdot \mathbf{q}) - I_2^2\right]^{1/2}\right\}}{|\mathbf{A}|^2}. \tag{7.40}$$

These expressions for the momenta are substituted into (7.33) to find the generating function S. As is often the case, the integrals involved need never be carried out, since we need S only to obtain the new coordinates Q^1 and Q^2. It is straightforward to find

$$Q^1 = \frac{\partial S}{\partial I_1} = \mp \frac{1}{|\mathbf{A}|^2}\left[2m|\mathbf{A}|^2(I_1 - \mathbf{A} \cdot \mathbf{q}) - I_2^2\right]^{1/2} + \text{const.}, \tag{7.41}$$

$$Q^2 = \frac{\partial S}{\partial I_2} = \frac{1}{|\mathbf{A}|^2}\left(\mathbf{A}' \cdot \mathbf{q} - \frac{I_2 Q^1}{m}\right). \tag{7.42}$$

Now since $H = I_1$, we find $\dot{Q}^1 = 1$ and $Q^1 = t +$ constant. Similarly, $Q^2 =$ constant. It is a matter of some straightforward algebra to show that this leads to the same form of the solution for q^1 and q^2 as given in Section 6.4; however, the constants involved are no longer initial values for $\mathbf{q}$ and $\dot{\mathbf{q}}$.

7.5. LIOUVILLE'S THEOREM

In Chapter 2 we introduced phase space as the space with coordinates and coordinate time derivatives as the variables. We recognize this space now as the tangent bundle manifold. In the general physics literature phase space uses "p_i's" and "q^i's" as coordinates and is the cotangent bundle. The Hamiltonian phase flow defined in Section 7.3 for $H(\mathbf{p}, \mathbf{q})$ refers to a one-parameter group of diffeomorphisms on the cotangent bundle: $g_H^t\colon T^*M \to T^*M$, where M is the configuration manifold for the mechanical system under study.

The phase flow has some interesting properties and one of the most interesting and useful is called Liouville's theorem. *Liouville's theorem* states that phase flows preserve volume in phase space. To prove this theorem we use a fluid analogy and the corresponding language. We think of the phase flow as a fluid flow and note the dynamical equation governing this flow can be written in the form

$$\frac{d\mathbf{x}}{dt} = \mathbf{v}(\mathbf{x}, t), \tag{7.43}$$

where $\mathbf{x} = (\mathbf{p}, \mathbf{q})$ and $\mathbf{v} = (\dot{\mathbf{p}}, \dot{\mathbf{q}})$ is given by Hamilton's equations. Equation (7.43) is an abbreviated form of (7.5) and (7.6). Recall that an incompressible fluid flow is characterized by the fact that the density does not change as one follows a fluid element in the flow. The density does not change in the flow iff $\nabla \cdot \mathbf{v} = 0$. See Problem 7.8. For such an incompressible flow we focus on an initially given volume of fluid. Follow these same fluid particles as the flow proceeds. Since the mass of this fixed collection of fluid particles does not change and further because the density at each fluid element is constant, we must have that the total volume occupied by the fluid elements is also unchanged. In other words, the volume is preserved under an incompressible flow. To see that the phase flow is incompressible we need only show $\nabla \cdot \mathbf{v} = 0$. The operator "$\nabla$" and the "velocity" $\mathbf{v}$ are in phase space.

$$\nabla \cdot \mathbf{v} = \frac{\partial \dot{q}^i}{\partial q^i} + \frac{\partial \dot{p}_i}{\partial p_i} = \frac{\partial}{\partial q^i}\left(\frac{\partial H}{\partial p_i}\right) - \frac{\partial}{\partial p_i}\left(\frac{\partial H}{\partial q^i}\right) = 0.$$

As a consequence of Hamilton's equations the *phase flow is incompressible* and hence it preserves volume. If D is some domain in T^*M, then we might state Liouville's theorem in the form

$$\text{volume of } \left(g_H^t D\right) = \text{volume of } D. \tag{7.44}$$

Liouville's theorem finds many important applications in statistical mechanics.

One of the interesting results that follows from knowing that the phase flow preserves volume is known as the *Poincaré recurrence theorem*. Following Arnold (1978), we let $g\colon M \to M$ be a volume-preserving, continuous, one-to-one mapping that maps a bounded region D of $\mathbb{R}^n$ onto itself: $gD = D$. Then in any neighborhood U of an arbitrary point x_0 in D there exists a point $x \in U$ that eventually returns to U, that is, $g^n x \in U$ for some $n > 0$. See Fig. 7.2. To prove this theorem we consider the images of the

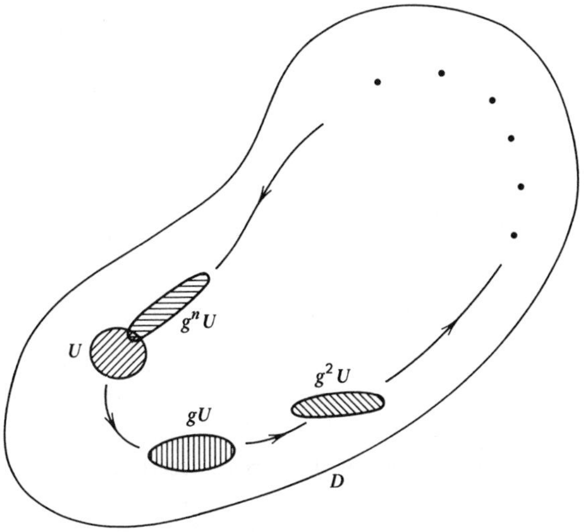

FIGURE 7.2. Eventual intersection of the neighborhoods implied by the Poincaré recurrence theorem.

neighborhood U under the mapping g. These are U, gU, $g^2U, \ldots$. All of these neighborhoods have the same volume, since by assumption g preserves volume. If they never intersected then D would have infinite volume, which is contrary to the bounded assumption we placed on D. Thus for some $k, l \geqslant 0$, where we assume $k > l$, we have $g^k U \cap g^l U \neq \varnothing$. Therefore $g^{k-l} U \cap U \neq \varnothing$. Let $k - l = n$ and then $g^n U \cap U \neq \varnothing$. Note again the conditions on g are that it preserve volume and map a bounded region onto itself.

As an example of the use of this theorem consider the two-dimensional torus with angular coordinates on it ϕ_1 and ϕ_2. See Fig. 7.3. Consider the system of ordinary differential equations on the torus: $\dot{\phi}_1 = \alpha_1$ and $\dot{\phi}_2 = \alpha_2$. Clearly, $\nabla \cdot \mathbf{v} = 0$ and the corresponding motion $(\phi_1, \phi_2) \to (\phi_1 + \alpha_1 t, \phi_2 + \alpha_2 t)$ preserves the volume $d\phi_1\, d\phi_2$. If α_1/α_2 is irrational a flowline on the torus will come arbitrarily close to every point on the torus.*

In this chapter we have learned how system motion is visualized in terms of trajectories on the phase space (cotangent bundle). Lagrangian mechanics is properly viewed on the tangent bundle and Hamiltonian mechanics on phase space. Poisson brackets proved to be a convenient way of formulating the total time derivative of phase space functions. The conditions for a canonical transformation were also defined in terms of Poisson brackets.

*In technical language, the flow is *dense* on the torus.

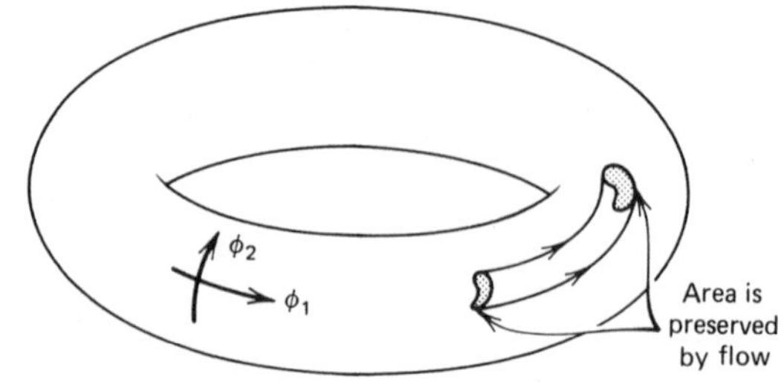

FIGURE 7.3. Area-preserving map on a torus.

Constants of Hamiltonian flow were discussed and their relationship to symmetries was compared with Noether's theorem. For a Hamiltoni ın system we have found that for n degrees of freedom, n first integrals are sufficient to solve the system completely. A complete set of such integrals does not always exist. Although not a first integral, phase-space volume is always an invariant of nondissipative, Hamiltonian phase flow.

PROBLEMS

7.1. Prove Euler's theorem for homogeneous functions: If $f: \mathbb{R}^m \to \mathbb{R}$ is such that $f(\lambda \mathbf{x}) = \lambda^n f(\mathbf{x})$, then $x^i f_{,i} = nf$ (λ is arbitrary). Hint: Consider a differentiation with respect to a λ.

7.2. Obtain the Hamiltonian and the canonical equations of motion for

(a) Kepler's problem. (Do not from the outset confine the motion to a plane.)

(b) Particle motion in an electromagnetic field.

7.3. Consider a particle moving in a plane that is under the influence of the generalized potential $U(r, \dot{r}) = k(1 + \dot{r}^2)/r$. Find the Hamiltonian and the canonical equations of motion. Reduce the problem to quadratures. Make a qualitative sketch of the trajectories on a suitable manifold.

7.4. Given a Hamiltonian, how does one find the Lagrangian? Show that Hamilton's equations follow from the Euler–Lagrange equations.

7.5. Consider the Hamiltonian of a system with one degree of freedom given below.

$$H = \frac{p^2}{2\alpha} - bqpe^{-\alpha t} + \frac{b\alpha q^2 e^{-\alpha t}(\alpha + be^{-\alpha t})}{2} + \frac{kq^2}{2}$$

where α, b, and k are constants.

(a) Find a Lagrangian corresponding to this Hamiltonian.

(b) Find an equivalent Lagrangian that is not explicitly dependent on time.

(c) What is the Hamiltonian corresponding to this new Lagrangian and what is the relationship between the two Hamiltonians?

7.6. Show that under the metric induced isomorphism of $T^*M_{\mathbf{x}}$ and $TM_{\mathbf{x}}$ that a tangent vector $\mathbf{v}$ with components v^i in a local coordinate system has its image in $T^*M_{\mathbf{x}}$ with components v_i.

7.7. Find the dual basis of $T^*SO(3)_R$ to the basis of the tangent space $TSO(3)_R$ given in Eqs. (6.17)–(6.19).

7.8. Show that if $\rho(\mathbf{x}, t)$ is the density of a fluid and $\mathbf{v}(\mathbf{x}, t)$ is the flow velocity of that fluid, then $(d\rho/dt) = 0$ if and only if $\nabla \cdot \mathbf{v} = 0$. Hint: Combine the continuity equation for the density and the formula for the convective derivative of a scalar.

7.9. Show that for a Hamiltonian system without explicit time dependence, it is impossible to have asymptotically stable equilibrium positions. See page 41 for a definition of asymptotic stability.

7.10. In studying the stability of a physical system to perturbations, the dynamical equations lead to the following differential equation for the perturbation $x(t)$.

$$\ddot{x} + 2A\dot{x} + Bx = 0,$$

where A and B are fixed parameters of the system. Solve the differential equation and in the (A, B) parameter plane ascertain the regions of stability and instability with special attention to the difference in the nature of the solutions separated by the parabola $A^2 = B$. Obtain a Lagrangian for this equation. Obtain the associated Hamiltonian. Discuss the relationship of the Hamiltonian to the energy of the system. Discuss the energy of the system in each of the important regions of the (A, B) parameter plane.

7.11. A mechanical system is described by the differential equations

$$\alpha - a = \frac{d^2}{dt^2}(1 + r^2)^{1/2}$$

$$\ddot{r} - r\dot{\phi}^2 + r = \frac{\alpha r}{(1 + r^2)^{1/2}}$$

$$r\ddot{\phi} + 2\dot{r}\dot{\phi} = 0$$

where a is a constant. Eliminate α and derive a Lagrangian for the remaining dependent variables; also obtain a Hamiltonian. Find two constants of the motion.

7.12. Formulate a result for Hamiltonians similar to the result (6.29). Find infinitesimal phase space mappings corresponding to the constants I_1 and I_2 given in Problem 6.3.

7.13. Prove the Jacobi relation (7.30) for Poisson brackets.

CHAPTER EIGHT

DYNAMICS ON PHASE SPACE

Mechanics is the paradise of the mathematical sciences because by means of it one comes to the fruits of mathematics.

LEONARDO DA VINCI

The previous chapter contained an introduction to Hamiltonian dynamics, and one important feature that emerged was the opportunity of treating momentum and configuration coordinates on an equal footing. For Hamiltonian mechanics, motions of mechanical systems are represented as trajectories on the cotangent bundle, that is, as curves in phase space. The natural coordinates on phase space are (p_i, q^i), the canonical momentum and its conjugate generalized coordinate. When the dichotomy between momentum coordinates and configuration coordinates is unnecessary, coordinates are denoted as (x^α), with Greek indices. Unless stated otherwise for specific examples, phase space as $2n$ dimensions, corresponding to n degrees of freedom for the dynamical system under study.

Our discussion of dynamics on phase space begins with a more detailed consideration of vector flows generated by real-valued functions. The Hamiltonian receives specific attention and is always assumed to have no explicit time dependence unless stated otherwise.

The algebraic properties of Poisson brackets motivate a brief study of differential k-forms. Poisson brackets determine a specific differential 2-form, which embodies fully the properties of Poisson brackets and is used to study in detail canonical transformations. Generating functions for canonical transformations are considered and lead to the Hamilton–Jacobi equation for the time evolution of mechanical systems.

8.1. FLOW-VECTOR FIELDS

In Section 7.3 we considered the flow map g_F^s: $T^*M \to T^*M$ associated with an arbitrary phase-space function $F(\mathbf{p}, \mathbf{q})$, where F: $T^*M \to \mathbb{R}$. We recall from (2.52) that the flow map is generated by the differential equation

$$\frac{d\mathbf{x}}{ds} = \mathbf{v}(\mathbf{x}), \tag{8.1}$$

where $\mathbf{x} = (\mathbf{p}, \mathbf{q})$. A *Hamiltonian flow* is defined to be a flow mapping given by (8.1), where the components of the vector $\mathbf{v}$ are given by

$$\left(\frac{dp_i}{ds}, \frac{dq^i}{ds}\right) = \left(-\frac{\partial F}{\partial q^i}, \frac{\partial F}{\partial p_i}\right). \tag{8.2}$$

When considering flows generated by real-valued functions on phase space, we restrict our attention exclusively to Hamiltonian flows.

In order to tie $\mathbf{v}$ notationally to F, we write $\mathbf{v}_F$ and call this the *flow vector associated with F*. Then (8.1) assumes the more specific form

$$\frac{d\mathbf{x}}{ds} = \mathbf{v}_F(\mathbf{x}). \tag{8.3}$$

For a phase-space function F that is differentiable everywhere, (8.2) and (8.3) define a vector field everywhere on the tangent bundle T^*M. We refer to the trajectories generated by (8.3) as a *congruence* of curves on T^*M.

At the point $\mathbf{x}_0$ in phase space, the vector $\mathbf{v}_F(\mathbf{x}_0)$ is tangent to the trajectory of (8.3) that passes through $\mathbf{x}_0$. In terms of the flow mapping

$$\left.\frac{d}{ds}\left(g_F^s \mathbf{x}_0\right)\right|_{s=0} = \mathbf{v}_F(\mathbf{x}_0). \tag{8.4}$$

The foregoing equation is entirely equivalent to (8.3), but (8.4) further emphasizes the equivalence of the flow mapping g_F^s, with its congruence curves, to the vector field $\mathbf{v}_F$. Given g_F^s or $\mathbf{v}_F$, one obtains the other through (8.3) or (8.4).

One function of immediate interest is a Hamiltonian $H(\mathbf{p}, \mathbf{q})$. In this case the curve parameter s becomes the time t. Equation (8.2) is exactly

Hamilton's equations (7.5), (7.6). The congruence of curves generated by $\mathbf{v}_H$ correspond to the trajectories, along which the system represented by H may evolve. A qualitative sketch of the congruences for two functions F and H is given in Fig. 8.1. The flow vector $\mathbf{v}_H(\mathbf{x})$ is a vector field on phase space wherever H is defined. The vector $\mathbf{v}_H(\mathbf{x}_0)$ is the tangent vector to the system trajectory passing through the phase space point $\mathbf{x}_0 = (\mathbf{p}_0, \mathbf{q}_0)$. Since this is true for every point, specifying the Hamiltonian flow-vector field $\mathbf{v}_H$ completely specifies the motion.

Another example of vectors and their associated flows is provided by the coordinate basis vectors. We denote the coordinate basis vectors as $\{\mathbf{e}_{p_i}, \mathbf{e}_{q^i}\}$ corresponding to the coordinates (p_i, q^i). Assuming that $\mathbf{v}_F = \mathbf{e}_{q^i}$ in (8.3), we find $dq^i/ds = 1$; all other derivatives vanish. Thus from (8.2) we find $\partial F/\partial p_i = 1$ and all other partial derivatives of F vanish. Thus $F = p_i$ generates the flow $\mathbf{v}_F = \mathbf{e}_{q^i}$. In an identical fashion we find $F = -q^i$ generates the flow $\mathbf{v}_F = \mathbf{e}_{p_i}$. We summarize these results as

$$F = p_i \qquad \text{gives} \quad \mathbf{v}_F = \mathbf{e}_{q^i}, \tag{8.5}$$

$$F = -q^i \quad \text{gives} \quad \mathbf{v}_F = \mathbf{e}_{p_i}. \tag{8.6}$$

Equations (8.5), (8.6) are valid for Hamiltonian flow. A definition for the flow-vector components, which is different from (8.2), would give different

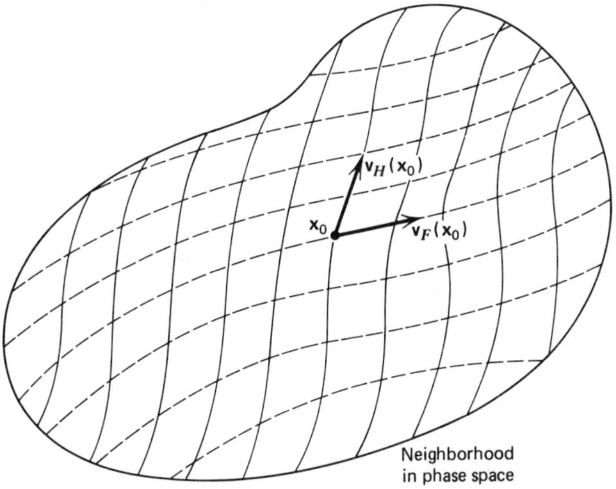

FIGURE 8.1. Sketch of some of the curves in the congruences associated with the functions F and H. At the point $\mathbf{x}_0$ the flow vectors $\mathbf{v}_F(\mathbf{x}_0)$ and $\mathbf{v}_H(\mathbf{x}_0)$ are indicated.

results for flow vectors corresponding to real-valued functions such as q^i and p_i.

As done in Section 6.1, we may interpret these flow vectors as operators. Analogous to the notation of (6.7), we define

$$\partial_{p_i} \equiv \frac{\partial}{\partial p_i} \quad \text{and} \quad \partial_{q^i} \equiv \frac{\partial}{\partial q^i}. \tag{8.7}$$

For

$$\mathbf{v}_F = \frac{\partial F}{\partial p_i}\mathbf{e}_{q^i} - \frac{\partial F}{\partial q^i}\mathbf{e}_{p_i} \quad \text{(Note sum on } i.) \tag{8.8}$$

we define the operator $\partial_{\mathbf{v}_F} \equiv \partial_F$. Then ∂_F may be written in the form

$$\partial_F = \frac{\partial F}{\partial p_i}\partial_{q^i} - \frac{\partial F}{\partial q^i}\partial_{p_i}. \quad \text{(Note sum on } i.) \tag{8.9}$$

The function F has also the usual expression for the corresponding 1-form:

$$\mathbf{d}F = \frac{\partial F}{\partial p_i}\mathbf{d}p_i + \frac{\partial F}{\partial q^i}\mathbf{d}q^i. \tag{8.10}$$

Comparing (8.8) and (8.10), we notice that for each differentiable 1-form $\mathbf{d}F$ there corresponds a flow-vector field $\mathbf{v}_F$. The correspondence in terms of components is given by

$$\left(-\frac{\partial F}{\partial q^i}, \frac{\partial F}{\partial p_i}\right) \leftrightarrow \left(\frac{\partial F}{\partial p_i}, \frac{\partial F}{\partial q^i}\right), \tag{8.11}$$

where the natural bases corresponding to the coordinates (p_i, q^i) are used for vectors and covectors.

This isomorphism between vector fields on phase space and differential 1-forms on phase space we denote as ω;

$$\omega(\mathbf{v}_F) = \mathbf{d}F, \tag{8.12}$$

where the isomorphism is defined by (8.11). The inverse mapping ω^{-1} goes from 1-forms to vector fields. One example would be the isomorphism between the Hamiltonian 1-form $\mathbf{d}H$ and the Hamiltonian flow vector $\mathbf{v}_H$.

The relationships (8.9) and (8.10) also provide several alternative expressions for the Poisson brackets of (7.21). For arbitrary phase-space functions

f and g, we have the alternative forms

$$\{f, g\} = \partial_g[f] = -\partial_f[g] = \mathbf{d}f[\mathbf{v}_g] = -\mathbf{d}g[\mathbf{v}_f]. \tag{8.13}$$

The skew symmetry and bilinearity of Poisson brackets are both manifestly evident in (8.13).

Equation (8.13) for the Poisson brackets suggest several interpretations. For example, we might interpret $\{f, g\}$ as the first-order change in the function f in the direction of the flow vector $\mathbf{v}_g$. Even somewhat more fruitful is to write (8.13) using the isomorphism in (8.12).

$$\{f, g\} = \omega(\mathbf{v}_f)[\mathbf{v}_g] = -\omega(\mathbf{v}_g)[\mathbf{v}_f]. \tag{8.14}$$

This form for the Poisson brackets is also skew symmetric and bilinear. Indeed, let us enumerate the properties of the composite symbol $\omega(\mathbf{v}_f)[\mathbf{v}_g]$. (1) It is a real-valued mapping on the tangent vectors at each point of phase space. (2) It is antisymmetric under interchange of $\mathbf{v}_f$ and $\mathbf{v}_g$. (3) It is bilinear in each argument. These properties 1–3 are those of a differential 2-form.

Thus motivated by this relationship between the Poisson bracket and the algebraic properties (1)–(3) for $\omega(\mathbf{v}_f)[\mathbf{v}_g]$, we turn to a brief study of differential k-forms.

8.2. DIFFERENTIAL FORMS AND THE EXTERIOR DERIVATIVE

Before launching into the definition of a differential k-form on a manifold M, let us review briefly those properties of differential 1-forms that we intend to generalize (cf. Section 7.2). A differential 1-form $\boldsymbol{\omega}^1$ is a covector field on a manifold M. This means that $\boldsymbol{\omega}^1(\mathbf{x})$, expanded in terms of a coordinate basis as

$$\boldsymbol{\omega}^1(\mathbf{x}) = \omega_i(\mathbf{x})\,\mathbf{d}x^i, \tag{8.15}$$

has components $\omega_i(\mathbf{x})$ that are differentiable to arbitrary order. At each point $\mathbf{x}$ in the manifold $\boldsymbol{\omega}^1(\mathbf{x})$ is a covector. This means in turn that $\boldsymbol{\omega}^1(\mathbf{x})$ at each $\mathbf{x}$ is a real-valued, linear mapping on the tangent vectors at $\mathbf{x}$. Let $\mathbf{v}(\mathbf{x})$ denote a vector field on the manifold M, so that $\mathbf{v}(\mathbf{x})$ is a tangent vector at each $\mathbf{x} \in M$, then a differential 1-form may be viewed as an operator on vector fields to give real-valued functions. This operator has one "slot," so to speak, in which to put the vector. Let $\boldsymbol{\omega}^1$ and $\mathbf{v}$ denote a differential 1-form and a vector field, respectively. Then $\boldsymbol{\omega}^1[\mathbf{v}]$ is a real-valued function on the manifold M. The number associated with the point $\mathbf{x} \in M$ is $\boldsymbol{\omega}^1(\mathbf{x})[\mathbf{v}(\mathbf{x})] \in \mathbb{R}$, that is, the number obtained by operating with the covector $\boldsymbol{\omega}^1(\mathbf{x}) \in T^*M_{\mathbf{x}}$ on the vector $\mathbf{v}(\mathbf{x}) \in TM_{\mathbf{x}}$.

It is perhaps helpful to work this out in terms of the components of $\boldsymbol{\omega}^1$ and $\mathbf{v}$.

$$\boldsymbol{\omega}^1(\mathbf{x})[\mathbf{v}(\mathbf{x})] = \omega_i(\mathbf{x})\,\mathbf{d}x^i\left[v^j(\mathbf{x})\mathbf{e}_j\right] = \omega_i(\mathbf{x})v^j(\mathbf{x})\,\mathbf{d}x^i[\mathbf{e}_j]$$

$$= \omega_i(\mathbf{x})v^j(\mathbf{x})\,\delta^i_j = \omega_i(\mathbf{x})v^i(\mathbf{x}) \in \mathbb{R} \text{ for all } \mathbf{x} \in M.$$

We first generalize the notation of a covector or 1-form to a k-form and subsequently consider differential k-forms. Consider the n-dimensional vector space $TM_{\mathbf{x}}$. A *k-form* ω^k on $TM_{\mathbf{x}}$ is a real-valued function $\boldsymbol{\omega}^k[\mathbf{v}_1, \mathbf{v}_2, \ldots, \mathbf{v}_k]$ of k different vectors from $TM_{\mathbf{x}}$, such that $\boldsymbol{\omega}^k$ is linear in each argument and antisymmetric under any pairwise interchange of arguments. If we let $V = TM_{\mathbf{x}} \times \cdots \times TM_{\mathbf{x}}$, where there are k factors, then we can state the requirements for a k-form as follows: $\boldsymbol{\omega}^k\colon V \to \mathbb{R}$, where

$$\boldsymbol{\omega}^k\left[\mathbf{v}_1, \ldots, a\mathbf{u}_j + b\mathbf{w}_j, \ldots, \mathbf{v}_k\right] = a\boldsymbol{\omega}^k\left[\mathbf{v}_1, \ldots, \mathbf{u}_j, \ldots, \mathbf{v}_k\right]$$
$$+ b\boldsymbol{\omega}^k\left[\mathbf{v}_1, \ldots, \mathbf{w}_j, \ldots, \mathbf{v}_k\right], \text{ (linearity)} \tag{8.16}$$

$$\boldsymbol{\omega}^k\left[\mathbf{v}_1, \ldots, \mathbf{v}_j, \ldots, \mathbf{v}_i, \ldots, \mathbf{v}_k\right] = -\boldsymbol{\omega}^k\left[\mathbf{v}_1, \ldots, \mathbf{v}_i, \ldots, \mathbf{v}_j, \ldots, \mathbf{v}_k\right],$$
$$\text{(skew symmetry)} \tag{8.17}$$

where $i, j = 1, \ldots, k$.

As was done in Section 7.2 for 1-forms or covectors, it is straightforward to demonstrate that k-forms constitute a vector space for each k, and this is left to the reader. For a k-form, we refer to k as the *rank* of the form.

Among k-forms there is an algebraic operation, "$\wedge$", called the exterior (wedge) product for making higher-rank forms out of ones with lower rank. For 1-forms $\boldsymbol{\omega}^1$, $\boldsymbol{\sigma}^1$ we define the 2-form called the *exterior product* $\boldsymbol{\omega}^1 \wedge \boldsymbol{\sigma}^1$ as follows:

$$\boldsymbol{\omega}^1 \wedge \boldsymbol{\sigma}^1[\mathbf{v}_1, \mathbf{v}_2] = \det\begin{bmatrix} \boldsymbol{\omega}^1(\mathbf{v}_1) & \boldsymbol{\sigma}^1(\mathbf{v}_1) \\ \boldsymbol{\omega}^1(\mathbf{v}_2) & \boldsymbol{\sigma}^1(\mathbf{v}_2) \end{bmatrix}, \tag{8.18}$$

where $\mathbf{v}_1$ and $\mathbf{v}_2$ are arbitrary vectors from $TM_{\mathbf{x}}$. With a moment's thought one confirms that $\boldsymbol{\omega}^1 \wedge \boldsymbol{\sigma}^1$ of (8.18) is a 2-form, that is, it is a real-valued mapping on $TM_{\mathbf{x}} \times TM_{\mathbf{x}}$, it is linear in each argument, and it changes sign when $\mathbf{v}_1$ and $\mathbf{v}_2$ are interchanged. From (8.18) we also see that

$$\boldsymbol{\omega}^1 \wedge \boldsymbol{\sigma}^1 = -\boldsymbol{\sigma}^1 \wedge \boldsymbol{\omega}^1. \tag{8.19}$$

A basis of 1-forms on $TM_{\mathbf{x}}$ is given by the set $\{\mathbf{d}x^i\}$ as shown in (7.15) and (7.16). By forming exterior products of this basis, we can form a basis for 2-forms. This we now demonstrate by letting $\boldsymbol{\omega}^2$ be an arbitrary 2-form and letting $\{\mathbf{e}_i\}$ be a basis of vectors on $TM_{\mathbf{x}}$ corresponding to the basis $\{\mathbf{d}x^i\}$ of $T^*M_{\mathbf{x}}$. Let $\omega_{ij} \equiv \boldsymbol{\omega}^2[\mathbf{e}_i, \mathbf{e}_j] = -\boldsymbol{\omega}^2[\mathbf{e}_j, \mathbf{e}_i] = -\omega_{ji}$. Also let $\mathbf{v} = v^i\mathbf{e}_i$ and $\mathbf{u} = u^i\mathbf{e}_i$ be arbitrary vectors in $TM_{\mathbf{x}}$. We consider the 2-form $(1/2)\omega_{ij}\mathbf{d}x^i \wedge \mathbf{d}x^j$ and show that it is identical to $\boldsymbol{\omega}^2$ by showing that it takes on exactly the same values for arbitrary $\mathbf{v}$ and $\mathbf{u}$.

$$\boldsymbol{\omega}^2[\mathbf{v},\mathbf{u}] = \boldsymbol{\omega}^2\left[v^i\mathbf{e}_i, u^j\mathbf{e}_j\right] = v^iu^j\boldsymbol{\omega}^2[\mathbf{e}_i,\mathbf{e}_j] = v^iu^j\omega_{ij}. \tag{8.20}$$

$$\begin{aligned}\tfrac{1}{2}\omega_{ij}\mathbf{d}x^i \wedge \mathbf{d}x^j[\mathbf{v},\mathbf{u}] &= \tfrac{1}{2}\omega_{ij}\left(\mathbf{d}x^i[\mathbf{v}]\,\mathbf{d}x^j[\mathbf{u}] - \mathbf{d}x^i[\mathbf{u}]\,\mathbf{d}x^j[\mathbf{v}]\right)\\ &= \tfrac{1}{2}\omega_{ij}\left(v^iu^j - u^iv^j\right) = \omega_{ij}v^iu^j.\end{aligned} \tag{8.21}$$

In the last computation we made use of (8.18) and the fact that $\omega_{ij} = -\omega_{ji}$.

We see that all 2-forms can be written in terms of the basis 1-forms $\mathbf{d}x^i \wedge \mathbf{d}x^j$. Equation (8.19) shows that $i \neq j$ in order that the 2-form $\mathbf{d}x^i \wedge \mathbf{d}x^j$ be nonzero. Also note that $\mathbf{d}x^i \wedge \mathbf{d}x^j$ and $\mathbf{d}x^j \wedge \mathbf{d}x^i$ are linearly dependent. Counting the linearly independent, nonzero possibilities, we see that the dimension of the vector space of 2-forms on $TM_{\mathbf{x}}$ is $n(n-1)/2$.

All that has been done for 2-forms can be extended in an obvious way for higher-rank forms. The higher-rank forms can all be constructed from exterior products of 1-forms, as was done for 2-forms, by using a determinant like (8.18). Leaving the details of this construction to the interested reader, it suffices us to note that in terms of basis 1-forms, an arbitrary k-form may be written as

$$\boldsymbol{\omega}^k = \frac{1}{k!}\omega_{i_1i_2\ldots i_k}\mathbf{d}x^{i_1} \wedge \mathbf{d}x^{i_2} \wedge \cdots \wedge \mathbf{d}x^{i_k}. \tag{8.22}$$

The components $\omega_{i_1i_2\ldots i_k}$ are antisymmetric under pairwise interchange. The factor $(1/k!)$ is included in (8.22) if the components of $\boldsymbol{\omega}^k$ are to be given by $\boldsymbol{\omega}^k[\mathbf{e}_{i_1},\ldots,\mathbf{e}_{i_k}] = \omega_{i_1\ldots i_k}$.

Just as little arrows provide a geometric model or mental image for vectors, it helps to have a geometric model for k-forms. This model must

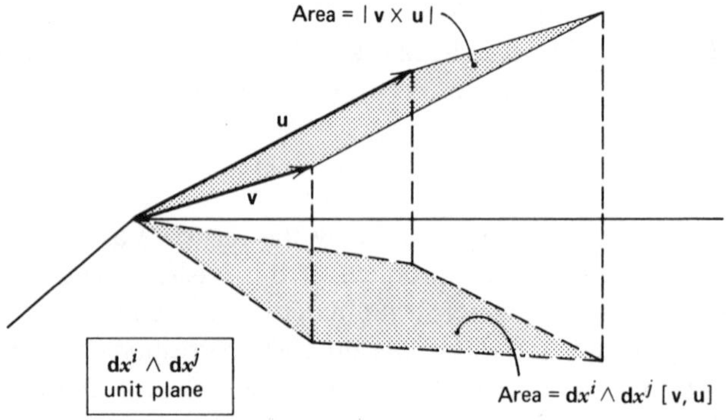

FIGURE 8.2. Sketch of the area $|\mathbf{v} \times \mathbf{u}|$ projected onto the (ij) coordinate plane suggesting the interpretation of basis 2-forms such as $\mathbf{d}x^i \wedge \mathbf{d}x^j$ as unit planes.

reflect the properties of the forms. One such geometric model is to think of $\mathbf{d}x^i$ as a line and refer to it as the (i) *unit line*. This is because $\mathbf{d}x^i[\mathbf{v}] = v^i$ gives the projection of the vector $\mathbf{v}$ on the ith coordinate axis. In the same vein, the geometric model of a 2-form $\mathbf{d}x^i \wedge \mathbf{d}x^j$ is a plane and we refer to it as the (ij) *unit plane*. Again, this conceptualization of $\mathbf{d}x^i \wedge \mathbf{d}x^j$ as a unit plane results because $\mathbf{d}x^i \wedge \mathbf{d}x^j[\mathbf{v},\mathbf{u}]$ is the area of the parallelogram formed by the vectors $\mathbf{v}$ and $\mathbf{u}$ projected onto the (i, j) coordinate plane. For a sketch of this situation see Fig. 8.2.

Continuing, we think of $\mathbf{d}x^i \wedge \mathbf{d}x^j \wedge \mathbf{d}x^k$ as the (ijk) *unit volume*, since $\mathbf{d}x^i \wedge \mathbf{d}x^j \wedge \mathbf{d}x^k[\mathbf{v},\mathbf{u},\mathbf{w}]$ is the projection of the volume represented by the three n-dimensional vectors $\mathbf{v},\mathbf{u},\mathbf{w}$ into the (ijk) coordinate volume. For higher-rank forms we picture them as unit k-volumes. For example, $\mathbf{d}x^i \wedge \mathbf{d}x^j \wedge \mathbf{d}x^k \wedge \mathbf{d}x^l$ is a $(ijkl)$ unit 4-volume. A sketch of these geometric models for the basis k-forms, $k = 1, 2, 3$, is given in Fig. 8.3.

Now we come to the issue of differential k-forms and generalize just as easily as was done for 1-forms. A *differential k-form* is a k-form whose components in (8.22) are functions of position, and these functions $\omega_{i_1 \dots i_k}(\mathbf{x})$ are differentiable to arbitrary order.

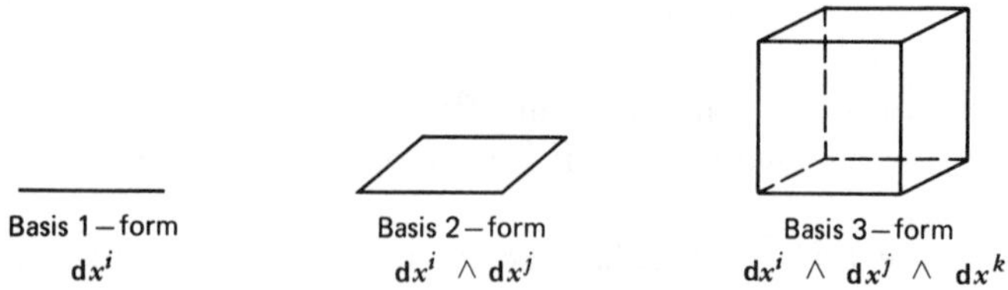

FIGURE 8.3. Sketch of the basis k-forms for $k = 1, 2, 3$.

Differential k-forms can be viewed as operators on vector fields. Let $\boldsymbol{\omega}^k(\mathbf{x})$ denote a differential k-form on the manifold M, and let $\mathbf{v}_1(\mathbf{x}),\ldots, \mathbf{v}_k(\mathbf{x})$ be k different vector fields. Then $\boldsymbol{\omega}^k(\mathbf{x})[\mathbf{v}_1(\mathbf{x}),\ldots, \mathbf{v}_k(\mathbf{x})]$ is a real-valued mapping on the manifold M in exactly the same way that $\boldsymbol{\omega}^1[\mathbf{v}(\mathbf{x})]$ is a real-valued mapping. To be specific, we again write out this real-valued function in terms of the real-valued functions that constitute the components of $\boldsymbol{\omega}^k$ and the vectors $\mathbf{v}_1,\ldots, \mathbf{v}_k$. If $\boldsymbol{\omega}^k = (1/k!)\omega_{i_1\cdots i_k}(\mathbf{x})\,\mathbf{d}x^{i_1} \wedge \cdots \wedge \mathbf{d}x^{i_k}$, where $\omega_{i_1\cdots i_k} = \boldsymbol{\omega}^k[\mathbf{e}_{i_1},\ldots, \mathbf{e}_{i_k}]$, then

$$\boldsymbol{\omega}^k[\mathbf{v}_1,\ldots, \mathbf{v}_2] = \boldsymbol{\omega}^k\left[v_1^{i_1}\mathbf{e}_{i_1},\ldots, v_k^{i_k}\mathbf{e}_{i_k}\right]$$

$$= v_1^{i_1} \cdots v_k^{i_k}\boldsymbol{\omega}^k\left[\mathbf{e}_{i_1},\ldots, \mathbf{e}_{i_k}\right] = \omega_{i_1\cdots i_k}(\mathbf{x})v_1^{i_1}(\mathbf{x}) \cdots v_k^{i_k}(\mathbf{x}). \tag{8.23}$$

As the point in the manifold changes then $\boldsymbol{\omega}^k[\mathbf{v}_1,\ldots, \mathbf{v}_k]$ takes on the real values obtained from the product of components in (8.23).

An example of a differential k-form for $k = 2$ is provided by the Poisson bracket expression given in (8.14).

$$\boldsymbol{\gamma}[\mathbf{v}_f, \mathbf{v}_g] \equiv \omega(\mathbf{v}_g)[\mathbf{v}_f]. \tag{8.24}$$

From the discussion of (8.14) we recognize that $\boldsymbol{\gamma}$ is antisymmetric under interchange of $\mathbf{v}_f$ and $\mathbf{v}_g$, it is a real-valued mapping on vector fields, and it is linear in each argument. Indeed this 2-form $\boldsymbol{\gamma}$ is the differential form of primary interest in subsequent sections of this chapter.

We have shown how to make higher-rank forms from ones with lower rank using the exterior product and this carries over in an obvious way to differential forms as well. However, there is yet a second way to make higher-rank differential forms from ones of lower rank. The process involves differentiation and is an extension of the process that produces the differential 1-form $\mathbf{d}f$ from the real-valued mapping f. Let $\boldsymbol{\omega}^k = \omega_{i_1\cdots i_i}\mathbf{d}x^{i_1} \wedge \cdots \wedge \mathbf{d}x^{i_k}$ be an arbitrary differential k-form. The *exterior derivative* of $\boldsymbol{\omega}^k$ is given by

$$\mathbf{d}\boldsymbol{\omega}^k = \left(\mathbf{d}\omega_{i_1\cdots i_k}\right)\mathbf{d}x^{i_1} \wedge \cdots \wedge \mathbf{d}x^{i_k} = \left(\omega_{i_1\cdots i_k,j}\right)\mathbf{d}x^{j} \wedge \mathbf{d}x^{i_1} \wedge \cdots \wedge \mathbf{d}x^{i_k}. \tag{8.25}$$

As an example of calculating an exterior derivative, we show one of its most important properties. Consider $\mathbf{d}(\mathbf{d}\boldsymbol{\omega}^k) \equiv \mathbf{d}^2\boldsymbol{\omega}^k$. Following (8.25)

$$\mathbf{d}^2\boldsymbol{\omega}^k = \omega_{i_1 \cdots i_k, j, l}\mathbf{d}x^l \wedge \mathbf{d}x^j \wedge \mathbf{d}x^{i_1} \wedge \cdots \wedge \mathbf{d}x^{i_k}.$$

Partial differentiation is commutative and the indices (jl) on $\omega_{i_1 \cdots i_k, j, l}$ are symmetric under interchange. But the indices (jl) on $\mathbf{d}x^l \wedge \mathbf{d}x^j$ are antisymmetric under interchange and thus the sum on (jl) gives zero. Hence $\mathbf{d}^2\boldsymbol{\omega}^k = 0$. We summarize this result with other properties of the exterior derivative but leave the remaining demonstrations to Problem 8.2. If $\boldsymbol{\omega}^k$ and $\boldsymbol{\sigma}^l$ are arbitrary differential forms, then

$$\mathbf{d}(\boldsymbol{\omega}^k + \boldsymbol{\sigma}^k) = \mathbf{d}\boldsymbol{\omega}^k + \mathbf{d}\boldsymbol{\sigma}^k \tag{8.26}$$

$$\mathbf{d}(\boldsymbol{\omega}^k \wedge \boldsymbol{\sigma}^l) = \mathbf{d}\boldsymbol{\omega}^k \wedge \boldsymbol{\sigma}^l + (-1)^k \boldsymbol{\omega}^k \wedge \mathbf{d}\boldsymbol{\sigma}^l \tag{8.27}$$

$$\mathbf{d}(\mathbf{d}\boldsymbol{\omega}^k) = 0 \tag{8.28}$$

Although the exterior derivative operator $\mathbf{d}$ may look strange, the familiar operators of vector calculus $\nabla, \nabla\cdot, \nabla\times$ can all be considered special cases of $\mathbf{d}$. We have already seen how the covariant components of ∇f are the components of $\mathbf{d}f$. The other operators $\nabla\cdot$ and $\nabla\times$ are studied in Problem 8.5 in terms of the exterior derivative $\mathbf{d}$.

The discussion of differential k-forms in this section is necessarily brief and suited to the particular needs of the present chapter. The interested reader can find many treatments that fully discuss differential forms. Excellent treatments with a physics orientation can be found in Arnold (1978), Choquet-Bruhat and DeWitt-Morette (1982), Schutz (1980), and Flanders (1963).

8.3. THE CANONICAL 2-FORM $\boldsymbol{\gamma}$

After this introduction to differential k-forms, we wish to focus attention on the 2-form obtained from Poisson brackets described earlier in (8.14) and (8.24). Let f and g be two real-valued functions on phase space and let $\mathbf{v}_f$ and $\mathbf{v}_g$ be their corresponding flow-vector fields. The *canonical 2-form* $\boldsymbol{\gamma}$ is defined according to

$$\boldsymbol{\gamma}[\mathbf{v}_f, \mathbf{v}_g] \equiv -\{f, g\}, \tag{8.29}$$

for arbitrary f and g. This definition in (8.29) is merely a restatement of

(8.24), together with (8.14). The choice of sign in (8.29) is made for later convenience.

Despite the definition of the canonical 2-form γ in terms of Poisson brackets, it is not necessary to go back to (8.29) each time we wish to consider the action of γ on arbitrary vector fields $\mathbf{v}, \mathbf{u}$. Indeed, it would be inconvenient if for each $\mathbf{v}(\mathbf{x})$ and $\mathbf{u}(\mathbf{x})$ it were necessary to find functions for use in Poisson brackets, which generate $\mathbf{v}(\mathbf{x})$ and $\mathbf{u}(\mathbf{x})$ as flows. Fortunately, a general expression in terms of basis 2-forms is obtainable for γ.

Let $\mathbf{v}(\mathbf{x})$ and $\mathbf{u}(\mathbf{x})$ be two arbitrary, differentiable vector fields, which we express in terms of the coordinate basis vectors:

$$\mathbf{v}(\mathbf{x}) = v^i(\mathbf{x})\mathbf{e}_{q^i} + v_i(\mathbf{x})\mathbf{e}_{p_i}$$

$$\mathbf{u}(\mathbf{x}) = u^j(\mathbf{x})\mathbf{e}_{q^j} + u_j(\mathbf{x})\mathbf{e}_{p_j}.$$

In these expressions for $\mathbf{v}$, and similarly for $\mathbf{u}$, we note that v^i and v_i are *not* contravariant and covariant counterparts, but correspond to different basis vectors. The index written as a superscript (subscript) corresponds to the basis vectors $\mathbf{e}_{q^i}$ ($\mathbf{e}_{p_i}$). Substituting $\mathbf{v}(\mathbf{x})$ and $\mathbf{u}(\mathbf{x})$ into γ, we find readily

$$\gamma[\mathbf{v}(\mathbf{x}), \mathbf{u}(\mathbf{x})] = v^i u^j \gamma[\mathbf{e}_{q^i}, \mathbf{e}_{q^j}] + v_i u_j \gamma[\mathbf{e}_{p_i}, \mathbf{e}_{p_j}]$$
$$+ \left(v_j u^i - v^i u_j\right)\gamma[\mathbf{e}_{p_j}, \mathbf{e}_{q^i}]. \tag{8.30}$$

Once the action of γ on the basis vector combinations occurring in (8.30) is known, then the 2-form γ is completely determined. We recall from (8.5) and (8.6) that $f = p_i$ generates the flow for $\mathbf{e}_{q^i}$ and $g = -q^i$ generates the flow for $\mathbf{e}_{p_i}$. Substituting this f and g into (8.29) gives

$$\gamma[\mathbf{e}_{q^i}, \mathbf{e}_{q^j}] = \{p_j, p_i\} = 0. \tag{8.31}$$

$$\gamma[\mathbf{e}_{p_i}, \mathbf{e}_{p_j}] = \{q^j, q^i\} = 0. \tag{8.32}$$

$$\gamma[\mathbf{e}_{p_j}, \mathbf{e}_{q^i}] = \{p_i, -q^j\} = \{q^j, p_i\} = \delta_i^j. \tag{8.33}$$

Any 2-form, and in particular γ, can be written as a sum of basis 2-forms. In the (p_i, q^i) chart the basis 2-forms are of three types: (1) $\mathbf{d}p_i \wedge \mathbf{d}p_j$, (2) $\mathbf{d}q^i \wedge \mathbf{d}q^j$, and (3) $\mathbf{d}p_i \wedge \mathbf{d}q^j$. Equation (8.32) shows that there are no basis 2-forms of type 1 in γ. Similarly, (8.31) shows that there are none of type 2. Equation (8.33) implies that only when $i = j$ in the basis 2-forms of type 3 do we obtain a nonzero coefficient and that coefficient must be 1. From the

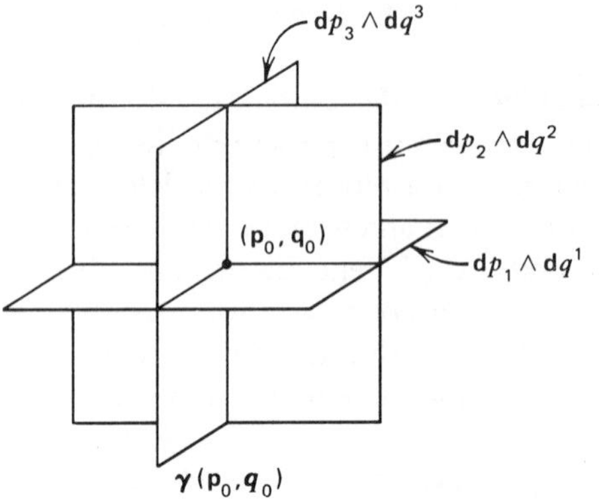

FIGURE 8.4. Sketch of the canonical 2-form as a sum of unit planes.

fundamental Poisson bracket relations (8.31)–(8.33), which are equivalent to (7.28), we obtain

$$\boldsymbol{\gamma} = \mathbf{d}p_i \wedge \mathbf{d}q^i. \quad \text{(Note sum on } i.\text{)} \tag{8.34}$$

With this result (8.30) becomes

$$\boldsymbol{\gamma}[\mathbf{v}(\mathbf{x}), \mathbf{u}(\mathbf{x})] = \left(v_j u^j - u_j v^j\right). \tag{8.35}$$

Form (8.34) for $\boldsymbol{\gamma}$ is valid in the entire neighborhood where the coordinates (p_i, q^i) are adequate. See Fig. 8.4 for a geometric model of $\boldsymbol{\gamma}$. Because all the coefficients turned out to be constant, $\mathbf{d}\boldsymbol{\gamma} = 0$. This is evident from (8.34) and (8.28). A form $\boldsymbol{\omega}^k$ such that $\mathbf{d}\boldsymbol{\omega}^k = 0$ is called *closed*. From (8.35) it is evident that if $\mathbf{v}(\mathbf{x})$ is not identically zero, then there always exists a $\mathbf{u}(\mathbf{x})$ such that $\boldsymbol{\gamma}[\mathbf{v}, \mathbf{u}] \neq 0$.*

The isomorphism given in (8.11) and (8.12) can now be given somewhat more directly using $\boldsymbol{\gamma}$. Let $\mathbf{v}$ be an arbitrary vector field. We define a 1-form corresponding to $\mathbf{v}$ by

$$\boldsymbol{\omega}^1{}_{\mathbf{v}} = \boldsymbol{\gamma}[\ ,\mathbf{v}] = \mathbf{d}q^i[\mathbf{v}]\,\mathbf{d}p_i - \mathbf{d}p_i[\mathbf{v}]\,\mathbf{d}q^i.$$

and assert that this is the same isomorphism as that given in (8.12). To verify this we let $\mathbf{v} = \mathbf{v}_F$, in which case (8.2) shows $\mathbf{d}q^i[\mathbf{v}] = \partial F/\partial p_i$ and

*A form with this property is termed *nondegenerate*. An even-dimensional manifold M^{2n}, together with a nondegenerate, closed 2-form defined on M^{2n}, is called a *symplectic* manifold. Phase space is a symplectic manifold.

$\mathbf{d}p_i[\mathbf{v}] = -\partial F/q^i$. Substitution into the preceding equation results in

$$\boldsymbol{\omega}^1_{\mathbf{v}_F} = \frac{\partial F}{\partial p_i}\,\mathbf{d}p_i + \frac{\partial F}{\partial q^i}\,\mathbf{d}p^i,$$

which we recognize immediately to be $\mathbf{d}F$.

8.4. CANONICAL TRANSFORMATIONS

Canonical transformations were defined in Section 7.3 as those transformations on phase space preserving the Poisson bracket relations in (7.28). Since the Poisson bracket relations determine the form of $\boldsymbol{\gamma}$ in the given coordinate system through (8.31)–(8.33), a canonical transformation preserves the form of $\boldsymbol{\gamma}$. To be specific, a coordinate transformation $(p_i, q^i) \rightarrow (P_i, Q^i)$ is a canonical transformation iff

$$\boldsymbol{\gamma} = \mathbf{d}p_i \wedge \mathbf{d}q^i = \mathbf{d}P_i \wedge \mathbf{d}Q^i. \tag{8.36}$$

Equation (8.36) implies that the sum of the (p_i, q^i) unit planes is equal to the sum of the (P_i, Q^i) unit planes.

The result contained in (8.36) is a very convenient way in which to characterize canonical transformations. Although we have already considered canonical transformations in terms of Poisson brackets, and subsequently will study yet other ways of viewing canonical transformations, the verification of (8.36) is perhaps the most efficient method for checking that a given transformation is canonical.

As an example let us consider the following transformation:

$$x = X\cos\lambda + \frac{P_Y \sin\lambda}{m\omega}; \qquad y = Y\cos\lambda + \frac{P_X \sin\lambda}{m\omega}$$

$$p_x = -m\omega Y\sin\lambda + P_X\cos\lambda; \qquad p_y = -m\omega X\sin\lambda + P_Y\cos\lambda$$

Using $\mathbf{d}f = f_{,i}\,\mathbf{d}x^i$, we find the corresponding 1-forms:

$$\mathbf{d}x = \cos\lambda\,\mathbf{d}X + \frac{1}{m\omega}\sin\lambda\,\mathbf{d}P_Y; \qquad \mathbf{d}y = \cos\lambda\,\mathbf{d}Y + \frac{1}{m\omega}\sin\lambda\,\mathbf{d}P_X;$$

$$\mathbf{d}p_x = -m\omega\sin\lambda\,\mathbf{d}Y + \cos\lambda\,\mathbf{d}P_X; \qquad \mathbf{d}p_y = -m\omega\sin\lambda\,\mathbf{d}X + \cos\lambda\,\mathbf{d}P_Y.$$

Taking exterior products,

$$\mathbf{d}p_x \wedge \mathbf{d}x = -m\omega \sin\lambda\cos\lambda\,\mathbf{d}Y \wedge \mathbf{d}X - \sin^2\lambda\,\mathbf{d}Y \wedge \mathbf{d}P_Y$$

$$+\cos^2\lambda\,\mathbf{d}P_X \wedge \mathbf{d}X + \frac{1}{m\omega}\sin\lambda\cos\lambda\,\mathbf{d}P_X \wedge \mathbf{d}P_Y;$$

$$\mathbf{d}p_y \wedge \mathbf{d}y = -m\omega \sin\lambda\cos\lambda\,\mathbf{d}X \wedge \mathbf{d}Y - \sin^2\lambda\,\mathbf{d}X \wedge \mathbf{d}P_X$$

$$+\cos^2\lambda\,\mathbf{d}P_Y \wedge \mathbf{d}Y + \frac{1}{m\omega}\sin\lambda\cos\lambda\,\mathbf{d}P_Y \wedge \mathbf{d}P_X.$$

The foregoing 2-forms are then added according to (8.34) to find $\mathbf{d}p_x \wedge \mathbf{d}x + \mathbf{d}p_y \wedge \mathbf{d}y = \mathbf{d}P_X \wedge \mathbf{d}X + \mathbf{d}P_Y \wedge \mathbf{d}Y$. Hence the preceding transformation is canonical.

The reader is invited to verify that the preceding transformation is canonical using the Poisson bracket relations of (7.28). A comparison of the labor involved shows that verification of (8.36) is considerably more efficient. Much of the labor involved in checking the Poisson brackets that are zero is eliminated. Nevertheless, canonical transformations come up in many ways, and alternative characterizations of canonical transformations are more convenient in other situations.

To consider a third, but related, view of canonical transformations, let the coordinates (p_i, q^i) be collectively denoted as $\{x^\alpha\}$ and similarly denote (P_i, Q^i) as $\{X^\alpha\}$. The coordinates are identified as $x^\alpha = p_i$ for $\alpha = i$ and $x^\alpha = q^i$ for $\alpha = n + i$, where $i = 1, \ldots, n$. A similar identification relates (P_i, Q^i) and $\{X^\alpha\}$. This notation forces us to give up the information contained in the positioning of an index on the "p's" and "q's". These index positions reflected the transformation properties under an arbitrary coordinate transformation in configuration space.

Let g: $T^*M \to T^*M$ be a transformation on phase space such that $x^\alpha \to X^\beta(x^\alpha)$ under the mapping g. The basis 1-forms are then related by

$$\mathbf{d}X^\beta = \frac{\partial X^\beta}{\partial x^\alpha}\,\mathbf{d}x^\alpha \tag{8.37}$$

and this allows us to relate the components of all forms. Let $\boldsymbol{\omega}^2$ be an arbitrary 2-form.

$$\boldsymbol{\omega} = \frac{1}{2!}\omega_{\alpha\beta}\,\mathbf{d}x^\alpha \wedge \mathbf{d}x^\beta = \frac{1}{2!}\bar{\omega}_{\alpha\beta}\,\mathbf{d}X^\alpha \wedge \mathbf{d}X^\beta, \tag{8.38}$$

where both $\omega_{\alpha\beta}$ and $\bar{\omega}_{\alpha\beta}$ are antisymmetric. Substituting from (8.37), we find

$\bar{\omega}_{\alpha\beta}$ and $\omega_{\alpha\beta}$ are related by

$$\omega_{\alpha\beta} = \frac{\partial X^\mu}{\partial x^\alpha}\frac{\partial X^\nu}{\partial x^\beta}\bar{\omega}_{\mu\nu}, \tag{8.39}$$

which is the usual transformation law relating the components of a 2-index tensor. Specializing to the canonical 2-form $\boldsymbol{\gamma}$,

$$\boldsymbol{\gamma} = \mathbf{d}p_i \wedge \mathbf{d}q^i = \frac{1}{2!}\gamma_{\alpha\beta}\mathbf{d}x^\alpha \wedge \mathbf{d}x^\beta. \tag{8.40}$$

From (8.40) we identify the components of $\gamma_{\alpha\beta}$ and find

$$\gamma_{\alpha\beta} = \begin{bmatrix} 0 & I \\ -I & 0 \end{bmatrix}, \tag{8.41}$$

where I is the $n \times n$ identify matrix.

If the transformation g is required to be canonical so that

$$\boldsymbol{\gamma} = \mathbf{d}P_i \wedge \mathbf{d}Q^i = \frac{1}{2!}\bar{\gamma}_{\alpha\beta}\mathbf{d}X^\alpha\mathbf{d}X^\beta \tag{8.42}$$

as well, then $\bar{\gamma}_{\alpha\beta}$ is given by the matrix in (8.41) also. Hence, for the components of $\boldsymbol{\gamma}$, (8.39) becomes

$$\gamma_{\alpha\beta} = \frac{\partial X^\mu}{\partial x^\alpha}\frac{\partial X^\nu}{\partial x^\beta}\gamma_{\mu\nu}. \tag{8.43}$$

Equation (8.43) is to be viewed as a restriction on the transformation $x^\alpha \to X^\beta(x^\alpha)$ in order that it be canonical.

Frequently, (8.43) is given in matrix form. For this purpose we use the convention discussed in the paragraph following Eq. (1.11) and let the Jacobian matrix J have components $J^\mu{}_\alpha = (\partial X^\mu / \partial x^\alpha)$. Then (8.43) becomes in matrix notation

$$\gamma = J^t \gamma J. \tag{8.44}$$

Equation (8.44) is sometimes made the definition of a canonical transformation.*

*The matrix equation (8.44) is also referred to as a *symplectic* transformation.

We observe that all the Poisson bracket relations in (7.21) and (7.28), respectively, are contained in the Poisson bracket relations

$$\{f, g\} = \gamma_{\alpha\beta}\frac{\partial f}{\partial x^\alpha}\frac{\partial g}{\partial x^\beta},$$
$$\{x^\alpha, x^\beta\} = \gamma_{\alpha\beta}. \tag{8.45}$$

To conclude this discussion of canonical transformations we consider a more abstract view, less oriented toward coordinates. We consider a mapping induced on forms by a general manifold map g: $M \to N$. Recall from Section 6.3 the definition of the derivative map g_*, which maps tangent vectors on M into tangent vectors on N. Analogous to g_*, we define the *pull-back* map on forms in N into forms on M. This map is denoted as g^* and defined for arbitrary k-forms by

$$(g^*\boldsymbol{\omega}^k)[\mathbf{v}_1,\ldots,\mathbf{v}_k] = \boldsymbol{\omega}^k[g_*\mathbf{v}_1,\ldots,g_*\mathbf{v}_k], \tag{8.46}$$

where $\mathbf{v}_1,\ldots,\mathbf{v}_k$ are arbitrary vectors in a tangent space on M. The pull-back map gets its name from the fact that g^* goes backward from the direction of g and g_*, that is, $g^*\boldsymbol{\omega}^k$ is a k-form on M where $\boldsymbol{\omega}^k$ is a k-form on N.

As an example, let us see how g^* works on 1-forms. Let $\{x^\alpha\}$ denote a coordinate system in M and $\{X^\alpha\}$ a coordinate system in N. In terms of basis 1-forms $\boldsymbol{\omega}^1 = \omega_\alpha \mathbf{d}X^\alpha$ and $g^*\boldsymbol{\omega}^1 = \bar{\omega}_\alpha \mathbf{d}x^\alpha$. Now let $\mathbf{v} = v^\alpha \mathbf{e}_\alpha$ be an arbitrary tangent vector on M and from Section 6.3 we have that $g_*\mathbf{v} = v^\alpha(\partial X^\beta/\partial x^\alpha)\mathbf{E}_\beta$, where $\{\mathbf{e}_\alpha\}$ and $\{\mathbf{E}_\alpha\}$ are basis vectors for the $\{x^\alpha\}$ and $\{X^\alpha\}$ coordinate systems, respectively. By (8.46) we have

$$g^*\boldsymbol{\omega}^1[\mathbf{v}] = \bar{\omega}_\alpha \mathbf{d}x^\alpha\left[v^\beta \mathbf{e}_\beta\right] = v^\alpha\bar{\omega}_\alpha$$
$$= \boldsymbol{\omega}^1[g_*\mathbf{v}] = \omega_\alpha \mathbf{d}X^\alpha\left[v^\gamma\left(\frac{\partial X^\beta}{\partial x^\gamma}\right)\mathbf{E}_\beta\right] = \omega_\alpha v^\gamma\left(\frac{\partial X^\alpha}{\partial x^\gamma}\right)$$

Since the vector $\mathbf{v}$ is arbitrary we have that

$$\bar{\omega}_\alpha = \frac{\partial X^\beta}{\partial x^\alpha}\omega_\beta, \tag{8.47}$$

which is nothing more than the usual tensor transformation law for covectors.

With regard to canonical transformations, we let the manifold map g be a mapping on phase space g: $T^*M \to T^*M$. Assume g is a canonical transformation. Then from (8.43) we see that g^* preserves the canonical 2-form $\boldsymbol{\gamma}$, that is,

$$g^*\boldsymbol{\gamma} = \boldsymbol{\gamma} \tag{8.48}$$

for a canonical transformation.

As an example, we assume (8.48) and show that Poisson brackets are preserved. Let F, G be arbitrary functions on phase space. Then using (8.29)

$$\{F, G\}_{(\mathbf{P},\mathbf{Q})} = -\boldsymbol{\gamma}[\mathbf{v}_F, \mathbf{v}_G] = -g^*\boldsymbol{\gamma}[\mathbf{v}_F, \mathbf{v}_G] = \{F, G\}_{(\mathbf{p},\mathbf{q})}. \tag{8.49}$$

Recall that canonical transformations, which can be viewed as manifold maps on phase space, are of interest because they preserve Hamilton's equations of motion. The Hamiltonian flow g_H^t is a manifold map on phase space. Also, Hamilton's equations are certainly preserved under the flow map g_H^t, since they indeed determine the mapping. Thus it seems obvious that g_H^t should be a canonical transformation. Nevertheless, as a further illustration, we demonstrate explicitly that a Hamiltonian flow is a canonical transformation. We show this initially for an infinitesimal portion of the flow.

The new coordinates are obtained from the old by

$$P_i = p_{0i} + \dot{p}_{0i}\,\Delta t + O\left[(\Delta t)^2\right] = p_{0i} - \Delta t\,\frac{\partial H}{\partial q_0^i} + O\left[(\Delta t)^2\right],$$

$$Q^i = q_0^i + \dot{q}_0^i\,\Delta t + O\left[(\Delta t)^2\right] = q_0^i + \Delta t\,\frac{\partial H}{\partial p_{0i}} + O\left[(\Delta t)^2\right].$$

Calculating the corresponding forms by $\mathbf{d}f = f_{,i}\,\mathbf{d}x^i$, we find

$$\mathbf{d}P_i = \mathbf{d}p_{0i} - \Delta t\,\frac{\partial^2 H}{\partial p_{0j}\,\partial q_0^i}\,\mathbf{d}p_{0j} - \Delta t\,\frac{\partial^2 H}{\partial q_0^j\,\partial q_0^i}\,\mathbf{d}q_0^j + O\left[(\Delta t)^2\right],$$

$$\mathbf{d}Q^i = \mathbf{d}q_0^i + \Delta t\,\frac{\partial^2 H}{\partial p_{0j}\,\partial p_{0i}}\,\mathbf{d}p_{0j} + \Delta t\,\frac{\partial^2 H}{\partial q_0^j\,\partial p_{0i}}\,\mathbf{d}q_0^j + O\left[(\Delta t)^2\right].$$

Using exterior products to form $\boldsymbol{\gamma}$,

$$\begin{aligned}\mathbf{d}P_i \wedge \mathbf{d}Q^i = {} & \mathbf{d}p_{0i} \wedge \mathbf{d}q_0^i \\ & + \Delta t\left[\frac{\partial^2 H}{\partial p_{0j}\,\partial p_{0i}}\,\mathbf{d}p_{0i} \wedge \mathbf{d}p_{0j} + \frac{\partial^2 H}{\partial q_0^j\,\partial p_{0i}}\,\mathbf{d}p_{0i} \wedge \mathbf{d}q_0^j \right. \\ & \left. - \frac{\partial^2 H}{\partial p_{0j}\,\partial q_0^i}\,\mathbf{d}p_{0j} \wedge \mathbf{d}q_0^i - \frac{\partial^2 H}{\partial q_0^j\,\partial q_0^i}\,\mathbf{d}q_0^j \wedge \mathbf{d}q_0^i\right] + O\left[(\Delta t)^2\right].\end{aligned}$$

The middle terms in the square brackets cancel each other. The first and the last terms are zero because the partial derivative factor is symmetric under index interchange and the exterior products are antisymmetric. Therefore $\mathbf{d}P_i \wedge \mathbf{d}Q^i = \mathbf{d}p_{0i} \wedge \mathbf{d}q_0^i$ and the infinitesimal transformation generated by the Hamilton phase flow is a canonical transformation.

The result of Problem 8.7 shows that for two manifold maps h and g, $(h \circ g)^* = g^* \circ h^*$. If both g and h are canonical transformations, then $h^*\gamma = \gamma$ and $g^*\gamma = \gamma$. Thus we find that $(h \circ g)^*\gamma = \gamma$. Consequently, the Hamiltonian flow g_H^t, for finite t, is canonical, since it can be built up as a sequence of infinitesimal flows, each of which is a canonical transformation.

The recognition of the Hamiltonian flow as a canonical transformation shows that a discovery of this canonical transformation would constitute an integration of the equations of motion. A solution technique for the problem of motion based on this observation is called the Hamilton–Jacobi method and is discussed in Section 8.7.

8.5. GENERATING FUNCTIONS

Let $(\mathbf{p}, \mathbf{q})$ and $(\mathbf{P}, \mathbf{Q})$ denote coordinates in phase space that are related by a canonical transformation. If we let ψ be the 1-form

$$\psi = p_i \mathbf{d}q^i + \mathbf{d}f, \tag{8.50}$$

then $\mathbf{d}\psi = \gamma$, since $\mathbf{d}^2 = 0$. In similar fashion the 1-form

$$\chi = P_i \mathbf{d}Q^i + \mathbf{d}F \tag{8.51}$$

gives $\mathbf{d}\chi = \gamma$. Hence ψ and χ can differ at most by an exact 1-form, that is,

$$\mathbf{d}S = p_i \mathbf{d}q^i - P_i \mathbf{d}Q^i, \tag{8.52}$$

since the exterior derivative of both ψ and χ gives γ. The phase space function S is called a *generating function*.

Generally speaking, an arbitrary transformation is given by $2n$ functions of $2n$ variables, $X^\alpha = X^\alpha(x^\beta)$. However, canonical transformations are determined by one function of $2n$ variables. This function is the generating function and it "generates" in the following way. Assume that near some point $(\mathbf{p}_0, \mathbf{q}_0)$ of T^*M, the functions $Q^i(\mathbf{p}, \mathbf{q})$ can be inverted to find $\mathbf{p} = \mathbf{p}(\mathbf{Q}, \mathbf{q})$. This assumption is embodied in the demand that the Jacobian of the map $(\mathbf{p}, \mathbf{q}) \to (\mathbf{Q}, \mathbf{q})$ not vanish.

$$\frac{\partial(\mathbf{Q}, \mathbf{q})}{\partial(\mathbf{p}, \mathbf{q})} = \det \frac{\partial \mathbf{Q}}{\partial \mathbf{p}} \neq 0. \tag{8.53}$$

Then $\mathbf{Q}$ and $\mathbf{q}$ can be regarded as the independent variables. The function S of (8.52) becomes $S(\mathbf{p}, \mathbf{q}) = S_1(\mathbf{Q}, \mathbf{q})$. Compute its differential

$$\mathbf{d}S = \frac{\partial S_1}{\partial q^i}\, \mathbf{d}q^i + \frac{\partial S_1}{\partial Q^i}\, \mathbf{d}Q^i \tag{8.54}$$

and compare with (8.52). Identifying coefficients, we obtain

$$p_i = \frac{\partial S_1}{\partial q^i} \quad \text{and} \quad P_i = -\frac{\partial S_1}{\partial Q^i}. \tag{8.55}$$

A generating function $S_1(\mathbf{Q}, \mathbf{q})$, with $\mathbf{Q}$ and $\mathbf{q}$ as the independent variables, is called a *free* or *Type 1* generating function. The necessary condition (8.53) can be stated in an alternative form. If $\det(\partial \mathbf{Q}/\partial \mathbf{p}) \neq 0$, then the inverse transformation also has a nonvanishing Jacobian and $\det(\partial \mathbf{p}/\partial \mathbf{Q}) \neq 0$. Substituting from (8.55), we obtain

$$\det\left(\frac{\partial p_j}{\partial Q^i}\right) = \det\left(\frac{\partial^2 S_1}{\partial Q^i\, \partial q^j}\right) \neq 0. \tag{8.56}$$

Consider a simple example with a single degree of freedom. Let $Q = p$ and $P = -q + 2p$. We see immediately that $\mathbf{d}P \wedge \mathbf{d}Q = \mathbf{d}p \wedge \mathbf{d}q$. Applying (8.55) gives

$$\frac{\partial S_1}{\partial q} = p = Q, \quad \text{and} \quad \frac{\partial S_1}{\partial Q} = -P = -(-q + 2p) = q - 2Q.$$

Integrating these equations gives $S_1(Q, q) = qQ - Q^2$, which is a generating function for the given canonical transformation. Condition (8.56) is also satisfied.

Thus we see how these functions "generate" the canonical transformations. The Type 1 generating function and (8.55) fully embody the information contained in the transformation equations themselves. A generating function is very much like a Lagrangian or a Hamiltonian in the sense that a single function S contains all the information for the canonical transformation, when it is coupled with (8.52), or equivalently (8.55) for Type 1 functions. A Lagrangian contains all the information about the motion of a system when coupled with Hamilton's principle or equivalently the Euler–Lagrange equations. In the case that the motion may be solved by finding the canonical transformation between the initial values and the state of the system at later times, as discussed in Section 8.7, the generating function contains *all* the information about the motion of the system. This information is contained in an explicit way, not implicitly as in the Lagrangian and Hamiltonian.

Not all canonical transformations can be obtained from Type 1 generating functions because (8.56) cannot always be satisfied. If, for example, $Q = q$, $P = P(q, p)$, then Q and q cannot serve as independent variables. This leads us naturally to consider other possibilities. There are three other possibilities for independent variables in a neighborhood of the point $(\mathbf{p}_0, \mathbf{q}_0)$. They are $(\mathbf{P}, \mathbf{q})$, $(\mathbf{p}, \mathbf{Q})$, and $(\mathbf{p}, \mathbf{P})$. The generating functions with these respective choices for independent variables are referred to as Type 2, Type 3, and Type 4 generating functions, respectively. To obtain the equations corresponding to (8.55) for a Type 2 generating function, we perform a Legendre transformation on (8.52). Subtracting $Q^i \mathbf{d}P_i$ from both sides gives

$$\mathbf{d}S - Q^i \mathbf{d}P_i = p_i \mathbf{d}q^i - P_i \mathbf{d}Q^i - Q^i \mathbf{d}P_i = p_i \mathbf{d}q^i - \mathbf{d}\left(Q^i P_i\right).$$

Transferring across the equal sign, we obtain then

$$\mathbf{d}S_2 = \mathbf{d}\left(S_1 + P_i Q^i\right) = p_i \mathbf{d}q^i + Q^i \mathbf{d}P_i, \tag{8.57}$$

where $S_2(\mathbf{P}, \mathbf{q}) = S_1 + P_i Q^i$. Identifying the coefficients of $\mathbf{d}q^i$ and $\mathbf{d}P_i$ in (8.57) with the partial derivatives of S_2, we obtain

$$p_i = \frac{\partial S_2}{\partial q^i} \quad \text{and} \quad Q^i = \frac{\partial S_2}{\partial P_i}, \tag{8.58}$$

which corresponds to (8.55).

Analogous results for Type 3 and Type 4 generating functions are left to Problem 8.8.

8.6. EXTENDED PHASE SPACE

Often we wish to consider Hamiltonians and/or canonical transformations that depend on time. For this case our geometrical characterization of mechanical motion is best carried out in *extended phase space*. Extended phase space is obtained by simply adding a time axis to the phase space manifold. Refer to Fig. 8.5. This extended phase space is now of dimension $2n + 1$, where n denotes the number of degrees of freedom. Just as in phase space, there is in extended phase space a flow determined by the Hamiltonian $H(\mathbf{p}, \mathbf{q}, t)$. This flow determines a local vector field in extended phase space, $\mathbf{v}_H$.

Let g_H^t denote this flow mapping as before, where on extended phase space $g_H^t(\mathbf{x}(t_0), t_0) = (\mathbf{x}(t_0 + t), t_0 + t)$. Differentiating with respect to time and using (8.4),

$$\frac{d}{dt}\left[g_H^t(\mathbf{x}(t_0), t_0)\right] = (\dot{\mathbf{x}}, 1) = \mathbf{v}_H. \tag{8.59}$$

We can write $\mathbf{v}_H$ in the form

$$\mathbf{v}_H = -\frac{\partial H}{\partial q^i}\mathbf{e}_{p_i} + \frac{\partial H}{\partial p_i}\mathbf{e}_{q^i} + \mathbf{e}_t, \tag{8.60}$$

or in operator form

$$\partial_{\mathbf{v}_H} = -\frac{\partial H}{\partial q^i}\partial_{p_i} + \frac{\partial H}{\partial p_i}\partial_{q^i} + \partial_t. \tag{8.61}$$

Aside from the Hamiltonian flow vector $\mathbf{v}_H$, the central object in the previous sections of this chapter has been the canonical 2-form γ. In extended phase space the 2-form that plays the central role is a generalization:

$$\Gamma = \gamma - \mathbf{d}H \wedge \mathbf{d}t = \mathbf{d}p_i \wedge \mathbf{d}q^i - \mathbf{d}H \wedge \mathbf{d}t. \tag{8.62}$$

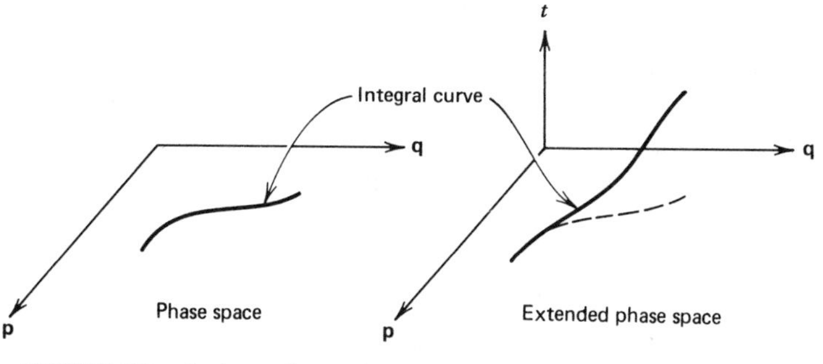

FIGURE 8.5. An integral curve in phase space and in extended phase space.

This is a closed 2-form and can be obtained as the exterior derivative of the 1-form

$$\Psi = p_i\,\mathbf{d}q^i - H\,\mathbf{d}t + \mathbf{d}S. \tag{8.63}$$

The 2-form Γ has a special relationship with the flow vector $\mathbf{v}_H$. This relationship is contained in the statement that $\mathbf{v}_H$ is the unique (to within a constant multiple) vector field such that

$$\Gamma[\mathbf{v}_H, \mathbf{u}] = 0, \quad \text{for all vector fields } \mathbf{u}. \tag{8.64}$$

A demonstration of (8.64) is straightforward and left to Problem 8.9.

To illuminate the meaning of this special relationship between Γ and $\mathbf{v}_H$, we note the following property of any 2-form on a $(2n + 1)$ − dimensional manifold. Any 2-form $\boldsymbol{\omega}^2$ on a $(2n + 1)$-dimensional manifold (e.g., extended phase space) has at least one vector $\mathbf{v}$ satisfying

$$\boldsymbol{\omega}^2[\mathbf{v}, \mathbf{u}] = 0, \text{ for all vectors } \mathbf{u}. \tag{8.65}$$

To prove this consider the matrix of coefficients $\omega_{\alpha\beta}$, where $\boldsymbol{\omega}^2 = (1/2!)\omega_{\alpha\beta}\,\mathbf{d}x^\alpha \wedge \mathbf{d}x^\beta$, $\alpha, \beta = 1, \ldots, 2n + 1$. The matrix ω, with coefficients $\omega_{\alpha\beta}$, is antisymmetric, that is, $\omega^t = -\omega$. From (1.34) it is evident that $\det(-\omega) = (-1)^{2n+1}\det(\omega)$. Since

$$\det \omega = \det \omega^t = \det(-\omega) = (-1)^{2n+1}\det \omega = -\det \omega,$$

we see that $\det \omega = 0$. Hence ω is a singular matrix and there exists at least one nonzero vector $\mathbf{v}$, such that $\omega_{\alpha\beta}v^\beta = 0$. Thus for every vector $\mathbf{u}$, $u^\alpha\omega_{\alpha\beta}v^\beta = 0$. Since $\boldsymbol{\omega}^2[\mathbf{v}, \mathbf{u}] = \omega_{\alpha\beta}v^\alpha u^\beta = 0$, Eq. (8.65) has been shown.

A 2-form that has only one such $\mathbf{v}$ (unique to within a constant multiple) satisfying (8.65) is said to be *nondegenerate*. For a nondegenerate differential 2-form, the flow lines generated by the vector field $\mathbf{v}$ satisfying (8.65) are called the *vortex lines* of $\boldsymbol{\omega}^2$.

Hence Γ, as given in (8.62), is that generalization of $\boldsymbol{\gamma}$ in extended phase space that has vortex lines corresponding to the Hamiltonian flow lines. The 2-form Γ is unique (Problem 8.9) and fully characterizes the Hamiltonian flow in extended phase space just as $\mathbf{v}_H$ does.

If in a new chart $(\mathbf{P}, \mathbf{Q}, T)$ for extended phase space

$$p_i\,\mathbf{d}q^i - H\,\mathbf{d}t = P_i\,\mathbf{d}Q^i - H'\,\mathbf{d}T + \mathbf{d}S,$$

then the 1-form $P_i\,\mathbf{d}Q^i - H'\,\mathbf{d}T + \mathbf{d}S$ will give exactly the same vortex lines. This leads to a more general definition of canonical transformation. A *canonical transformation* on extended phase space is a transformation that leaves invariant the flow lines (integral curves) of $\mathbf{v}_H$, or equivalently leaves invariant the vortex lines of Γ. For such a transformation, which may be time dependent, we contemplate only one time $t = T$ and thus

$$p_i \mathbf{d}q^i - H\mathbf{d}t = P_i \mathbf{d}Q^i - H'\mathbf{d}t + \mathbf{d}S. \tag{8.66}$$

The vortex lines of $\mathbf{d}(P_i \mathbf{d}Q^i - H'\mathbf{d}t) = \Gamma$ satisfy Hamilton's equations and thus $\dot{P}_i = -(\partial H'/\partial Q^i)$ and $\dot{Q}^i = (\partial H'/\partial P_i)$. Writing (8.66) in the equivalent form

$$p_i \mathbf{d}q^i - P_i \mathbf{d}Q^i + (H' - H)\mathbf{d}t = \mathbf{d}S \tag{8.67}$$

enables us readily to identify the partial derivatives of the generating function S. We conclude

$$p_i = \frac{\partial S}{\partial q^i}, \quad P_i = -\frac{\partial S}{\partial Q^i}, \quad H' = H + \frac{\partial S}{\partial t}. \tag{8.68}$$

The foregoing results are for a Type 1 generating function and similar results hold for other types as well. The Hamiltonians H and H' are related as in (8.68) for all types.

8.7. THE HAMILTON–JACOBI EQUATION

In this section we develop a method of integrating the equations of motion that is quite different in spirit from the Newtonian, Lagrangian, or Hamiltonian methods discussed previously. The Hamilton–Jacobi method is based on the following observations. (1) The Hamiltonian flow is a canonical transformation and thus the flow map that takes $(\mathbf{p}(t_0), \mathbf{q}(t_0), t_0) \to (\mathbf{p}(t + t_0), \mathbf{q}(t + t_0), t + t_0)$ can be obtained from a generating function S. (2) The value of the Hamiltonian at the initial time is just a number and may be chosen to be zero.

In light of these observations we seek a Type 1 generating function $S(\mathbf{q}, \mathbf{Q}, t)$ as in (8.68). The new coordinates are almost the initial coordinates $\mathbf{q}(t_0)$, but since we choose $H' = 0$, the $\mathbf{Q}$ will be constants, but functions of the initial $\mathbf{q}(t_0)$. Likewise, the new momenta $\mathbf{P}$ obtained from (8.68) are constants that are functions of the initial values. Summarizing we state the Hamilton–Jacobi theorem. The generating function $S(\mathbf{q}, \mathbf{Q}, t)$ satisfying

$$\frac{\partial S}{\partial t} + H\left(\frac{\partial S}{\partial \mathbf{q}}, \mathbf{q}, t\right) = 0 \tag{8.69}$$

and depending on n constants $Q^1, \ldots, Q^n$, such that $\det(\partial^2 S/\partial Q^i\, \partial q^j) \neq 0$ represents a complete solution to the problem of motion. For completeness we also give here the relations for the momenta:

$$p_i = \frac{\partial S}{\partial q^i} \quad \text{and} \quad P_i = -\frac{\partial S}{\partial Q^i}, \tag{8.70}$$

where the P_i are constants. Equations (8.70) define the canonical transformation from $2n$ constants of the motion (P_i, Q^i) to the (p_i, q^i), which are the general time-dependent motions of the system. It can be shown directly (Problem 8.13) that a solution to the Hamilton–Jacobi equation (8.69) with (8.70) ensures that Hamilton's equations for $(\mathbf{p}, \mathbf{q})$ are satisfied.

It is important to recognize that there is nothing special in choosing a Type 1 generating function. A completely analogous Hamilton–Jacobi theorem may be given for $S_2(\mathbf{P}, \mathbf{q})$ in which (8.69) is the same but (8.70) is replaced by the relations in (8.58). The determinant condition is also appropriately modified.

It is somewhat remarkable that an integration of (8.69) should prove to be an effective solution technique. Equation (8.69) is nearly always nonlinear because H is almost always quadratic in the momenta p_i, and consequently (8.69) is quadratic in $(\partial S/\partial q^i)$. Furthermore, (8.69) is a partial differential equation as opposed to Hamilton's equations, which are ordinary differential equations.

The difficulties mentioned in the previous paragraph are partially offset by the great number of systems for which (8.69) turns out to be separable. If $(\partial H/\partial t) = 0$, then a partial integration of (8.69) is immediate

$$S(\mathbf{q}, \mathbf{Q}, t) = -tE + W(\mathbf{q}, \mathbf{Q}). \tag{8.71}$$

Then (8.69) reduces to

$$H\left(\frac{\partial W}{\partial \mathbf{q}}, \mathbf{q}\right) = E. \tag{8.72}$$

The first constant coordinate Q^1 is just E. The notation chosen is to suggest that E is usually the energy.

A further integration of (8.72) usually proceeds by a separation of variables. If the assumption

$$W(\mathbf{q}, \mathbf{Q}) = W^1(q^1, \mathbf{Q}) + W^2(q^2, \mathbf{Q}) + \cdots = \sum_{i=1}^{n} W^i(q^i, \mathbf{Q}) \tag{8.73}$$

leads to a separation of variables, then (8.72) reduces to a set of n ordinary differential equations and the "Q's" are separation constants.

As an illustration of the Hamilton–Jacobi method we consider the simple problem of a one-dimensional harmonic oscillator.

$$H = \frac{p^2}{2} + \frac{q^2}{2}.$$

Since H is constant, we have the situation described by (8.71) and (8.72). The first constant coordinate is $Q = E$, the first separation constant. Equation (8.72) takes the specific form for this Hamiltonian

$$\frac{1}{2}\left(\frac{\partial W}{\partial q}\right)^2 + \frac{q^2}{2} = E$$

and we find

$$W = \int (2E - q^2)^{1/2}\, dq.$$

The integration is carried out by the trigonometric substitution $q = (2E)^{1/2}\sin\theta$. We find

$$W = E\sin^{-1}\left(\frac{q}{(2E)^{1/2}}\right) + q\left(\frac{E}{2}\right)^{1/2}\left(1 - \frac{q^2}{2E}\right)^{1/2}.$$

Thus the Hamilton–Jacobi function is

$$S = -Et + E\sin^{-1}\left(\frac{q}{(2E)^{1/2}}\right) + \frac{q}{2}(2E - q^2)^{1/2}$$

and we find

$$-P = \frac{\partial S}{\partial E} = -t + \sin^{-1}\left(\frac{q}{(2E)^{1/2}}\right)$$

and

$$p = \frac{\partial S}{\partial q} = (2E)^{1/2}\left(1 - \frac{q^2}{2E}\right)^{1/2}.$$

These results can be written in the form $q = (2E)^{1/2}\sin(t - P)$ and $p = (2E)^{1/2}\cos(t - P)$. Note that in this case $P, Q = E$ are not the initial values for p and q.

In this chapter we have considered a very geometrical view of Hamiltonian dynamics. The evolution of a mechanical system produces flow lines or system trajectories in phase space. These flow lines are the integral curves of the flow vector $\mathbf{v}_H$. Differential forms were introduced and the canonical 2-form $\boldsymbol{\gamma}$, or its generalization for extended phase space Γ, played the central role in characterizing canonical transformation. We considered in detail how generating functions lead to canonical transformations. The Hamiltonian–Jacobi theory leading to a solution of the problem of motion by finding a generating function for the Hamilton flow was also considered.

PROBLEMS

8.1. Let $\boldsymbol{\omega}^k$ be a differential k-form on the manifold M that is of dimension n. Show that $\boldsymbol{\omega}^k = 0$ for $k > n$.

8.2. Derive Eqs. (8.26) and (8.27).

8.3. Let the magnetic field components comprise the components of the differential 2-form

$$\omega_{\mathbf{B}} = B_x\, dy \wedge dz + B_y\, dz \wedge dx + B_z\, dx \wedge dy.$$

Show that $\mathbf{d}\omega_{\mathbf{B}} = 0$.

8.4. Show that the vector field in the plane given by $\mathbf{B} = c\hat{\phi}$ generates the flow $g^t(r_0, \phi_0) = (r_0, \phi_0 + ct/r)$.

8.5. Let $\mathbf{A}$ be an arbitrary vector field in $\mathbb{R}^3$, where $\mathbf{A} = A^i\mathbf{e}_i$. Consider the differential forms $\boldsymbol{\omega}^1{}_A = A_i\,\mathbf{d}x^i$ and $\boldsymbol{\omega}^2{}_A = \frac{1}{2}A^i\varepsilon_{ijk}\,\mathbf{d}x^j \wedge \mathbf{d}x^k$. Show that

$$\mathbf{d}\boldsymbol{\omega}^1{}_A = (\nabla \times \mathbf{A})^i\boldsymbol{\Sigma}_i \qquad \text{and}$$

$$\mathbf{d}\boldsymbol{\omega}^2{}_A = (\nabla \cdot \mathbf{A})\boldsymbol{\Omega},$$

where

$$\boldsymbol{\Sigma}_i = \tfrac{1}{2}\varepsilon_{ijk}\,\mathbf{d}x^j \wedge \mathbf{d}x^k \quad \text{and} \quad \boldsymbol{\Omega} = (g)^{1/2}\,\mathbf{d}x^1 \wedge \mathbf{d}x^2 \wedge \mathbf{d}x^3.$$

The quantity $g = \det(g_{ij})$ as in Problem 1.5, and the forms $\boldsymbol{\Sigma}_i$ and $\boldsymbol{\Omega}$ are called the *surface 2-form* and the *volume 3-form*, respectively. Derive the usual expressions for $\nabla \cdot \mathbf{A}$ and $\nabla \times \mathbf{A}$ in cylindrical and spherical coordinates using these results.

8.6. Prove the following transformation is canonical three different ways: by (7.28), by (8.36), and by (8.44).

$$q^1 = \left(\sqrt{(2P_1)}\sin Q^1 + P_2\right)/\sqrt{(m\omega)}\,;$$

$$q^2 = \left(\sqrt{(2P_1)}\cos Q^1 + Q^2\right)/\sqrt{(m\omega)}$$

$$p_1 = \sqrt{(m\omega)}\left(\sqrt{(2P_1)}\cos Q^1 - Q^2\right)/2;$$

$$p_2 = \sqrt{(m\omega)}\left(-\sqrt{(2P_1)}\sin Q^1 + P_2\right)/2.$$

8.7. Show that the pull-back map satisfies $(h \circ g)^* = g^* \circ h^*$ and that $g^*(\omega^k \wedge \sigma^l) = g^*\omega^k \wedge g^*\sigma^l$.

8.8. Obtain the following formulas for the generating functions of Types 3 and 4.

(a) $S_3(\mathbf{p}, \mathbf{Q}) = S_1(\mathbf{q}, \mathbf{Q}) - p_i q^i$ where $q^i = q^i(\mathbf{p}, \mathbf{Q}, t)$.

$$P_i = -\frac{\partial S_3}{\partial Q^i} \quad \text{and} \quad q^i = -\frac{\partial S_3}{\partial p_i} \text{ where } \det\frac{\partial^2 S_3}{\partial Q^i\,\partial p_j} \neq 0.$$

(b) $S_4(\mathbf{p}, \mathbf{P}) = S_1(\mathbf{q}, \mathbf{Q}) - p_i q^i + P_i Q^i$ where $q^i = q^i(\mathbf{p}, \mathbf{P}, t)$ and $Q^i = Q^i(\mathbf{p}, \mathbf{P}, t)$ with

$$Q^i = \frac{\partial S_4}{\partial P_i} \quad \text{and} \quad q^i = -\frac{\partial S_4}{\partial p_i} \text{ and } \det\frac{\partial^2 S_4}{\partial P_i\,\partial p_j} \neq 0.$$

8.9. Show that $\mathbf{v}_H$ of (8.60) satisfies (8.64) and that $\mathbf{v}_H$ is unique to within a constant multiple. Given $\mathbf{v}_H$, to what degree is the 2-form Γ satisfying (8.64) unique?

8.10. Consider the Lagrangian for a particle of charge e, mass m, in an electromagnetic field with potentials $\Phi(t, \mathbf{x})$ and $\mathbf{A}(t, \mathbf{x})$:

$$L = \frac{m\dot{x}^2}{2} - e\Phi(t, \mathbf{x}) + \frac{e}{c}\dot{\mathbf{x}} \cdot \mathbf{A}(t, \mathbf{x}).$$

(a) Obtain the Hamiltonian for this system.

(b) Consider the following gauge transformation:

$$\mathbf{A}'(t,\mathbf{x}) = \mathbf{A}(t,\mathbf{x}) + \nabla\Lambda;\ \Phi'(t,\mathbf{x}) = \Phi(t,\mathbf{x}) - \frac{1}{c}\frac{\partial\Lambda}{\partial t},$$

where $\Lambda = \Lambda(t,\mathbf{x})$ is the gauge function. Is such a gauge transformation a canonical transformation and if so find a generating function of appropriate type.

8.11. Solve the problem of the motion of a point mass under the influence of a uniform gravitational field, *using the Hamilton–Jacobi method*. Find both the path and the dependence of the coordinates on time. Compare your results with those obtained by more familiar methods.

8.12. The Lagrangian of a particular system is given by

$$L = \frac{\dot{x}^2}{2(ax+b)} + \frac{1}{2}y^2\dot{y}^2 - (c+dy),$$

where a, b, c, d are all constants. Write out the Hamiltonian for this system. Solve the Hamilton–Jacobi equation and use the solution to construct a transformation that makes the Hamiltonian cyclic. Can you solve this problem directly using the canonical equations?

8.13. Demonstrate directly that a solution to the Hamilton–Jacobi equation will ensure that Hamilton's canonical equations are satisfied.

8.14. Consider the so-called *3-particle Toda lattice* of three particles moving on a ring with exponential forces between them. See Fig. 8.6. The Hamiltonian can be written as

$$H = \tfrac{1}{2}\left(p_1^{\,2} + p_2^{\,2} + p_3^{\,2}\right) + e^{-(q^3-q^2)} + e^{-(q^2-q^1)} + e^{-(q^1-q^3)} - 3.$$

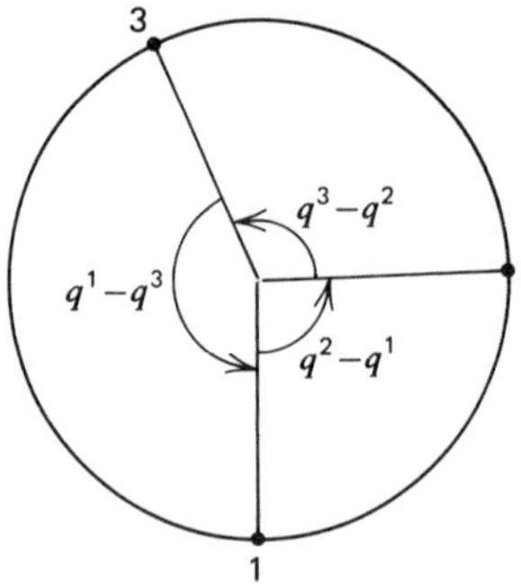

FIGURE 8.6. The three-particle Toda lattice for Problem 8.14.

Show that a canonical transformation can be made, with some possible rescaling, so that a mathematically equivalent dynamic problem is determined by the Hamiltonian

$$\overline{H} = \tfrac{1}{2}\left(p_x^2 + p_y^2\right) + \tfrac{1}{24}\left(e^{-2x-2y\sqrt{3}} + e^{-2x+2y\sqrt{3}} + e^{4x}\right) - \tfrac{1}{8}.$$

Hint: Examine the limiting small-oscillations problem for a suggested transformation:

8.15. Consider a particle of unit mass moving in one dimension under the influence of the potential $V = (1/2)\tan^2(x/l)$. Find the solution to the problem of motion using the Hamilton–Jacobi method.

8.16. Show that the following transformation is canonical:

$$P = p + q^2 + pq^2; \qquad Q = \tan^{-1} q.$$

Construct a generating function of appropriate type.

8.17. Given the Hamiltonian $H = (p_1 - kq_1)^2/2m + p_2{}^2/2m$ solve for the motion using the Hamilton–Jacobi theorem.

8.18. (a) Show the transformation

$$Q = m^{1/2}\left(q + (q-p)^2\right); \qquad P = (m)^{-1/2}\left(p + (q-p)^2\right)$$

to be canonical.

(b) Find a generating function for this transformation.

(c) If the Hamiltonian of a particle is given by

$$H = \frac{1}{2m}\left(p^2 + p(1-2q) - q^2\right),$$

use a canonical transformation to find the motion.

(d) Now solve for the motion of this system by using the Hamilton–Jacobi method.

CHAPTER NINE

ACTION-ANGLE VARIABLES

Although I intend to leave the description of this empire to a particular treatise, yet in the meantime I am content to gratify the curious reader with some general ideas.

SWIFT

In this chapter we define and discuss action-angle variables and construct these variables for some simple but typical systems. We introduce Stokes' theorem and the notion of integration on forms. Stokes' theorem for forms is used to obtain several integral invariants of Hamiltonian flow. We discuss canonical perturbation theory and simple nonlinear resonances. The preservation of invariant tori under perturbation is discussed and numerical examples of the Toda and Henon–Hiles potentials are used to illustrate the KAM theorem.

9.1. INVARIANT TORI AND ACTION-ANGLE VARIABLES

In Section 7.4 we considered completely integrable systems. Recall that the central requirement for complete integrability was the existence of n first integrals I_i, $i = 1, \ldots, n$. These first integrals must be linearly independent, single-valued functions of phase space coordinates $(\mathbf{p}, \mathbf{q})$, and be in involution, that is, $\{I_i, I_j\} = 0$ for all i and j. When these conditions are fulfilled, the system was shown to be completely integrable.

For conservative systems we have often used the fact that the motion in phase space must take place on the energy surface. When additional constants of the motion exist, then the motion is further restricted. For a completely integrable system the motion is restricted to an n-dimensional submanifold $M_{\mathbf{I}_0}$ of the phase space manifold M. The points of $M_{\mathbf{I}_0}$ are given by

$$M_{\mathbf{I}_0} = \{(\mathbf{p}, \mathbf{q}) \in M \mid I_i(\mathbf{p}, \mathbf{q}) = I_{i0} = \text{constant}, \qquad i = 1, \ldots, n\}.$$

For different choices of the constants $\mathbf{I}_0 = (I_{10}, I_{20}, \ldots, I_{n0})$ we get different submanifolds.

The first integrals $I_i(\mathbf{p}, \mathbf{q})$ are functions on phase space and consequently generate flow vectors according to (8.2). We denote these flow vectors as $\mathbf{v}_i$, where i is the label distinguishing the invariants $I_i(\mathbf{p}, \mathbf{q})$, $i = 1, \ldots, n$. The corresponding operators are denoted $\boldsymbol{\partial}_i$. From (8.13) and the involutive constraint, $\boldsymbol{\partial}_i[I_j] = \{I_j, I_i\} = 0$, and thus none of the I_j change along the flow lines of $\mathbf{v}_i$. Hence, the flow lines of $\mathbf{v}_i$ must lie entirely in the submanifold $M_{\mathbf{I}_0}$. Since this result is valid for any $i = 1, \ldots, n$, *all* $\mathbf{v}_i$ have flow lines that are entirely within the submanifold $M_{\mathbf{I}_0}$.

Denote the path parameter along the flow lines of $\mathbf{v}_i$ as ξ^i, $i = 1, \ldots, n$. Then the operators $\boldsymbol{\partial}_i$ are the partial derivative operators $(\partial/\partial\xi^i)$. The flow lines generated by the vector fields can be used as a coordinate system for the submanifold $M_{\mathbf{I}_0}$ if two conditions are met. (1) The vector fields $\mathbf{v}_i$ must be linearly independent. (2) As operators, the vector fields must commute. Condition 1 is familiar and 2 is a restatement of the condition

$$\frac{\partial^2 f}{\partial\xi^i\,\partial\xi^j} - \frac{\partial^2 f}{\partial\xi^j\,\partial\xi^i} = 0$$

for an arbitrary, single-valued function f. Condition 2 may be written in operator form:

$$\boldsymbol{\partial}_i(\boldsymbol{\partial}_j f) - \boldsymbol{\partial}_j(\boldsymbol{\partial}_i f) = 0,$$

and therefore the operators $\boldsymbol{\partial}_i$ and $\boldsymbol{\partial}_j$ must commute for $i, j = 1, \ldots, n$. We verify below that these conditions are met by the flow vectors $\mathbf{v}_i$, corresponding to the first integrals I_i.

Since by assumption the functions $I_i(\mathbf{p}, \mathbf{q})$ are linearly independent, it follows that the 1-forms $\mathbf{d}I_i$ are linearly independent. Since the $\mathbf{d}I_i$ are linearly independent and $\mathbf{d}I_i[\mathbf{v}_f] = \partial_i[f]$ for an arbitrary function f, the $\mathbf{v}_i$ are also linearly independent.

Consider the commutivity of the vector fields. From (8.13) we see

$$\partial_i[\partial_j[f]] - \partial_j[\partial_i[f]] = \{\partial_j[f], I_i\} - \{\partial_i[f], I_j\}$$

$$= \{\{f, I_j\}, I_i\} - \{\{f, I_i\}, I_j\} = \{\{I_j, I_i\}, f\} = 0,$$

where the Jacobi identity (7.30) and the involutive constraint have been used.

All conditions are met everywhere on $M_{\mathbf{I}_0}$ and thus the flow lines of the $\mathbf{v}_i$ serve as a global coordinate mesh.

If in addition the submanifold is compact, that is, the motion is bounded, then the submanifold $M_{\mathbf{I}_0}$ must have the topology of an n-dimensional torus. This topological result for a compact, n-dimensional manifold, on which n vector fields giving a basis are defined everywhere, is easily visualized in the case $n = 2$. Figure 9.1 contrasts the compact manifolds S^2 and T^2. There is always some point on the sphere where the conditions for the vector fields to form a coordinate basis are *not* satisfied. This result is often made graphic by contrasting hair combed on a torus with hair combed on a sphere. On a sphere, there will always be a "crown" or "swirl," some place where the scalp shows through. On a torus, the hair can be combed to fully cover the surface without "crowns" or "swirls."

These n-tori or submanifolds $M_{\mathbf{I}_0}$ (labeled by specific values of the I_i, or J_i to be discussed shortly) are referred to as *invariant tori*. The word "invariant" is part of the name and reflects the fact that the various values

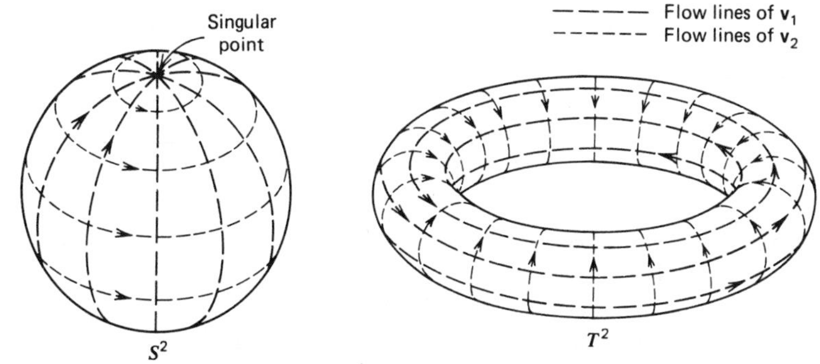

FIGURE 9.1. Flow lines on the surface of a sphere and a torus showing that only the torus has on it two vector fields that are everywhere linearly independent, commute, and hence give a good coordinate mesh everywhere.

of the invariants I_i label the different tori. The name "invariant tori" is not intended to connote some invariance property of the tori themselves.

This picture of bounded motion taking place on an n-dimensional torus is very intuitive and suggests the use of angle variables θ^i, one for each independent direction for going around the torus. The θ^i are different from the coordinates Q^i, which are conjugate to the I_i and discussed in Section 7.4. The θ^i taken together with the I_i do not form a canonically conjugate set of coordinates. The geometric picture of an n-torus with angle coordinates θ^i is very appealing, but to use the θ^i as a set of canonical coordinates, a set of canonically conjugate momenta is needed. We seek a new set of momentum coordinates J_i, which are functions of the I_i only, and corresponding angle coordinates θ^i such that $\boldsymbol{\gamma} = \mathbf{d}p_i \wedge \mathbf{d}q^i = \mathbf{d}I_i \wedge \mathbf{d}Q^i = \mathbf{d}J_i \wedge \mathbf{d}\theta^i$ [cf. (8.36)].

It was stated earlier that the functions $I_i(\mathbf{p}, \mathbf{q})$ are single-valued and hence may be inverted to find $p_i(\mathbf{I}, \mathbf{q})$. Even though the relationship between the p's and the I's is one-to-one, the same is not true of the relationship between the p's and q's. Generally, the $p_i(\mathbf{I}, \mathbf{q})$ are multivalued functions of $\mathbf{q}$. Even the simplest of systems may have several values of the momenta for a given $\mathbf{q}$. (For example, a simple pendulum may be moving to the right or the left at a given instantaneous position.) Consequently, on $M_{\mathbf{I}_0}$ there may be many values of $\mathbf{p}$ for a given $\mathbf{q}$. So the generating function of (7.33)

$$S(\mathbf{q}, \mathbf{I}) = \int_{\mathbf{q}_0}^{\mathbf{q}} p_j(\mathbf{I}, \mathbf{q}')\, dq'^j \tag{9.1}$$

will also be multivalued. In other words, the value of $S(\mathbf{q}_0, \mathbf{I})$, computed from (9.1) around a closed path, need not be zero, depending on the closed path taken. On an n-dimensional torus, we can choose at least, but no more than, n closed paths that cannot be deformed into one another or to a point. Let $\{\gamma_i\}$ denote a set of n such paths on the n-torus $M_{\mathbf{I}_0}$. We define the *action momentum* J_i by

$$J_i = \frac{1}{2\pi} \oint_{\gamma_i} p_j(\mathbf{I}, \mathbf{q})\, dq^j. \tag{9.2}$$

The action J_i is the change in the generating function of (9.1) over the closed curve γ_i, divided by 2π, that is, $J_i = (\Delta_{\gamma_i} S)/2\pi$.

The quantity J_i does not depend on the curve γ_i. To show this, we consider two curves γ_i and γ_i', which can be deformed into one another. Referring to Fig. 9.2, we form a new curve Γ by connecting γ_i and γ_i' with an interconnecting two-way link as shown in Fig. 9.2b. Since $p_j\, dq^j = dS$, the integral around Γ is zero because it can be deformed to a point. The

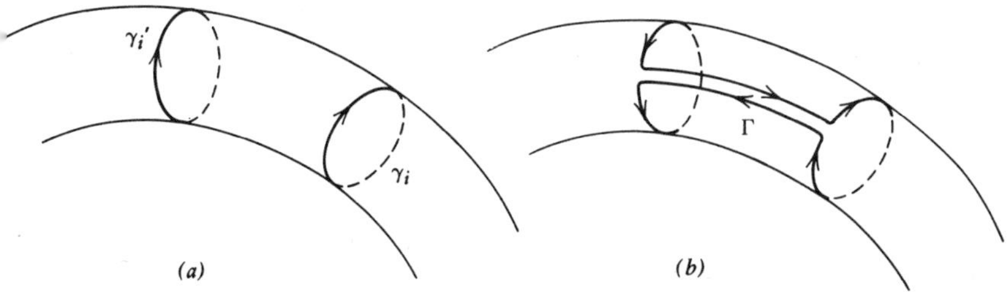

FIGURE 9.2. (*a*) Two closed paths γ_i and γ_i', which may be deformed into one another. (*b*) Path formed by connecting γ_i and γ_i' with two-way link.

contribution to the line integral along the connecting link is zero because it is traversed twice in opposite directions. The integral along Γ traverses γ_i' in the opposite sense from γ_i. Thus we have

$$0 = \oint_\Gamma p_i\, dq^i = \oint_{\gamma_i} p_j\, dq^j - \oint_{\gamma_i'} p_j\, dq^j$$

which shows that the line integral of $p_j\, dq^j$ is the same around γ_i or γ_i'. Since the action J_i does not depend on which γ_i is chosen for the line integral, for actual calculation one always attempts to chose a path for which only one term in the sum $p_j\, dq^j$ contributes, that is, a path for which all dq^j except one are zero.

In definition (9.2) all dependence on the coordinates $\mathbf{q}$ has been integrated out and so J_i depends only on the values of the invariants I_i, which characterize the manifold in question; that is, $J_i = J_i(\mathbf{I})$. Similarly, the I_i can now be considered as functions of the J_i; that is, $I_i = I_i(\mathbf{J})$. So we may consider the generating function of (9.1) as a function of the q^i and J_i, a Type 2 function.

$$S(\mathbf{q}, \mathbf{J}) = \int_{\mathbf{q}_0}^{\mathbf{q}} p_j(\mathbf{J}, \mathbf{q})\, dq^j \tag{9.3}$$

Having defined the new momenta, the new coordinates are calculated in standard fashion according to (8.58).

$$\theta^i = \frac{\partial S}{\partial J_i}. \tag{9.4}$$

It can be shown explicitly that the action-angle coordinates $(\mathbf{J}, \boldsymbol{\theta})$ form a canonical set (Problem 9.5).

Our motivation for considering action momenta was to obtain conjugate coordinates to angles in various directions around the tori. However, once the action J_i are defined, we obtain the coordinates from (9.4). Are the coordinates of (9.4) angles? To verify this we calculate how much θ^i changes in one complete circuit around the closed curve γ_j.

$$\Delta_{\gamma_j}\theta^i = \Delta_{\gamma_j}\frac{\partial S}{\partial J_i} = \frac{\partial}{\partial J_i}\Delta_{\gamma_j}S = \frac{\partial}{\partial J_i}2\pi J_j = 2\pi\delta^i_j. \tag{9.5}$$

Thus θ^i changes by 2π for one complete circuit around γ_i and does not change around the other γ_j. The θ^i are therefore appropriately identified as angles on the n-tori.

In the coordinates $(\mathbf{I}, \mathbf{q})$ the Hamiltonian H is only a function of the I_i (cf. Section 7.4). Consequently, H is only a function of the J_i, and Hamilton's canonical equations imply

$$\frac{dJ_i}{dt} = 0; \qquad \frac{d\theta^i}{dt} = \frac{\partial H}{\partial J_i} \equiv \omega^i(\mathbf{J}), \tag{9.6}$$

with the solution

$$J_i = \text{constant} \quad \text{and} \quad \theta^i(t) = \omega^i(\mathbf{J})t + \theta^i_0. \tag{9.7}$$

Once again we have conditionally periodic motion and the phase curve will densely cover the n-torus, specified by $\mathbf{J}$, unless the frequencies $\omega^i(\mathbf{J})$ are commensurable. By definition, the frequencies are commensurable, if there exists some time T and integers n_i, $i = 1,\ldots, n$, such that for each i, $\omega^i T = 2\pi n_i$. Thus $(\omega^i/\omega^j) = (n_i/n_j)$, or

$$\omega^i n_j - \omega^j n_i = 0, \qquad i, j = 1,\ldots, n. \tag{9.8}$$

When the frequencies are commensurable, the phase path on the torus will close on itself. The frequencies $\omega^i(\mathbf{J})$ are functions of the action, as the notation indicates, and in general change as $\mathbf{J}$ changes. A different $\mathbf{J}$ implies a change in the invariant n-torus under consideration. The tori corresponding to $\mathbf{J}$ values for which the $\omega^i(\mathbf{J})$ are commensurable are referred to as *rational* surfaces or *resonant* surfaces. The names are used interchangeably. In Fig. 9.3 a phase curve on a rational surface is compared to a curve on a surface where the $\omega^i(\mathbf{J})$ are not commensurable.

In action-angle variables we have a very geometrical picture for the phase space of a completely integrable system undergoing bounded motion. This picture consists of nested n-tori with motion in each independent direction given by a single angle coordinate θ^i measured around a circle of radius J_i. See Fig. 9.4 for a sketch of different 2-tori corresponding to different $\mathbf{J}$.

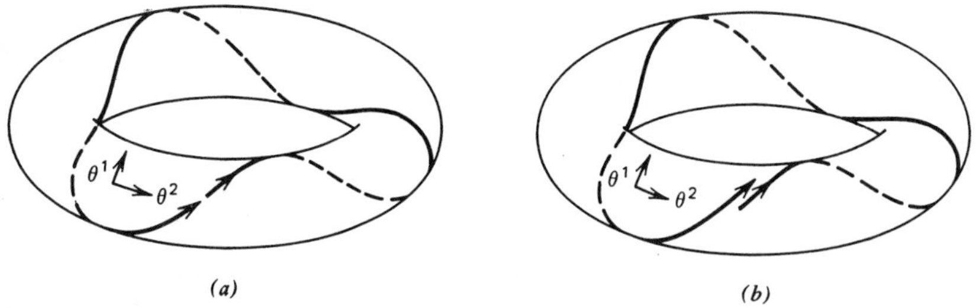

FIGURE 9.3. Comparison of phase curves on 2-tori. The torus in (a) is a rational surface where $\omega^1/\omega^2 = 3$. The torus in (b) has slightly different $\mathbf{J}$ values from (a) and $\omega^1(\mathbf{J})$ and $\omega^2(\mathbf{J})$ are not commensurable. The phase curve in (b) eventually covers the surface of the torus densely.

Example 1. As a simple but instructive example, we consider a simple harmonic oscillator.

$$H = \frac{p^2}{2} + \frac{q^2}{2}, \tag{9.9}$$

where the time scale has been chosen to make the frequency equal 1. We find $p = (2H - q^2)^{1/2}$ and the phase curves are circles in the (p, q) plane of radius $(2H)^{1/2}$. See Fig. 9.5.

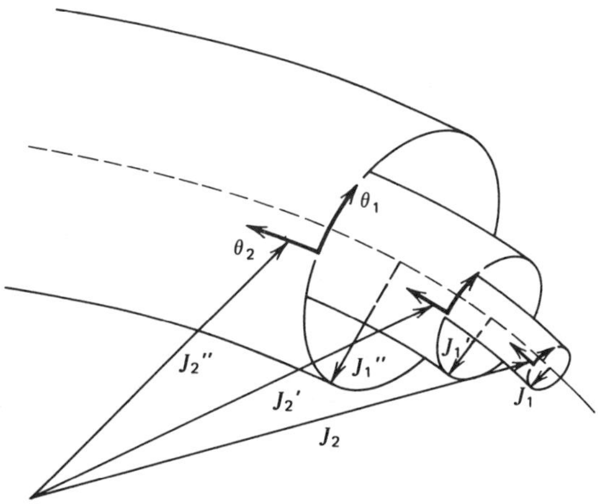

FIGURE 9.4. Nested toroids corresponding to constant values for the actions J_i for 2 degrees of freedom. The vectors $\mathbf{J} = (J_1, J_2)$, $\mathbf{J}' = (J_1', J_2')$, and $\mathbf{J}'' = (J_1'', J_2'')$ label the different tori. Angle coordinates (θ^1, θ^2) on the tori are also indicated.

The coordinates (p, q) can take on all possible real values and thus phase space is $\mathbb{R}^2$. The nested "toroids" consist of circles centered about the origin with each circle labeled by a different energy. From (9.2) the action momentum is

$$J = \frac{1}{2\pi}\oint p\,dq = \frac{1}{2\pi}\int_{\text{circle}} (2H - q^2)^{1/2}\,dq = \frac{\text{area of circle}}{2\pi} = H. \tag{9.10}$$

The action angle is obtained from (9.4).

$$\theta = \frac{\partial}{\partial J}\int_0^q (2J - q'^2)^{1/2}\,dq' = \sin^{-1}\left(\frac{q}{(2J)^{1/2}}\right), \tag{9.11}$$

or

$$(2J)^{1/2}\sin\theta = q. \tag{9.12}$$

The phase space variables (J, θ) are also shown in Fig. 9.5. For this simple system the change of canonical coordinates from (p, q) to (J, θ) is equivalent to a change from cartesian to polar coordinates in the plane $\mathbb{R}^2$.

Example 2. For a second, somewhat more interesting example consider a simple pendulum:

$$H = \frac{p^2}{2} - \cos\phi, \tag{9.13}$$

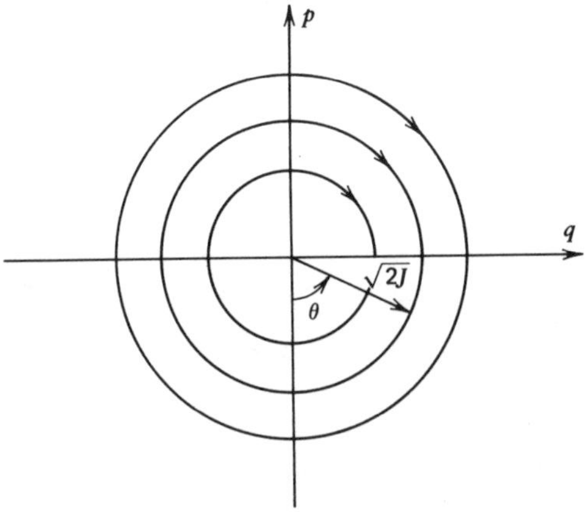

FIGURE 9.5. Phase curves of a one-dimensional, simple harmonic oscillator.

where variables have been appropriately scaled to be dimensionless. Again, one first integral is immediate, $E = H$. The invariant tori are given by (9.13) with $E = H$ held constant. The angle ϕ is chosen to range between π and $-\pi$. The phase space manifold for this system is sketched in Fig. 9.6. The phase curve (drawn in bold, dark lines in Fig. 9.6) separating the phase curves encircling the cylinder from those that do not is called the *separatrix*. If the cylinder in Fig. 9.6 is cut at $\phi = \pi$ and flattened out, the phase curve sketch of Fig. 2.3 is obtained. The phase curves on the cylinder are the invariant tori for this system.

We turn to a construction of the action-angle variables for the simple pendulum. The tori consist of simple curves and so there is only one possible circuit γ. However, we must respect the dependence on energy, which determines whether the curve γ is inside the separatrix or not.

For a phase curve inside the separatrix, the limiting angle $\phi_0 = \cos^{-1}(E)$ is less than π; that is, $\phi_0 < \pi$. Substituting into (9.2)

$$J = \frac{\sqrt{2}}{\pi}\int_{-\phi_0}^{\phi_0} (E + \cos\phi)^{1/2}\, d\phi = \frac{2\sqrt{2}}{\pi}\int_0^{\phi_0}(E + \cos\phi)^{1/2}\, d\phi. \tag{9.14}$$

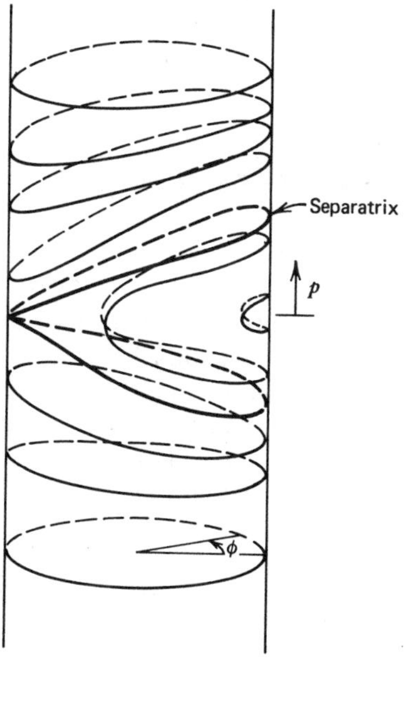

FIGURE 9.6. Phase space manifold for a simple pendulum. This manifold has the topology of a cylinder, $\mathbb{R}^1 \times S^1$. The invariant tori (curves) are sketched and the separatrix between bounded oscillations and free rotation is indicated. See Fig. 2.3 for a sketch of the phase curves (tori) mapped into the plane $\mathbb{R}^2$.

By defining $k = [2/(E+1)]^{1/2}$ and changing the integration variable to $\phi/2$, J can be written in terms of elliptic functions.

$$J = \frac{8}{\pi k} E(\phi_0/2, k), \tag{9.15}$$

where $E(\alpha, k)$ is the elliptic integral of the second kind.

For the motion on phase curves outside the separatrix, $\phi_0/2 \to \pi/2$ in (9.15) and J is given in terms of the complete elliptic integral of the second kind.

In the case of the simple pendulum it is not so easy to express the Hamiltonian as a function of J, as it was for the simple harmonic oscillator. For $\phi_0 < \pi$, the dependence of J on $E = H$ is through both k and ϕ_0 in (9.15). When $\phi_0 = \pi$, J depends on the energy E only through k. Consequently, the resulting expression for the frequency is more complicated, but can also be given in terms of elliptic integrals.

$$\omega(J) = \frac{\pi}{2k} [F(\phi_0/2, k)]^{-1}, \tag{9.16}$$

where $k = [2/(E(J)+1)]^{1/2}$ and $F(\alpha, k)$ is the elliptic integral of the first kind. The action angle is given by

$$\theta = k\omega(J) F(\phi/2, k). \tag{9.17}$$

Other examples are considered in the problems.

9.2. INTEGRALS OF FORMS AND STOKES' FORMULA

In doing the example of a simple harmonic oscillator in the previous section, we found that J was given by the area enclosed in the phase curve, divided by 2π. From (9.14) it is also clear that (9.15) represents the area enclosed in the phase curve. This result has a generality that extends beyond the one-dimensional case and provides one reason for considering the integration of forms. We remark, however, that integration of forms on manifolds is not a topic that can be fully studied in a short section. For the reader wishing a more complete account we recommend the following as giving readable introductions: Flanders (1963), Choquet-Bruhat and DeWitt-Morette (1982), and Arnold (1978). See also Misner, Thorne, and Wheeler (1973) for an excellent discussion, similar in spirit to the presentation of this section.

Unquestionably, the most fundamental integral theorem for physics, with wide-ranging applications, is called *Stokes' theorem*. This is not the usual Stokes' theorem, which relates line integrals to surface integrals, but a generalization of it that includes the usual divergence theorem as well. The content of the theorem is contained in the formula

$$\int_{\partial V} \omega = \int_V \mathbf{d}\omega, \tag{9.18}$$

where we must explain what the symbols in this formula mean. The quantity ω is a k-form and $\mathbf{d}\omega$ is its exterior derivative, a $(k+1)$-form. V is a $(k+1)$-dimensional "volume" and ∂V denotes its boundary, a k-dimensional "volume."

What does an integral over a form mean? In concept, this question is easily answered by referring back to the geometrical pictures of k-forms as discussed in Section 8.2. The basis k-forms were pictured as unit "volumes": a 1-form as a unit line, a 2-form as a unit area, and so on. The integral over a k-form is an integral over these unit k-volumes. Indeed, integrals provide a major motivation for studying differential k-forms. Rather than going carefully into all the details one would find in a mathematics text, we give a prescription of how one is to proceed with an evaluation of the integral over a form for simple "volumes." Complicated "volumes" may require a partition into pieces for which the following prescription is applicable. The integral of the k-form ω over the k-volume V

$$\int_V \omega$$

is given by the following rules.

1. A chart or coordinate system must be chosen in order to have an explicit representation of the form and the volume of integration in terms of this coordinate system.

2. The orientation of the volume of integration must be properly taken into account. In physical applications we refer to the "right-hand rule" as determining the positive circulation around a line integral, or the "outward pointing normal" as determining the positive direction for a surface integral. Both of these integration "volumes" have only two possible orientations, either clockwise or counterclockwise, either outward or inward. More general volumes also have only two possible orientations, which are determined by similar rules. To choose an orientation means: Choose an ordering for the basis vectors $(\mathbf{e}_1, \mathbf{e}_2, \ldots, \mathbf{e}_n)$ and then consider other basis sets as

belonging to the same orientation if they can be obtained from the first by a linear transformation with positive determinant. As a familiar example the "right"- and "left"-handed coordinate systems in $\mathbb{R}^3$ have their basis vectors related by a matrix with a negative determinant. An integral over a "volume" with the opposite orientation is simply the negative of the integral over the "volume" with the original orientation. The positive or correct orientation is a matter of convention.

3. In this local coordinate system choose a set of basis vectors for the volume V, $\{\mathbf{e}_1, \mathbf{e}_2, \ldots, \mathbf{e}_n\}$, ordered for the correct orientation as described under rule 2. This choice and ordering may serve as the designation of the orientation if one has not been fixed previously.

4. Construct the infinitesimal "volume" element from the set of vectors $\{\mathbf{e}_1\, dx^1, \mathbf{e}_2\, dx^2, \ldots, \mathbf{e}_n\, dx^n\}$. (Note that dx^i is the ordinary differential of integral calculus on $\mathbb{R}^n$, not $\mathbf{d}x^i$.) Evaluate the given form on the vectors making up the volume element in the correct order corresponding to the chosen orientation. Perform the integration of the resulting scalar over the designated "volume" V in local coordinates as usual.

We illustrate this prescription with some examples below, but first, what is to be gained by considering integrals over forms? Since a form is a geometric object, with meaning that transcends any particular coordinate system, the same is true of their integrals. Integrals of forms over surfaces and volumes are quantities that have meaning independent of any particular coordinatization that must be made in order to evaluate them. In any theory such as Hamiltonian dynamics where the coordinates can be of many kinds it is important to focus on quantities and properties that are independent of any particular coordinate choice. In this chapter alone we have gone from $(\mathbf{p}, \mathbf{q}) \to (\mathbf{I}, \mathbf{Q}) \to (\mathbf{J}, \boldsymbol{\theta})$ and so a coordinate-independent formulation is useful. In the next section we consider a number of integral invariants that are most easily understood from a geometric, coordinate independent viewpoint.

The foregoing prescription becomes clear by considering some examples. To be specific, consider in $\mathbb{R}^3$ the 2-form

$$\boldsymbol{\Phi} = \frac{1}{2!} B^i \varepsilon_{ijk}\, \mathbf{d}x^i \wedge \mathbf{d}x^j, \tag{9.19}$$

which we call the *flux 2-form* for the vector $\mathbf{B}$. In terms of the surface 2-form of Problem 8.5, (9.19) can be written in the form

$$\boldsymbol{\Phi} = B^i \Sigma_i. \tag{9.20}$$

We have in mind here the magnetic field for $\mathbf{B}$, but the flux 2-form can be constructed for any vector field using (9.20). Let the "volume" V be the area of the plane pictured in Fig. 9.7. We use the usual orientation in $\mathbb{R}^3$ with basis vectors $(\mathbf{e}_1, \mathbf{e}_2, \mathbf{e}_3)$. Let α denote the angle between V and the (x^1, x^2) plane. Then the infinitesimal area of V has vectors $\mathbf{v} = \mathbf{e}_1\cos\alpha\, dx^1 + \mathbf{e}_3\sin\alpha\, dx^3$ and $\mathbf{u} = \mathbf{e}_2\, dx^2$ as sides. The orientation of V is given by the "right-hand rule" or according to $\mathbf{v} \times \mathbf{u}$, if $\mathbf{v}$ and $\mathbf{u}$ are entered as the arguments of $\mathbf{\Phi}$ in that order. Having specified the components of the prescription, we calculate

$$\int_V \mathbf{\Phi} = \int_V \mathbf{\Phi}[\mathbf{v},\mathbf{u}] = \int_V \frac{1}{2!} B^i \varepsilon_{ijk}\, \mathbf{d}x^j \wedge dx^k[\mathbf{v},\mathbf{u}]$$

$$= \int_V \tfrac{1}{2} B^i \varepsilon_{ijk}\{\mathbf{d}x^j \wedge \mathbf{d}x^k\left[\mathbf{e}_1\cos\alpha\, dx^1, \mathbf{e}_2\, dx^2\right]$$

$$+\mathbf{d}x^j \wedge \mathbf{d}x^k\left[\mathbf{e}_3\sin\alpha\, dx^3, \mathbf{e}_2\, dx^2\right]\}$$

$$= \int_V \tfrac{1}{2} B^i \varepsilon_{ijk}\{\cos\alpha\, dx^1\, dx^2\left(\delta^j_1\delta^k_2 - \delta^j_2\delta^k_1\right) + \sin\alpha\, dx^3\, dx^2\left(\delta^j_3\delta^k_2 - \delta^j_2\delta^k_3\right)\}$$

$$= \int_V \tfrac{1}{2} B^i\{\cos\alpha\, dx^1\, dx^2\left(\varepsilon_{i12} - \varepsilon_{i21}\right) + \sin\alpha\, dx^3\, dx^2\left(\varepsilon_{i32} - \varepsilon_{i23}\right)\}$$

$$= \int_V \{B^3\cos\alpha\, dx^1\, dx^2 - B^1\sin\alpha\, dx^3\, dx^2\}$$

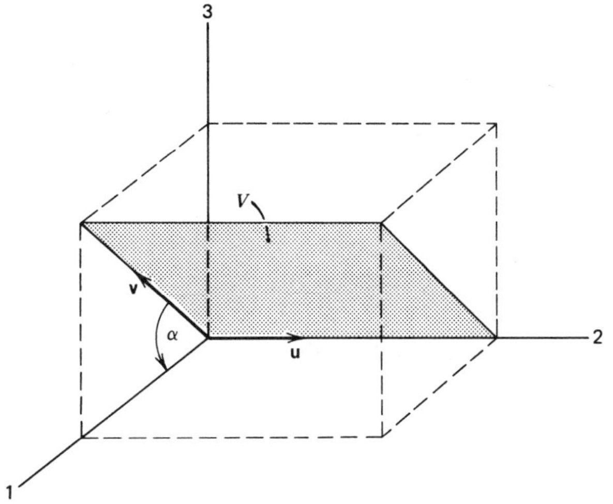

FIGURE 9.7. The area V over which the flux 2-form B of (9.22) is integrated. Positive orientation is given by $\mathbf{v} \times \mathbf{u}$.

This is the expected result for the flux of the vector $B^i\mathbf{e}_i = \mathbf{B}$ through the area V.

We return to Stokes' formula (9.18) and rather than giving the details of a proof, we examine how (9.18) reduces to familiar results for which the elementary proofs are also familiar. Suppose that V is some (open) surface with a boundary in $\mathbb{R}^3$. The boundary of V, ∂V, is then the closed curve around the outside edge of this surface. The positive orientation for V is given by the right-hand rule, that is, if one imagines traversing ∂V in the counterclockwise direction, then V is at the left. Since ∂V is a curve, it is appropriate to consider the integration of a 1-form over ∂V. The infinitesimal element of ∂V is $d\mathbf{l} = \mathbf{e}_j\, dx^j$. Let the 1-form $\boldsymbol{\omega}$ be given by $\boldsymbol{\omega} = A_i\,\mathbf{d}x^i$. Then the left-hand side of Stokes' formula (9.18) is

$$\int_{\partial V} \boldsymbol{\omega} = \int_{\partial V} A_i\,\mathbf{d}x^i\left[\mathbf{e}_j\, dx^j\right] = \int_{\partial V} \mathbf{A} \cdot d\mathbf{l}, \tag{9.21}$$

where the final integral is written in the familiar way using vectors.

Consider now the right-hand side of (9.18). The following calculation essentially supplies the details to part of Problem 8.5. We use the results of Problems 1.4 and 1.5 along with the definition of the surface 2-form $\boldsymbol{\Sigma}_i$ given in Problem 8.5. Using the definition (8.25) we compute

$$\begin{aligned}
\mathbf{d}\boldsymbol{\omega} &= A_{i,j}\,\mathbf{d}x^j \wedge \mathbf{d}x^i = \frac{1}{2}\left(A_{i,j} - A_{j,i}\right)\mathbf{d}x^j \wedge \mathbf{d}x^i \\
&= \frac{1}{2}A_{l,m}\left[\delta_i^l\delta_j^m - \delta_j^l\delta_i^m\right]\mathbf{d}x^j \wedge \mathbf{d}x^i \\
&= \frac{1}{2}A_{l,m}\varepsilon^{klm}\varepsilon_{kij}\,\mathbf{d}x^j \wedge \mathbf{d}x^i = (\nabla \times \mathbf{A})^k \frac{1}{2!}\varepsilon_{kji}\,\mathbf{d}x^j \wedge \mathbf{d}x^i \\
&\equiv (\nabla \times \mathbf{A})^k \boldsymbol{\Sigma}_k,
\end{aligned} \tag{9.22}$$

The final expression in (9.22) shows $\mathbf{d}\boldsymbol{\omega} = (\nabla \times \mathbf{A})^k\boldsymbol{\Sigma}_k$, which is the flux 2-form for the vector $\nabla \times \mathbf{A}$. Consequently, for this specific example (9.18) is equivalent to the familiar Stokes' theorem of vector calculus

$$\int_S (\nabla \times \mathbf{A}) \cdot \hat{\mathbf{n}}\, da = \int_C \mathbf{A} \cdot d\mathbf{l}.$$

The divergence theorem of vector calculus is also contained in (9.18). Let V denote a closed volume in $\mathbb{R}^3$ and ∂V be the closed surface of this volume.

The form $\boldsymbol{\omega}$ on the left-hand side of (9.18) must be a 2-form, $\boldsymbol{\omega} = \frac{1}{2}\omega_{ij}\mathbf{d}x^i \wedge \mathbf{d}x^j$. Since ω_{ij} is antisymmetric under index interchange, it can be written in terms of the components of an equivalent vector $\mathbf{A}$ as $\omega_{ij} = \varepsilon_{ijk}A^k$. The arbitrary 2-form $\boldsymbol{\omega}$ can then be written as a flux 2-form $\boldsymbol{\omega} = \frac{1}{2}A^k\varepsilon_{kij}\mathbf{d}x^i \wedge \mathbf{d}x^j$. Arbitrary 2-forms in $\mathbb{R}^3$ can always be written as flux 2-forms. Problem 8.5 shows that $\mathbf{d}\boldsymbol{\omega} = (\nabla \cdot \mathbf{A})\boldsymbol{\Omega}$, where $\boldsymbol{\Omega}$ is the volume 3-form, $\boldsymbol{\Omega} = (g)^{1/2}\mathbf{d}x^1 \wedge \mathbf{d}x^2 \wedge \mathbf{d}x^3$. The infinitesimal volume for V is given by the vectors $(\mathbf{e}_1\,dx^1, \mathbf{e}_2\,dx^2, \mathbf{e}_3\,dx^3)$, in that order for positive orientation. Thus

$$\int_V \mathbf{d}\boldsymbol{\omega} = \int_V (\nabla \cdot \mathbf{A})\boldsymbol{\Omega}\left[\mathbf{e}_1\,dx^1, \mathbf{e}_2\,dx^2, \mathbf{e}_3\,dx^3\right]$$

$$= \int_V (\nabla \cdot \mathbf{A})(g)^{1/2}\,dx^1\,dx^2\,dx^3 = \int_V (\nabla \cdot \mathbf{A})\,dv. \qquad (9.23)$$

If the normal to the surface area ∂V is in the $\mathbf{e}_1$ direction, then the infinitesimal of surface area with positive orientation is $(\mathbf{e}_2\,dx^2, \mathbf{e}_3\,dx^3)$. In general, if $\hat{\mathbf{n}}$ is the normal to the infinitesimal surface area and $\mathbf{f}_1$ and $\mathbf{f}_2$ are vectors making up the infinitesimal sides of this same area, then the order of $\mathbf{f}_1$ and $\mathbf{f}_2$ must be such that $(\hat{\mathbf{n}}, \mathbf{f}_1, \mathbf{f}_2)$ has the same orientation as $(\mathbf{e}_1, \mathbf{e}_2, \mathbf{e}_3)$, that is, can be obtained from $(\mathbf{e}_1, \mathbf{e}_2, \mathbf{e}_3)$ by a matrix transformation with positive determinant. For the infinitesimal area given by $(\mathbf{e}_2\,dx^2, \mathbf{e}_3 dx^3)$

$$\int_{\partial V} \boldsymbol{\omega} = \int_{\partial V} \tfrac{1}{2}A^k\varepsilon_{kij}\mathbf{d}x^i \wedge \mathbf{d}x^j\left[\mathbf{e}_2\,dx^2, \mathbf{e}_3\,dx^3\right] = \int_{\partial V} A^1\,dx^2\,dx^3.$$

More generally,

$$\int_{\partial V} \boldsymbol{\omega} = \int_{\partial V} \mathbf{A} \cdot \hat{\mathbf{n}}\,da, \qquad (9.24)$$

and (9.18) reduces to the familiar divergence theorem

$$\int_{\partial V} \mathbf{A} \cdot \hat{\mathbf{n}}\,da = \int_V (\nabla \cdot \mathbf{A})\,dv.$$

With this background in evaluating integrals of forms we look once again at (9.2), which defines the action momenta. First write it in terms of forms and then apply Stokes' formula to this equation:

$$J_i = \frac{1}{2\pi}\oint_{\gamma_i} p_j\,dq^j = \frac{1}{2\pi}\oint_{\gamma_i} p_j\,\mathbf{d}q^j = \frac{1}{2\pi}\int_{S_i} \mathbf{d}p_j \wedge \mathbf{d}q^j = \frac{1}{2\pi}\int_{S_i} \boldsymbol{\gamma}. \qquad (9.25)$$

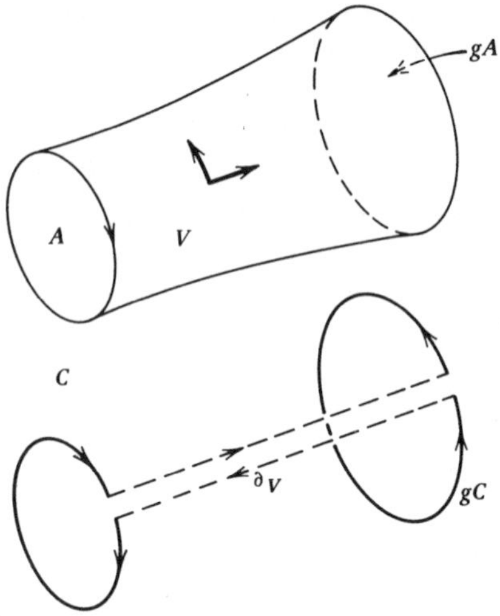

FIGURE 9.8. The tube of Hamiltonian flow lines generated by the curve C in extended phase space under the flow map g. The curves g and gC form the boundary of the volume V with the orientation as shown.

Therefore $2\pi J_i$ is the integral of the fundamental canonical 2-form γ over a surface S_i bounded by the closed curve γ_i. The quantity $2\pi J_i$ is interpreted as the sum of the area projections of the surface S_i on the coordinate planes $(p_1, q^1), (p_2, q^2), \ldots, (p_n, q^n)$ that constitute γ. Thus we have come full cycle and see why the simple harmonic oscillator gave $J = H$ in (9.10).

9.3. INTEGRAL INVARIANTS

Before leaving integrals over forms, we wish to show the existence of a number of integral invariants. It has already been shown that a volume in phase space is preserved under a Hamiltonian flow. This result is called Liouville's theorem and was discussed in Section 7.5.

Consider a "volume" in extended phase space that is made of some curve C swept forward in time as the system evolves. The "sides" of this "volume" are made up of flow lines of $\mathbf{v}_H$ given in (8.60). Let us denote the curve that C gets mapped into under g_H^t as gC, that is, $gC = g_H^t C$. As C is swept forward in time, a tube is created in phase space with C the bounding curve around one end and gC the bounding curve around the other end. These curves are then the boundary of the "area" represented by the sides of the tube. In terms of the notation used for Stokes' theorem, V is the surface "area" of the tube, excluding the ends, and ∂V is made up of C and gC. See Fig. 9.8 for a sketch showing the volume V, the boundary ∂V, and the orientations. The flow vector field $\mathbf{v}_H$ establishes one direction for the orientation on V, and the other vectors making up the basis are chosen in an arbitrary way. In Fig. 9.8 only one other basis vector besides $\mathbf{v}_H$ can be pictured.

Consider an integration over ∂V of the 1-form $\Psi = p_i\,\mathbf{d}q^i - H\,\mathbf{d}t$, introduced in Section 8.6. By Stokes' theorem (9.18)

$$\int_{\partial V} \Psi = \int_V \mathbf{d}\Psi = \int_V \Gamma, \tag{9.26}$$

where Γ is given in (8.62) Since the volume V is generated by the flow lines of $\mathbf{v}_H$, one of the vectors involved in evaluating Γ over V is always the Hamiltonian flow vector $\mathbf{v}_H$. But the vortex lines of Γ are the flow lines of $\mathbf{v}_H$ and $\Gamma[\mathbf{v}_H, \mathbf{u}] = 0$ from (8.64). Thus the right-hand side of (9.26) vanishes and we have

$$\int_{\partial V} \Psi = 0. \tag{9.27}$$

From Fig. 9.8 we see that the integral around ∂V is given by the line integral around C in the positive direction plus the line integral around gC in the negative direction. Hence from (9.27) we obtain

$$\oint_C p_i\,\mathbf{d}q^i - H\,\mathbf{d}t = \oint_{gC} p_i\,\mathbf{d}q^i - H\,\mathbf{d}t. \tag{9.28}$$

This integral invariant is called the *Poincaré–Cartan integral invariant.*

Since the curve C is arbitrary, several interesting and important results follow from (9.28). Choose C and gC to lie entirely in phase space. They are still ends of a flow tube, but such that $\mathbf{d}t = 0$ on C and gC. Then (9.28) becomes

$$\oint_C p_i\,\mathbf{d}q^i = \oint_{gC} p_i\,\mathbf{d}q^i. \tag{9.29}$$

Rather than interpreting C and gC as the boundary curve of V in Fig. 9.8, these curves may be viewed as the boundary of A and gA, as shown in the same figure. Applying Stokes' theorem again, (9.29) becomes

$$\int_A \gamma = \int_{gA} \gamma. \tag{9.30}$$

From definition (8.46) for the pull-back map g^* as applied to infinitesimal "volumes," that is since $g^*\omega^k[\mathbf{e}_1\,dx^1,\ldots,\mathbf{e}_k\,dx^k] = \omega^k[g_*\mathbf{e}_1\,dx^1,\ldots, g_*\mathbf{e}_k\,dx^k]$, the following relation for integrals of forms is obtained

$$\int_V g^*\omega = \int_{gC}\omega. \tag{9.31}$$

Apply (9.31) to (9.30) and obtain

$$\int_A \gamma = \int_{gA}\gamma = \int_A g^*\gamma. \tag{9.32}$$

Hence, we find once again that Hamiltonian flow preserves the canonical 2-form, that is, $g^*\gamma = \gamma$, since A is arbitrary.

Since $g^*\gamma = \gamma$, and hence (9.32) is valid for canonical transformations that are not necessarily Hamiltonian flows, Stokes' theorem implies

$$\oint_C p_i\,\mathbf{d}q^i = \int_A \gamma = \int_A g^*\gamma = \oint_C P_i\,\mathbf{d}Q^i. \tag{9.33}$$

The line integral in (9.33) is known as *Poincaré's integral invariant*. Comparing (9.25) with (9.33) we see once again the action momenta with a deeper appreciation for why they are invariants.

Besides the integrals occurring in (9.32) and (9.33), there are additional integral invariants, which we now obtain.

In Problem 8.7 the result is obtained that $g^*(\omega^k \wedge \omega^l) = g^*\omega^k \wedge g^*\omega^l$ for arbitrary forms ω^k and ω^l. We can apply this to exterior products of γ with itself. Let g be any canonical transformation, not necessarily a phase flow; then

$$g^*(\gamma \wedge \gamma) = g^*\gamma \wedge g^*\gamma = \gamma \wedge \gamma. \tag{9.34}$$

We see that the 4-form $\gamma \wedge \gamma$ is also preserved under a canonical transformation. This can also be extended to the higher rank forms defined by $(\gamma)^k \equiv \gamma \wedge \cdots \wedge \gamma$ (k factors). The proof is the same calculation as in (9.34) and we obtain

$$g^*(\gamma)^k = (\gamma)^k. \tag{9.35}$$

When $k = n$, $(\gamma)^n$ is a $2n$-form and is proportional to the unit $2n$-volume form on phase space. The invariance of $(\gamma)^n$ under Hamiltonian flow is a restatement of Liouville's theorem. Indeed, each of the $(\gamma)^k$ gives an integral invariant under the phase flow since

$$\int_A (\boldsymbol{\gamma})^k = \int_A g^*(\boldsymbol{\gamma})^k = \int_{gA} (\boldsymbol{\gamma})^k, \tag{9.36}$$

where we have used (9.35) and (9.31).

As an illustration of an invariant $(\boldsymbol{\gamma})^k$ consider the example of a canonical transformation following (8.36).

$$\boldsymbol{\gamma} = \mathbf{d}p_x \wedge \mathbf{d}x + \mathbf{d}p_y \wedge \mathbf{d}y = \mathbf{d}P_x \wedge \mathbf{d}X + \mathbf{d}P_y \wedge \mathbf{d}Y$$

We find for

$$\boldsymbol{\gamma} \wedge \boldsymbol{\gamma} = (\boldsymbol{\gamma})^2 = -2\mathbf{d}p_x \wedge \mathbf{d}p_y \wedge \mathbf{d}x \wedge \mathbf{d}y = -2\mathbf{d}P_X \wedge \mathbf{d}P_Y \wedge \mathbf{d}X \wedge \mathbf{d}Y.$$

Thus $(\boldsymbol{\gamma})^2$ is invariant under a canonical transformation and in fact $(\boldsymbol{\gamma})^2$ is proportional to the unit 4-volume in this four-dimensional phase space.

9.4. CANONICAL PERTURBATION THEORY

As shown in Section 9.1 action-angle variables are convenient for analyzing integrable systems. But what about systems that do not have a complete set of involutive invariants? Poincaré (1892) has proved that most Hamiltonians have only the energy as a first integral of the motion. We might well expect that even if we start with an integrable Hamiltonian, the smallest of perturbations would destroy the invariant tori, and phase curves would wander throughout the energy surface. This is not the case, but the elucidation of the actual situation followed Poincaré's work by many decades.

As a step toward studying more general Hamiltonians, consider a Hamiltonian that can be partitioned into a completely integrable piece and a perturbation, that is

$$H(\mathbf{J}, \boldsymbol{\theta}) = H_0(\mathbf{J}) + \varepsilon H_1(\mathbf{J}, \boldsymbol{\theta}), \tag{9.37}$$

The Hamiltonian $H(\mathbf{J}, \boldsymbol{\theta})$ may be completely integrable but the action-angle variables $(\mathbf{J}, \boldsymbol{\theta})$ are from solving the Hamiltonian H_0. We consider the partition (9.37) to be valid when the system point in phase space is near the invariant torus of H_0 with label $\mathbf{J}$. The phase space is the same for H as it is for H_0 and $\mathbf{d}p_i \wedge \mathbf{d}q^i = \mathbf{d}J_i \wedge \mathbf{d}\theta^i$. The coordinates $(\mathbf{J}, \boldsymbol{\theta})$ still represent a set of canonical coordinates even though they do not represent an action-angle set for H. H depends on $\boldsymbol{\theta}$ as well as $\mathbf{J}$.

If the perturbation is small, that is, $\varepsilon \ll 1$, then we might expect H to represent an integrable system. Should that be the case, there exists a set of action-angle coordinates $(\mathbf{J}', \boldsymbol{\theta}')$ such that $H(\mathbf{J}, \boldsymbol{\theta}) = H(\mathbf{J}')$. One way of finding the $(\mathbf{J}', \boldsymbol{\theta}')$ is to find the canonical transformation relating $(\mathbf{J}, \boldsymbol{\theta})$ and $(\mathbf{J}', \boldsymbol{\theta}')$. Thus it is natural to consider an expansion for the generating function of this desired transformation. In terms of the small parameter ε occurring in (9.37),

$$S(\mathbf{J}', \boldsymbol{\theta}) = \theta^i J_i' + \varepsilon S_1(\mathbf{J}', \boldsymbol{\theta}) + \cdots, \tag{9.38}$$

Applying (8.58) to $S(\mathbf{J}', \boldsymbol{\theta})$ for a Type 2 generating function

$$\theta'^i = \theta^i + \varepsilon \frac{\partial S_1}{\partial J_i'} + \cdots \quad \text{and} \quad J_i = J_i' + \varepsilon \frac{\partial S_1}{\partial \theta^i} + \cdots. \tag{9.39}$$

Since the known function $H_1(\mathbf{J}, \boldsymbol{\theta})$ is periodic in the action angles, it can be expanded in a Fourier expansion.

$$H_1(\mathbf{J}, \boldsymbol{\theta}) = \sum_{\mathbf{m}} H_{1\mathbf{m}}(\mathbf{J}) e^{i\mathbf{m}\cdot\boldsymbol{\theta}}, \tag{9.40}$$

where $\mathbf{m} = (m_1, m_2, \ldots, m_n)$, $m_i = 0, \pm 1, \pm 2, \ldots$. The perturbation function $S_1(\mathbf{J}', \boldsymbol{\theta})$ can also be expanded.

$$S_1(\mathbf{J}', \boldsymbol{\theta}) = \sum_{\mathbf{m} \neq 0} S_{1\mathbf{m}}(\mathbf{J}') e^{i\mathbf{m}\cdot\boldsymbol{\theta}}. \tag{9.41}$$

In (9.41) we have required $\mathbf{m} \neq 0$ to exclude a constant term and make the Fourier expansion consistent with (9.38). To express H_0 in terms of $\mathbf{J}'$, we make a Taylor's expansion in ε and use (9.39).

$$H_0(\mathbf{J}) = H_0(\mathbf{J}') + \varepsilon \left.\frac{\partial H_0}{\partial J_j}\right|_{\mathbf{J}'} \frac{\partial S_1}{\partial \theta^j} + \cdots. \tag{9.42}$$

To lowest order in ε, $(\partial H_0/\partial J_j)|_{\mathbf{J}'} = (\partial H_0/\partial J_j)|_{\mathbf{J}}$, where $\dot{\theta}^i = \omega^j(\mathbf{J}) = (\partial H_0/\partial J_j)$. Substitute in (9.42) to find

$$H_0(\mathbf{J}) = H_0(\mathbf{J}') + \varepsilon \omega^j(\mathbf{J}) \frac{\partial S_1}{\partial \theta^j} + \cdots. \tag{9.43}$$

Using the Fourier expansion (9.41)

$$\frac{\partial S_1}{\partial \theta^j} = \sum_{\mathbf{m} \neq 0} i m_j S_{1\mathbf{m}}(\mathbf{J}') e^{i\mathbf{m}\cdot\boldsymbol{\theta}}. \tag{9.44}$$

With (9.44) in (9.43), we substitute into (9.37) and order the terms in ε.

$$H(\mathbf{J}') = H_0(\mathbf{J}') + \varepsilon\left\{ \sum_{\mathbf{m}\neq 0} i\omega^j m_j S_{1\mathbf{m}}(\mathbf{J}')e^{i\mathbf{m}\cdot\boldsymbol{\theta}} + \sum_{\mathbf{m}} H_{1\mathbf{m}}(\mathbf{J})e^{i\mathbf{m}\cdot\boldsymbol{\theta}} \right\} + \cdots .$$

In the foregoing expansion for $H(\mathbf{J}')$, we equate to zero the coefficient of $e^{i\mathbf{m}\cdot\boldsymbol{\theta}}$ for each choice of $\mathbf{m}$.

$$H(\mathbf{J}') = H_0(\mathbf{J}') + \varepsilon H_{10}(\mathbf{J}') + \cdots$$

$$i(\omega(\mathbf{J})\cdot\mathbf{m})S_{1\mathbf{m}}(\mathbf{J}') + H_{1\mathbf{m}}(\mathbf{J}) = 0, \qquad \mathbf{m}\neq 0.$$

The terms constant in θ give the first of the foregoing equations. In the second solve for the Fourier coefficient of the generating function to find

$$S_{1\mathbf{m}}(\mathbf{J}') = \frac{iH_{1\mathbf{m}}(\mathbf{J})}{\boldsymbol{\omega}(\mathbf{J})\cdot\mathbf{m}}. \tag{9.45}$$

Equation (9.45) gives the coefficients for the expansion of the generating function in terms of the Fourier coefficients of the known perturbation Hamiltonian.

Substituting (9.45) into (9.41) for the first-order contribution to the generating function we find

$$S_1(\mathbf{J}',\boldsymbol{\theta}) = \sum_{\mathbf{m}\neq 0} \frac{iH_{1\mathbf{m}}(\mathbf{J})}{\boldsymbol{\omega}(\mathbf{J})\cdot\mathbf{m}} e^{i\mathbf{m}\cdot\boldsymbol{\theta}}. \tag{9.46}$$

However, in this expression there is a potentially serious problem. There may be $\mathbf{m} = (m_1, m_2, \ldots, m_n)$ for which $\boldsymbol{\omega}\cdot\mathbf{m} \simeq 0$. Equation (9.8) is the condition for the frequencies $\omega^j(\mathbf{J})$ to be commensurable, or in other words for the invariant torus labeled by $\mathbf{J}$ to be a rational or "resonant" surface. Since the relation between the frequencies in (9.8) holds for each i and j, a sum of the equations, $\omega^j n_i - \omega^i n_j = 0$, for all independent combinations of i and j, results in an equation of the form $\mathbf{k}\cdot\boldsymbol{\omega}(\mathbf{J}) = 0$, where $\mathbf{k} = (k_1, \ldots, k_n)$ is an n-vector of integers, which is not unique. Consequently, if the invariant torus, where (9.37) is assumed valid, is a resonant surface, then there is some $\mathbf{m}$ for which $\boldsymbol{\omega}\cdot\mathbf{m} \equiv 0$, unless perhaps, the Fourier sum in (9.40) is finite. When the Fourier sum is infinite and is over all integer vectors $\mathbf{m}$, it is not necessary that $\mathbf{J}$ correspond to a rational surface; there will always exist $\mathbf{m}$ that can make the scalar product $\boldsymbol{\omega}\cdot\mathbf{m}$ small. Thus we are faced with two series for which the convergence is in question: the Fourier series may not converge because of the "small denominators" problem, and it is not clear that the series in ε converges. We see in Section 9.5 that for "sufficiently irrational" $\boldsymbol{\omega}(\mathbf{J})$ there is a convergent perturbation series leading to a new action-angle set $(\mathbf{J}', \boldsymbol{\theta}')$ and new invariant tori. This result is known as the KAM theorem.

Alerted to the potential for resonant denominators in (9.46), we focus our attention on a particular resonant torus $\mathbf{J}^0$, and examine the behavior of the phase curves of the system as the perturbation is turned on. For a given set of commensurable frequencies $\omega^i(\mathbf{J}^0)$, there are many $\mathbf{m}_0$ for which $\mathbf{m}_0 \cdot \boldsymbol{\omega}(\mathbf{J}^0) = 0$. Denote $\mathbf{k}$ as the n-vector of integers, with $k_1 > 0$, such that $\|\mathbf{m}_0\|^2 = \Sigma m_i^2$ is a minimum. The averaging effect of the higher frequency oscillations $e^{i\mathbf{m}\cdot\boldsymbol{\theta}}$, in comparison to $e^{i\mathbf{k}\cdot\boldsymbol{\theta}}$, makes the terms $\mathbf{m} = \pm\mathbf{k}$ the dominant resonance in the Fourier sum. We then concern ourselves only with a single term in the sum (9.40) and write the Hamiltonian in the form:

$$H(\mathbf{J}, \boldsymbol{\theta}) = H_0(\mathbf{J}) - \varepsilon H_1 \cos(\mathbf{k} \cdot \boldsymbol{\theta}). \tag{9.47}$$

Let us choose a new set of canonical coordinates that are particularly suited to the motion of the phase point near this resonant surface. To carry out this change of coordinates we select a set of n vectors $(\mathbf{m}^1, \mathbf{m}^2, \ldots, \mathbf{m}^n)$ with integer entries, such that $\mathbf{m}^1 \equiv \mathbf{k}$. The other $(n-1)$ vectors are chosen to be a linearly independent set so that the $(n \times n)$ matrix $m^j{}_i$ is nonsingular with inverse $(m^{-1})^i{}_j$. Then make the following transformations, where $\mathbf{J}^0$ is fixed:

$$\psi^j = \mathbf{m}^j \cdot \boldsymbol{\theta}, \tag{9.48}$$

$$I_j = (m^{-1})^i{}_j (J_i - J_i^0). \text{ (Note sum on } i.) \tag{9.49}$$

We find $\mathbf{d}I_j \wedge \mathbf{d}\psi^j = \mathbf{d}J_j \wedge \mathbf{d}\theta^j$, and so the $(\mathbf{I}, \boldsymbol{\psi})$ form a set of canonical coordinates. The $\mathbf{J}^0$ resonant surface corresponds to $\mathbf{I} = 0$ in these new canonical coordinates. The Hamiltonian of (9.47) takes the form

$$H(\mathbf{I}, \boldsymbol{\psi}) = H_0(\mathbf{I}) - \varepsilon H_1(\mathbf{I}) \cos \psi^1. \tag{9.50}$$

All the coordinates $\psi^2, \ldots, \psi^n$ are cyclic and, hence, $I_2, \ldots, I_n$ are constant.

$H_0(\mathbf{I})$ is conveniently approximated in the neighborhood of the $\mathbf{J}^0$, $\mathbf{I} = 0$ resonance surface

$$H_0(\mathbf{I}) = H_0(0) + I_i \left(\frac{\partial H_0}{\partial I_i} \right)_0 + \tfrac{1}{2} I_i I_j \left(\frac{\partial^2 H_0}{\partial I_i \, \partial I_j} \right)_0 + \cdots .$$

Since $I_2, \ldots, I_n$ are constant and equal zero,

$$H_0(\mathbf{I}) = I_1\left(\frac{\partial H_0}{\partial I_1}\right)_0 + \tfrac{1}{2}(I_1)^2\left(\frac{\partial^2 H_0}{\partial {I_1}^2}\right)_0 + \cdots, \tag{9.51}$$

where the constant term $H_0(0)$ has been discarded. Evaluate the first term by the chain rule to obtain

$$\frac{\partial H_0}{\partial I_1} = \left(\frac{\partial H_0}{\partial J_i}\right)_0 \frac{\partial J_i}{\partial I_1} = \omega^i(\mathbf{J}^0) m^1{}_i = \boldsymbol{\omega}(\mathbf{J}^0)\cdot\mathbf{k} = 0.$$

Thus (9.50) becomes

$$H(\mathbf{I}, \boldsymbol{\psi}) \simeq \frac{1}{2}\left(\frac{\partial^2 H_0}{\partial {I_1}^2}\right)_0 (I_1)^2 - \varepsilon H_1(0)\cos\psi^1, \tag{9.52}$$

where H_1 has been evaluated on the resonant surface. By a trivial rescaling of the momentum variable and the Hamiltonian, (9.52) can be written in the form (9.13) appropriate for the simple pendulum. Figure 9.9 is a sketch of the phase curves mapped into $\mathbb{R}^2$ (cf. Fig. 9.6 and Fig. 2.3). The *resonance width* W_I is defined and noted in this figure. By evaluating (9.52) at $I_1 = 0$, $\psi^1 = \pi$, we obtain $H = \varepsilon H_1(0)$ on the separatrix. Thus the width of the

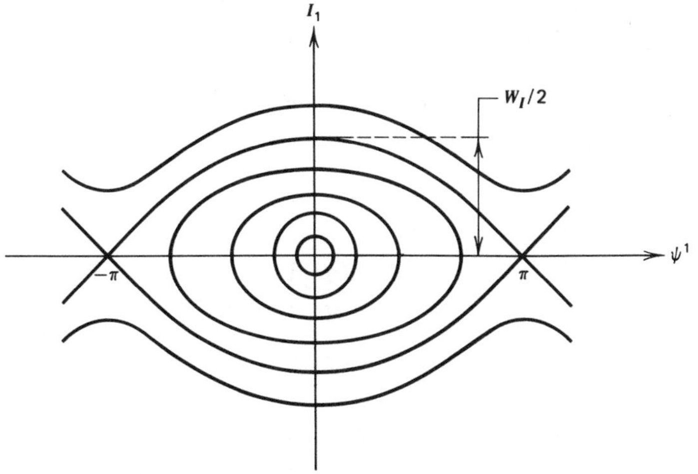

FIGURE 9.9. Phase curves mapped into $\mathbb{R}^2$ for a resonant surface. The resonance width W_I is indicated.

resonance is

$$W_I = 4\left[\varepsilon H_1(0)/\left(\frac{\partial^2 H_0}{\partial I_1{}^2}\right)_0\right]^{1/2}. \tag{9.53}$$

As noted earlier in Section 9.1 with respect to Fig. 9.6, the separatrix passing through $\pm\pi$ divides phase space into qualitatively different regions and for motion near the separatrix we can expect the system to be very sensitive to the small terms we have neglected in obtaining (9.52). However, we have treated a completely general system and thus the simple pendulum Hamiltonian of (9.52) may be considered as the archetype of a resonance for nonlinear systems.

It is instructive to return to the original action-angle coordinates and see how the invariant tori have been distorted by the perturbation. We cannot focus on all dimensions and so we focus on what is called a *Poincaré section*. A Poincaré section consists of choosing a two-dimensional phase plane, in which one plots intersection points of the phase trajectories with the plane. In such a Poincaré section for (J_1, θ^1), we look at the phase curves of Fig. 9.9, given in the (I_1, ψ^1) coordinates. From (9.48) and (9.49) we infer that the periodicity in the θ^i coordinate is given by $2\pi/k_i$ and the width of the resonance in the J_i direction is given by $k_i W_I$, where W_I is obtained in (9.53) and illustrated in Fig. 9.9. In Fig. 9.10 we sketch a Poincaré section and a piece of a distorted torus. The composite system of a series of "X" and "0" points with the separatrix and the tori within is often referred to as an *island chain*.

For a perturbation to an integrable system, there is not just one resonant surface as sketched in Fig. 9.10 but a whole host of such resonant surfaces nested within the invariant tori. To examine their proximity to one another, focus on a case with two degrees of freedom.

Recall that I_2 in (9.49) is constant and does not change. For brevity let $I_1 \equiv I$, $\psi^1 \equiv \psi$, and define $q(I) = \omega^1(I)/\omega^2(I)$. Since J_1 and J_2 are then only functions of I, $\boldsymbol{\omega} = (\dot\theta^1, \dot\theta^2) = \boldsymbol{\omega}(I)$. From (9.52) and Hamilton's equations, we obtain

$$\dot\psi = I\left(\frac{\partial^2 H_0}{\partial I^2}\right)_0. \tag{9.54}$$

Also from $\psi = \mathbf{k}\cdot\boldsymbol{\theta} = k_1\theta^1 + k_2\theta^2$ we find

$$\begin{aligned} d\psi &= \left[k_2 + k_1\frac{d\theta^1}{d\theta^2}\right] d\theta^2 = \left[k_2 + k_1\frac{\dot\theta^1}{\dot\theta^2}\right] d\theta^2 \\ &= \left[k_2 + k_1\frac{\omega^1}{\omega^2}\right] d\theta^2 = [k_2 + k_1 q]\, d\theta^2, \end{aligned} \tag{9.55}$$

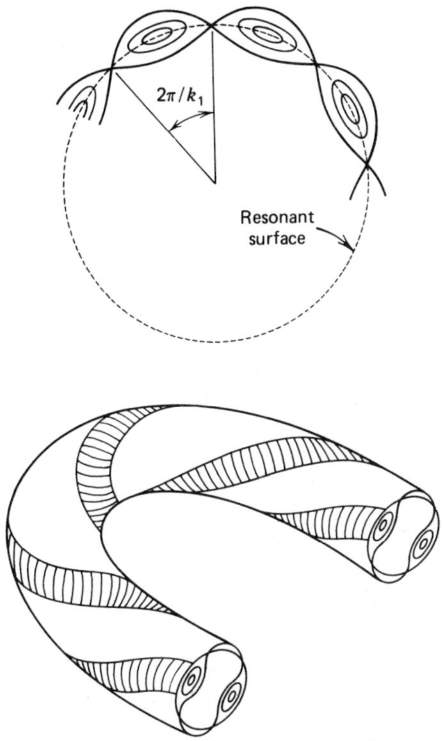

FIGURE 9.10. The island chain into which the resonant surface is perturbed. A portion of the distorted surface is shown in a Poincaré section where the spacing between "X" points is indicated. The manner in which such a chain of small tori twist around the original resonant surface is also sketched.

with $q(I)$ defined above. On the resonant surface $\omega^1(0)k_1 + \omega^2(0)k_2 = 0$ and

$$q(0) = \frac{\omega^1(0)}{\omega^2(0)} = -\frac{k_2}{k_1}. \tag{9.56}$$

Let $q' = (dq/dI)_{I=0}$, and expand $(d\psi/d\theta^2)$ near the resonant surface. Keeping only the lowest-order term,

$$d\psi \simeq [k_1 I q']\, d\theta^2, \tag{9.57}$$

and we obtain

$$\dot{\psi} = [k_1 I q'] \frac{d\theta^2}{dt} = k_1 I q' \omega^2. \tag{9.58}$$

Comparing (9.58) and (9.54) gives $(\partial^2 H_0/\partial I^2)_0 \simeq \omega^2 k_1 q'$. From (9.53) we see that $W_I \propto (1/q')^{1/2}$. Let ΔI denote the separation of two resonant surfaces. Then $\Delta I \simeq \Delta q/q' \propto (1/q')$, and

$$\frac{\Delta I}{W_I} \propto \left(\frac{1}{q'}\right)^{1/2}. \tag{9.59}$$

For q' an increasing function of I, one can expect to encounter resonant surfaces that will overlap.

Because phase flow lines satisfy $\nabla \cdot \mathbf{v} = 0$ in phase space (cf. Section 7.5) and the magnetic field satisfies $\nabla \cdot \mathbf{B} = 0$, magnetic field lines may be viewed as lines of phase flow. In a fusion device the invariant tori correspond to surfaces of constant pressure, on which the magnetic field is intended to provide the confinement for a hot plasma. Resonance overlap destroys the magnetic surfaces (invariant tori) and leads to a loss of confinement. Resonance overlap is an important issue in the design of magnetic fusion experiments.

9.5. NUMERICAL EXAMPLES AND THE KAM THEOREM

In a situation where resonant surfaces do indeed overlap, we can surely expect the foregoing simple analysis to be inadequate. In such a situation we turn to numerical methods to further elucidate system behavior. When resonant layers are sufficiently close, or the perturbation sufficiently large that the island widths overlap, we anticipate the destruction of the invariant tori. This behavior has been demonstrated in several numerical experiments that show in striking fashion the overlapping of resonances to produce stochastic or chaotic behavior. By stochastic we do not mean that there are any random forces operating. By *stochastic* we mean that the final positions of two phase points with nearly identical initial conditions can differ dramatically after finite time. Stated in other words, the final position in phase space of a system phase point gives no information about the final position of a second phase point of the system, when started with neighboring initial conditions and allowed to evolve.

As an example that makes the foregoing ideas clear, we consider the reduced Hamiltonian for the Toda lattice (Ford, 1975) as obtained in Problem 8.14,

$$H = \frac{p^2}{2} + V = \frac{1}{2}\left(p_x^2 + p_y^2\right) + \frac{1}{24}\left(e^{2x-2y\sqrt{3}} + e^{2x+2y\sqrt{3}} + e^{-4x}\right) - \frac{1}{8}. \tag{9.60}$$

For small values of the coordinates (9.60) is the Hamiltonian for a simple harmonic oscillator in two dimensions. In the system of (9.60), we may think of the energy itself as the small parameter. The larger the constant H is, then the larger will be the values for the coordinates x and y when the momenta p_x and p_y are zero. Thus the more the system will deviate from the harmonic oscillator approximation and be influenced by the nonlinear terms. Figure 9.11 shows a contour plot of the potential in (9.60). For small values of x and y, the equipotential contours are nearly circular. For larger values of $V(x, y)$, the deviation from the harmonic oscillator potential is significant.

Let us make a section plot of the phase curves that constitute the flow generated by the Hamiltonian (9.60). Since there is no explicit time dependence, the Hamiltonian itself is a constant of the motion. So, for a given energy, the phase flow can be expected to fill out a three-dimensional volume corresponding to the constant energy surface in phase space. We examine a Poincaré section of this volume. Specifically, we look at the flow

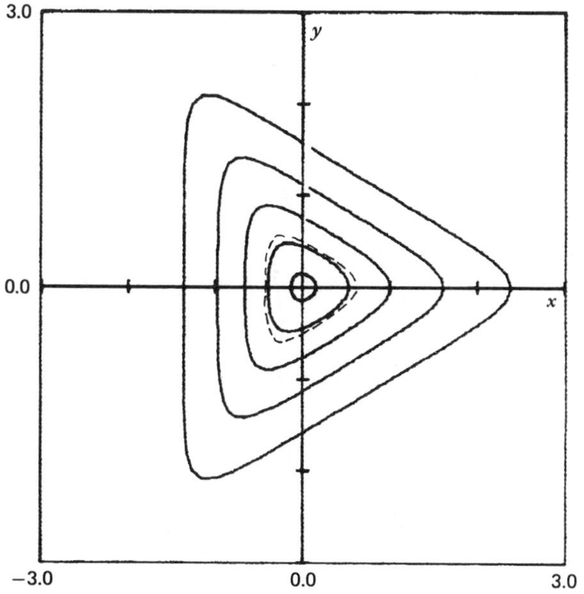

FIGURE 9.11. Equipotential contour plots of the Toda potential for the values 0.01, 0.12, 0.5, 2.0, 10.0 are represented by the solid curves with the largest value corresponding to the outermost curve. The dotted curve is for the Henon–Hiles potential in (9.65) with $V = 0.12$. For $V = 0.01$ the Henon–Hiles contour and the Toda contour cannot be distinguished in the scale used.

intersections with the (p_x, x) plane. Hamilton's equations give

$$\dot{x} = \frac{\partial H}{\partial p_x} = p_x; \qquad \dot{y} = \frac{\partial H}{\partial p_y} = p_y. \tag{9.61}$$

$$\dot{p}_x = -\frac{\partial H}{\partial x} = \frac{1}{6}\left(e^{-4x} - e^{2x}\cosh(2y\sqrt{3})\right). \tag{9.62}$$

$$\dot{p}_y = -\frac{\partial H}{\partial y} = -\frac{e^{2x}}{2\sqrt{3}}\sinh(2y\sqrt{3}). \tag{9.63}$$

Initial values for the variables in the (p_x, x) plane are supplied and then (9.61)–(9.63) are integrated forward in time numerically. Every time y passes through zero with $p_y > 0$ we place a point at the corresponding (p_x, x) coordinate. Figure 9.12 ($E = 0.01$) and Fig. 9.13 ($E = 0.12$) show the intersection points made in the (p_x, x) plane by intersections of phase curves. Because the intersection points, determined by a single phase curve specified by its initial conditions, are confined to closed, nested curves,

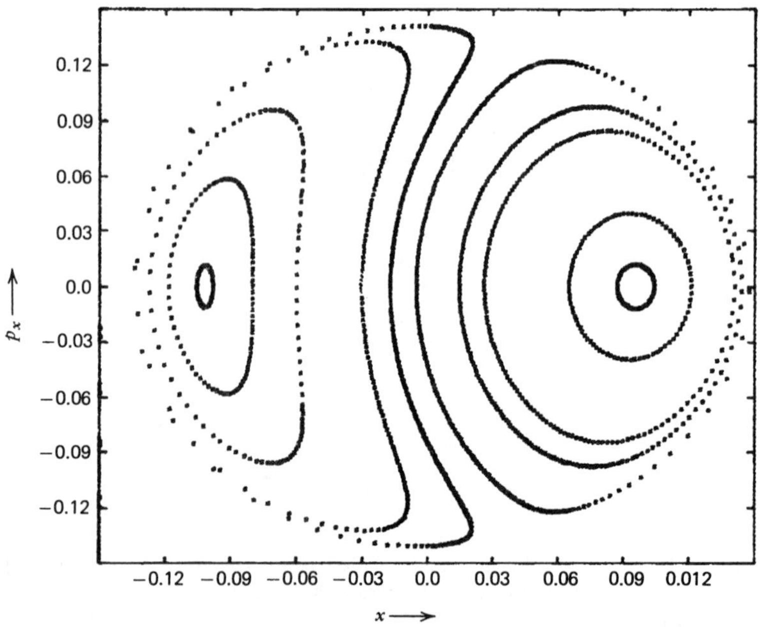

FIGURE 9.12. Poincaré section plots for the Toda system of (9.61)–(9.63) for $E = 0.01$. Each of the nested curves is determined by the intersections of a single-phase flow line. Different curves correspond to different initial values.

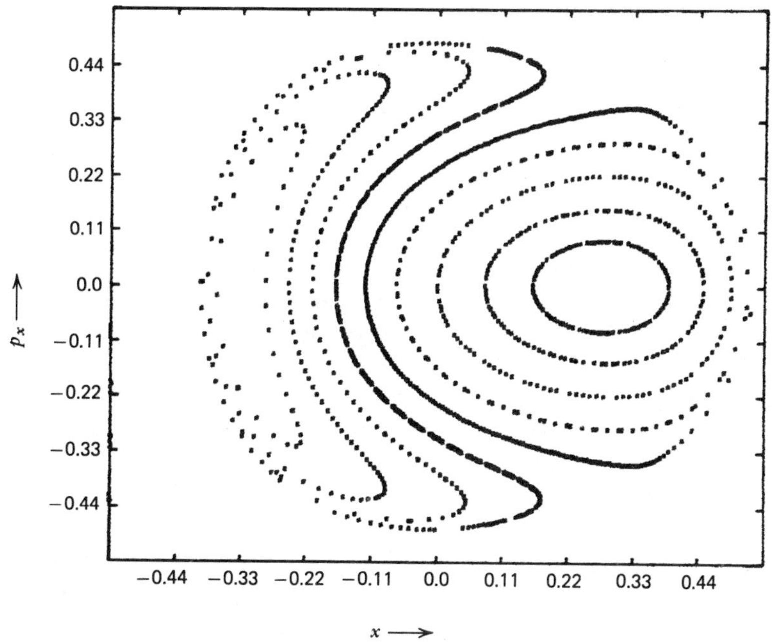

FIGURE 9.13. Poincaré section plots for Toda system of (9.61)–(9.63) for $E = 0.12$. Each of the nested curves is determined by the intersections of a single phase flow line. Different curves correspond to different initial values.

rather than dispersing throughout the Poincaré section, it is apparant that invariant tori exist. These tori are two-dimensional surfaces in the three-dimensional volume of the energy surface. It is evident that there is another constant of the motion. It was found by Henon (1969) to be

$$I = 8\dot{y}(\dot{y}^2 - 3\dot{x}^2) + (\dot{y} + \dot{x}\sqrt{3})e^{2x-2y\sqrt{3}} + (\dot{y} - \dot{x}\sqrt{3})e^{2x+2y\sqrt{3}} - 2\dot{y}e^{-4x}. \tag{9.64}$$

As given in (9.64), the constant I is not so easy to discover, but it is straightforward to check that it is indeed a constant of the motion. The closed curves in Figs. 9.12 and 9.13, which are given pointwise by the phase curve intersections, could be obtained easily as the curves for constant H and I from Eqs. (9.60) and (9.64), respectively.

Let us consider what, at first sight, looks like it will be a simpler system. We take the potential of (9.60), expand it for small values of the coordinates (x, y), and only keep the cubic additions to the harmonic oscillator potential. This gives the following Hamiltonian system:

$$H = \frac{p^2}{2} + V = \frac{1}{2}\left(p_x^2 + p_y^2\right) + \frac{1}{2}(x^2 + y^2) + y^2 x - \frac{1}{3}x^3. \quad (9.65)$$

The cubic terms constitute a perturbation away from the two-dimensional harmonic oscillator potential. We refer to the system with Hamiltonian (9.65) as the *Henon–Hiles system* after the investigators who first explored in detail the properties of (9.65) in a classic paper (Henon and Hiles, 1964).

The Henon–Hiles system has a "disassocation" energy in that for $H > \frac{1}{6}$ the values of the coordinates (x, y) are unbounded. The equipotential surfaces are not very different from those of the Toda potential. In Fig. 9.11 the dotted curve is the Henon–Hiles curve for the same energy as the solid curve for the Toda lattice that it touches. For the value of the energy corresponding to the innermost curve of Fig. 9.11, one cannot distinguish between the curve for the Toda potential and the curve for the Henon–Hiles potential. In general, the equipotential curves for the two systems are very similar.

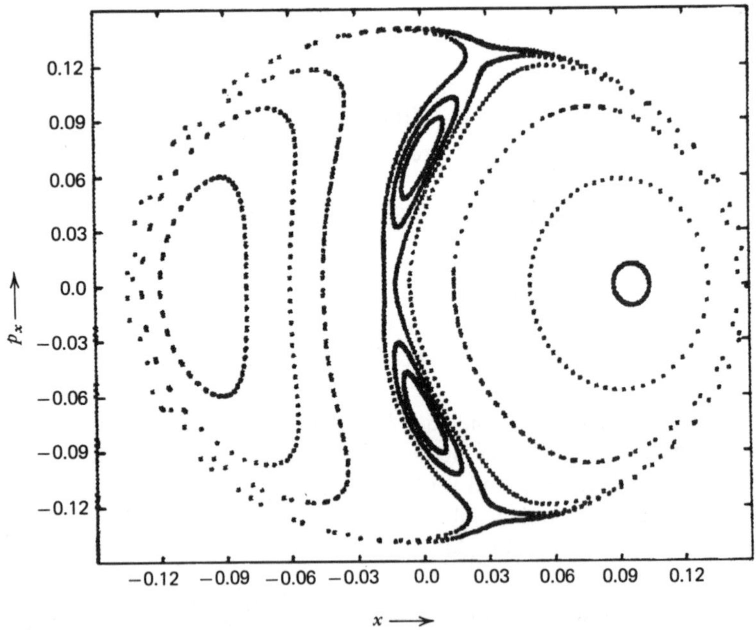

FIGURE 9.14. Poincaré section plot for the Henon–Hiles system with $E = 0.01$.

One might expect the curves and surfaces of a Poincaré section plot to be similar as well, but not so! As Figs. 9.14 and 9.15 show, things are very different indeed! For low values of the energy there are good surfaces, as shown in Fig. 9.14. For larger values of the energy, the phase point may visit regions of phase space where the nonlinear cubic terms become increasingly important.

Figure 9.15 shows clearly a case where the nonlinear terms are having a significant effect in large regions of accessible phase space. In Fig. 9.15 nested tori persist in the neighborhood of "0" points. In the transition regions between, island chains associated with nonlinear resonances can be distinguished. Several island structures are clearly visible around the "0" points in Fig. 9.15. In between regions where island structures and closed curves are distinguishable, there is a region of stochasticity, where the phase curve intersections appear in a random fashion. The intersection points in the stochastic region of Fig. 9.15 were made by 1000 intersections of a single phase curve!

From the discussion in Section 9.4, we have seen how a resonant surface breaks up into a sequence of alternating "X" and "0" points. This is an example of the Poincaré–Birkhoff fixed point theorem. The primary conclusion of this theorem is that a rational or resonant surface will break up into

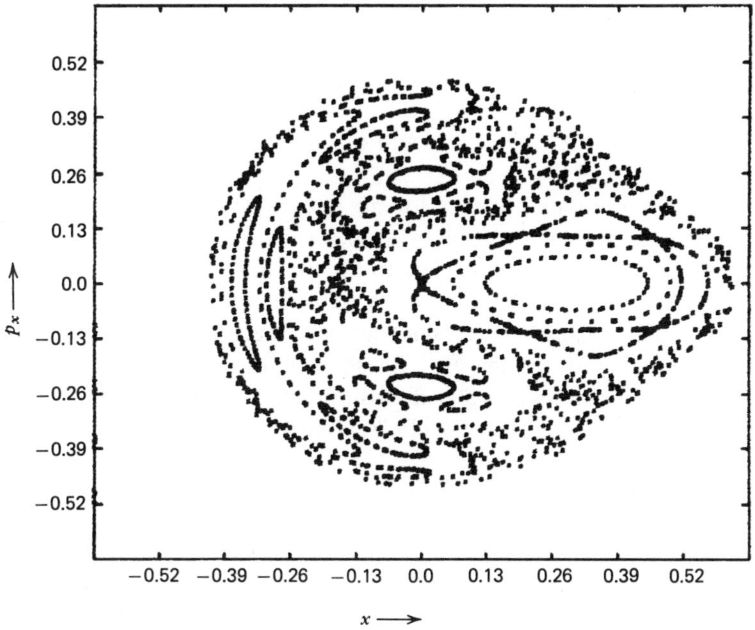

FIGURE 9.15. Poincaré section plot for the Henon–Hiles system with $E = 0.12$.

an even number of fixed points, half of them "X" points and half of them "0" points. We can see several such island chains in Fig. 9.15.

The rational surfaces are dense in the irrational surfaces, and one might well inquire as to the fate of such irrational surfaces. Indeed, we might ask whether the picture of nested toroids for an integrable system is completely destroyed by perturbations. Numerically, we have seen that this is not the case and that even for strong perturbations some invariant tori do persist. This result is the subject of the celebrated KAM theorem, after Kolmogoroff, Arnold, and Moser. Even a precise statement of this mathematical theorem would take us too far afield, but the interested reader can find a physicist's introduction to the mathematical details in Thirring (1978). The KAM theorem is probably the most important result in classical mechanics in this century. In qualitative terms the KAM theorem says that for a small perturbation a large number of the invariant surfaces will be preserved; they may be significantly distorted but still preserved. In essence, a convergent (so-called superconvergent) perturbation sequence was detailed in the proof of the KAM theorem that has a faster rate of convergence than the scheme sketched in Section 9.4. This scheme worked in those cases for which the unperturbed frequencies $\omega^i(\mathbf{J})$ satisfy certain conditions:

1. They are nonresonant.
2. They satisfy the inequalities

$$|\mathbf{m}\cdot\boldsymbol{\omega}(\mathbf{J})| > c|\mathbf{m}|^{-\gamma} \quad \text{for all integer vectors } \mathbf{m} = (m_1, m_2, \ldots, m_n),$$

where $|\mathbf{m}| = \sum_{i=1}^{n}|m_i| > 0$ and c, γ are fixed positive constants.

When these conditions are met for an integrable system, most tori will persist when the system is altered by a small perturbation. "Most" is used in a measure-theoretic sense and refers to the tori that are not resonant.

The details relating to overlap of resonant surfaces to produce stochastic regions of phase space, how resonant surfaces break up, applications of the KAM theorem, and so on, are all areas of ongoing research and the interested reader can find further discussion and references in the following reviews: Whiteman (1977), Berry (1978), Chirikov (1979).

PROBLEMS

9.1. Find action-angle variables for the system described in Problem 8.16.

9.2. Consider the system of two uncoupled simple harmonic oscillators given by the Hamiltonian

$$H = \frac{p_x^2}{2} + \frac{p_y^2}{2} + \frac{\omega_x^2 x^2}{2} + \frac{\omega_y^2 y^2}{2}.$$

Find action-angle variables for this system.

9.3. Consider the Hamiltonian for a harmonic oscillator, $H = p^2/2 + q^2/2$, and the curve in phase space defined by the equation $(q - a)^2 + p^2 = R^2$, where R is a constant. Calculate Poincaré's integral invariant for this curve. Consider the transformations

(a) $P = (q + p)/\sqrt{2}$; $Q = (q - p)/\sqrt{2}$

(b) $P = (p + q)/\sqrt{2}$; $Q = (p - q)/\sqrt{2}$.

Verify that Poincaré's integral invariant is indeed invariant under the preceding transformation that is canonical and that it is not invariant for the one that is not canonical. What role is the Hamiltonian playing in all of this?

9.4. Consider the Hamiltonian of Problem 9.2 to be perturbed by the following perturbation Hamiltonian

$$H_1 = \varepsilon\left(xy^2 - \tfrac{1}{3}x^3\right).$$

Write the total Hamiltonian in terms of the action-angle variables of the unperturbed system. Write out H_1 in Fourier expansion in terms of action angles. Are there any resonant denominators in the Fourier expansion of the generating function? How do the resonances appear? With $\varepsilon = 1$, this is again the Henon–Hiles system.

9.5. Show explicitly that action-angle coordinates $(\mathbf{J}, \boldsymbol{\theta})$ given by (9.2)–(9.4) are canonical coordinates.

CHAPTER TEN

ALGEBRAIC ASPECTS OF MOTION

Behold, the former things are come to pass, and new things do I declare.
ISAIAH 42:9

In the preceding chapter we focused on the geometric aspects of motion as it took place in phase space. In this chapter we continue to focus on motion in phase space but we are particularly interested in examining the algebraic properties of phase-space vectors and functions, which we always assume to be differentiable to all orders necessary. In this chapter we cease to distinguish notationally between a vector $\mathbf{v}$ and its corresponding operator $\partial_{\mathbf{v}}$. We think of the symbol $\mathbf{v}$ simultaneously as a tangent vector to a curve and an operator on functions. This practice is more consistent with the literature on this subject. We do continue the practice of enclosing in square brackets [ ] the argument of the operator $\mathbf{v}$.

10.1. THE LIE ALGEBRA OF VECTOR FIELDS

In Section 8.1 we introduced Hamiltonian vector fields associated with real-valued, differentiable functions on phase space. Recall that if $F: T^*M \to \mathbb{R}$ is a real-valued function on phase space, then the associated 1-form and vector fields are given respectively by

$$\mathbf{d}F = \frac{\partial F}{\partial q^i}\mathbf{d}q^i + \frac{\partial F}{\partial p_i}\mathbf{d}p_i, \tag{10.1}$$

$$\mathbf{v}_F = \frac{\partial F}{\partial p_i}\frac{\partial}{\partial q_i} - \frac{\partial F}{\partial q^i}\frac{\partial}{\partial p_i}, \tag{10.2}$$

where (p_i, q^i) is a local coordinate system on phase space. It is not necessary that F be the Hamiltonian of a physical system for the Hamiltonian flow vector $\mathbf{v}_F$ of (10.2) to be defined. However, if F is a Hamiltonian of some physical system, then the flow lines of $\mathbf{v}_F$ are the phase trajectories of the system.

Let F, G, H be arbitrary, differentiable, real-valued functions on phase space and recall (8.13)

$$\{F, G\} = -\{G, F\} = \mathbf{v}_G[F] = -\mathbf{v}_F[G] = \mathbf{d}F[\mathbf{v}_G] = -\mathbf{d}G[\mathbf{v}_F]. \tag{10.3}$$

The Poisson bracket of two such functions, F and G, is also a real-valued, differentiable function on phase space. Associated with this function is a flow vector field and it is natural to ask what the relationship is of $\mathbf{v}_{\{F,G\}}$ to $\mathbf{v}_F$ and $\mathbf{v}_G$. Using the Jacobi identity (7.30) we calculate

$$\begin{aligned} \mathbf{v}_{\{F,G\}}[H] &= \{H, \{F, G\}\} = \{\{H, F\}, G\} - \{\{H, G\}, F\} \\ &= \mathbf{v}_G[\mathbf{v}_F[H]] - \mathbf{v}_F[\mathbf{v}_G[H]] \\ &\equiv [\mathbf{v}_G, \mathbf{v}_F][H]. \end{aligned} \tag{10.4}$$

The quantity $[\mathbf{v}_G, \mathbf{v}_F]$ is called the *commutator* of the vector fields $\mathbf{v}_G$ and $\mathbf{v}_F$. The flow field associated with the Poisson bracket $\{F, G\}$ is the commutator $[\mathbf{v}_G, \mathbf{v}_F]$. It would be pleasant if the order of the "F" and the "G" were the same in the Poisson brackets and in the commutator, but the order is determined by the choice of sign in (8.8) so that $\mathbf{v}_F = (\dot{\mathbf{p}}, \dot{\mathbf{q}})$ for phase trajectories.

The commutator is often referred to as the *Lie bracket* and is defined for all differentiable vector fields, not just those obtained as vector flows from differentiable functions. Let $\mathbf{v}, \mathbf{u}, \mathbf{w}$ be differentiable vector fields on T^*M, and let $\{x^i\}$ denote a local coordinate system. The $\{x^i\}$ coordinates correspond to the usual (p_i, q^i) in that $x^i = p_i$, $x^{i+n} = q^i$, for $i = 1, \ldots, n$. If F is a real-valued, differentiable function on T^*M, then

$$[\mathbf{v},\mathbf{u}][F] = v^i \frac{\partial}{\partial x^i}\left(u^j \frac{\partial F}{\partial x^j}\right) - u^i \frac{\partial}{\partial x^i}\left(v^j \frac{\partial F}{\partial x^j}\right)$$

$$= \left(v^i \frac{\partial u^j}{\partial x^i} - u^i \frac{\partial v^j}{\partial x^i}\right)\frac{\partial F}{\partial x^j}$$

$$= \left(v^i u^j{}_{,i} - u^i v^j{}_{,i}\right)\frac{\partial}{\partial x^j}[F].$$

In component form the commutator vector field is given by

$$[\mathbf{v},\mathbf{u}] = \left(v^i u^j{}_{,i} - u^i v^j{}_{,i}\right)\frac{\partial}{\partial x^j} = \left(\mathbf{v}[u^j] - \mathbf{u}[v^j]\right)\frac{\partial}{\partial x^j}. \qquad (10.5)$$

From (10.5) it is evident that the Lie bracket is antisymmetric; that is,

$$[\mathbf{v},\mathbf{u}] = -[\mathbf{u},\mathbf{v}]. \qquad (10.6)$$

As is the case for the Poisson bracket, relations (10.3) and results (10.4)–(10.6) apply at each point of phase space where the relevant quantities are defined.

The algebraic properties induced on phase-space functions by the Poisson bracket, and on vector fields by the Lie bracket, are typical of what is called a Lie algebra. A *Lie algebra* is a vector space S with a binary operation defined on the elements of S. This binary operation is usually denoted as $[\ ,\ ]: S \times S \to S$, which satisfies

$$[a\mathbf{u} + b\mathbf{v}, \mathbf{w}] = a[\mathbf{u},\mathbf{w}] + b[\mathbf{v},\mathbf{w}] \qquad \text{(linear)} \qquad (10.7)$$

$$[\mathbf{u},\mathbf{v}] = -[\mathbf{v},\mathbf{u}] \qquad \text{(antisymmetric)} \qquad (10.8)$$

$$[\mathbf{u},[\mathbf{v},\mathbf{w}]] + [\mathbf{v},[\mathbf{w},\mathbf{u}]] + [\mathbf{w},[\mathbf{u},\mathbf{v}]] = 0. \qquad \text{(Jacobi identity)} \qquad (10.9)$$

Properties (10.7)–(10.9) have all been established for Poisson brackets. For vector fields linearity follows from the linearity of differentiation and is immediately inferred from (10.5). Antisymmetry for the Lie bracket is contained in (10.6) and the Jacobi identity is established in Problem 10.1. An even more familiar example of a Lie algebra is provided by vectors in $\mathbb{R}^3$, with the cross product "$\times$" operation. Vectors formed from $\mathbf{A} \times \mathbf{B}$, where $\mathbf{A}$ and $\mathbf{B}$ are arbitrary vectors in $\mathbb{R}^3$, satisfy (10.7)–(10.9).

On phase space there is yet another interesting Lie algebra. Let $\mathbf{u}, \mathbf{v}, \mathbf{w}$ be differentiable vector fields. For each such vector field we define a corresponding adjoint operator. For arbitrary $\mathbf{u}$ the *adjoint operator* $\tilde{\mathbf{u}}$ on an arbitrary vector field $\mathbf{v}$ is defined by

$$\tilde{\mathbf{u}}[\mathbf{v}] = [\mathbf{u}, \mathbf{v}], \tag{10.10}$$

The quantity $\tilde{\mathbf{u}}[\mathbf{v}]$ is another vector field. The elements of the algebra are $\tilde{\mathbf{u}}, \tilde{\mathbf{v}}, \tilde{\mathbf{w}}, \ldots$, and so on. For the binary operation on these elements take the difference of two adjoints applied in reverse order, that is, $[\tilde{\mathbf{u}}, \tilde{\mathbf{v}}]$, where

$$\begin{aligned}
[\tilde{\mathbf{u}}, \tilde{\mathbf{v}}][\mathbf{w}] &= \tilde{\mathbf{u}}(\tilde{\mathbf{v}}[\mathbf{w}]) - \tilde{\mathbf{v}}(\tilde{\mathbf{u}}[\mathbf{w}]) = [\mathbf{u}, \tilde{\mathbf{v}}[\mathbf{w}]] - [\mathbf{v}, \tilde{\mathbf{u}}[\mathbf{w}]] \\
&= [\mathbf{u}, [\mathbf{v}, \mathbf{w}]] - [\mathbf{v}, [\mathbf{u}, \mathbf{w}]] = [[\mathbf{u}, \mathbf{v}], \mathbf{w}] = [\widetilde{\mathbf{u}, \mathbf{v}}][\mathbf{w}].
\end{aligned} \tag{10.11}$$

We have used the Jacobi identity for vector fields. The adjoint operators form a Lie algebra, and the ˜ operation, which gives adjoint operators from vector fields, bears a similar relationship to the operation of obtaining vector fields $\mathbf{v}_F$ from functions. This similarity is more evident if we display (10.10) and (10.3) in similar fashion:

$$\begin{aligned}
\mathbf{v}_F[G] &= \{G, F\}, \\
\tilde{\mathbf{v}}[\mathbf{u}] &= [\mathbf{v}, \mathbf{u}].
\end{aligned} \tag{10.12}$$

The relationships displayed in (10.12) define a correspondence or mapping between elements in one algebra to elements in another. An *algebra homomorphism* is defined as a linear mapping that preserves the bracket operation. More specifically, if A, B denote two Lie algebras, then ϕ: $A \to B$ is an algebra homomorphism if $\phi(\lambda_1\mathbf{a}_1 + \lambda_2\mathbf{a}_2) = \lambda_1\phi(\mathbf{a}_1) + \lambda_2\phi(\mathbf{a}_2)$ and $\phi([\mathbf{a}_1, \mathbf{a}_2]) = [\phi(\mathbf{a}_1), \phi(\mathbf{a}_2)]$, where λ_1, λ_2 are scalars and $\mathbf{a}_1, \mathbf{a}_2 \in A$. If such a map is also a bijection, then it is called an *isomorphism*.

From (10.12) we define the mappings by $F \to -\mathbf{v}_F$ and $\mathbf{v} \to \tilde{\mathbf{v}}$, which are easily shown to be Lie algebra homomorphisms. These foregoing maps are not always isomorphisms, and to illustrate in a specific case, we do an example. Consider the two-dimensional phase space with coordinates (p, q). On this phase space we consider the simple functions p and q. The Poisson bracket $\{q, p\} = 1$, and hence the Lie algebra generated by the Poisson brackets on these functions consists of the elements $(1, p, q)$. Any of these functions $(1, p, q)$ can play the role of F in the mapping $F \to -\mathbf{v}_F$. All brackets are zero except $\{q, p\} = 1$. For the vector fields, the $\mathbf{v}_F$, we have from (8.5), (8.6), (8.9) that $\mathbf{v}_q = -\partial/\partial p$, $\mathbf{v}_p = \partial/\partial q$, and $\mathbf{v}_1 = 0$. The Lie algebra $(1, q, p)$ with Poisson brackets maps homomorphically onto the Lie algebra $(\partial/\partial q, \partial/\partial p)$ with commutator brackets. The mapping is not an isomorphism because $\partial/\partial q$ and $\partial/\partial p$ commute whereas q and p do not (in terms of Poisson brackets).

The Lie algebra property is of interest because of the intimate connection of Poisson brackets with dynamical evolution through (7.24). When a dynamical variable has no explicit time dependence, then the Poisson brackets with the Hamiltonian contain all information about the time evolution of that particular variable. In Section 10.3 Lie algebraic techniques are used to study canonical transformations.

10.2. LIE DERIVATIVE

Consider an arbitrary vector field $\mathbf{v}$ on an n-dimensional manifold M, which in some neighborhood U generates a flow on M, that is, a congruence of nonintersecting curves (cf. Section 8.1) Figure 8.1 shows congruences of curves for two independent vector fields. Assume that the flow of $\mathbf{v}$ is parameterized by the parameter λ, with λ = constant surfaces corresponding to connected submanifolds of M. The curves in the congruence or flow are labeled with some set of $(n - 1)$ parameters, which when taken together with λ, form an appropriate coordinate system in the neighborhood U. Denote the flow mapping as $g_{\mathbf{v}}^{\lambda}$, where $g_{\mathbf{v}}^{0}$ is the identity map. The tangent to each of these curves is the vector field $\mathbf{v}$ generating the flow, as defined in (8.4). The flow mapping $g_{\mathbf{v}}^{\lambda}$ is a manifold map, that is, $g_{\mathbf{v}}^{\lambda}: M \to M$, and consequently the derivative map $(g_{\mathbf{v}}^{\lambda})_*$ and the pull-back map $(g_{\mathbf{v}}^{\lambda})^*$, have both been defined. The mappings $(g_{\mathbf{v}}^{\lambda})_*$ and $(g_{\mathbf{v}}^{\lambda})^*$ relate vectors and forms, in short all geometric objects, at points connected by the flow map. The process of relating geometric objects along the flow by the derivative and pull-back maps is called *dragging along*. We also define a dragged-along function. This is done in a way consistent with the view of real-valued functions as 0-forms. If F is an arbitrary function defined in U, then the dragged-along function F^* is defined by

$$F^*(\mathbf{x}_0) = F\big(g_{\mathbf{v}}^{\lambda}(\mathbf{x}_0)\big) = F(\mathbf{x}(\lambda)) \tag{10.13}$$

If we view functions as 0-forms, then this is the pull-back map of (8.46).

We have frequently discussed invariants of the motion, that is, invariants of Hamiltonian flow. The very question of invariance implies a comparison along the flow. A way to quantify a comparison of geometric objects along a flow is through the *Lie derivative*.

Consider first the difference between $F^*(\mathbf{x}_0)$ and $F(\mathbf{x}_0)$. We define the Lie derivative of the function F at $\mathbf{x}_0 \in M$:

$$L_{\mathbf{v}}[F]|_{x_0} = \lim_{\lambda \to 0} \frac{1}{\lambda}[F^*(\mathbf{x}_0) - F(\mathbf{x}_0)] \tag{10.14}$$

Despite its appearance, the Lie derivative of a function is a familiar quantity, as the following calculation shows.

$$F^*(\mathbf{x}_0) = F(\mathbf{x}(\lambda)) = F(\mathbf{x}_0) + \lambda \left.\frac{\partial F}{\partial x^i}\right|_{x_0} \left.\frac{\partial x^i}{\partial \lambda}\right|_{x_0} + \cdots$$

Substituting into (10.14) and then taking the limit we find

$$L_{\mathbf{v}}[F]|_{\mathbf{x}_0} = \left.\frac{\partial F}{\partial x^i}\right|_{\mathbf{x}_0} \left.\frac{dx^i}{d\lambda}\right|_{\mathbf{x}_0}$$

Recall $v^i(\mathbf{x}_0) = (dx^i/d\lambda)|_{\mathbf{x}_0}$, which gives

$$L_{\mathbf{v}}[F]|_{\mathbf{x}_0} = \left.\frac{\partial F}{\partial x^i}\right|_{\mathbf{x}_0} v^i(\mathbf{x}_0) = \mathbf{v}(\mathbf{x}_0)[F].$$

Since the point $\mathbf{x}_0$ is arbitrary, we have

$$L_{\mathbf{v}}[F] = \mathbf{v}[F], \tag{10.15}$$

and indeed, we see that the Lie derivative of a function F, with respect to a vector field $\mathbf{v}$, is the same thing as $\mathbf{v}$ (as a directional derivative operator) operating on F.

If F is to be invariant under the dragging of $\mathbf{v}$, then $F^*(\mathbf{x}_0) = F(\mathbf{x}_0)$ and $L_{\mathbf{v}}[F] = 0$. This gives yet another way of characterizing constants of the motion. They are those functions such that $L_{\mathbf{v}_H}[F] = 0$. The vector field $\mathbf{v}_H$ is the Hamiltonian flow vector. We have for an invariant F that $L_{\mathbf{v}_H}[F] = \mathbf{v}_H[F] = \{F, H\} = 0$.

Having considered the changes in scalar functions under a flow, let us consider vector fields as well. Let $\mathbf{u}$ denote a second vector field in U with associated flow $g_{\mathbf{u}}^{\mu}$. The the Lie derivative of $\mathbf{u}$ with respect to $\mathbf{v}$ is given by

$$L_{\mathbf{v}}[\mathbf{u}]|_{\mathbf{x}_0} = \lim_{\lambda \to 0} \frac{1}{\lambda}\{(g_{\mathbf{v}}^{-\lambda})_*(\mathbf{u}(\mathbf{x}(\lambda))) - \mathbf{u}(\mathbf{x}_0)\}. \tag{10.16}$$

This Lie derivative considers the difference between $\mathbf{u}(\mathbf{x}_0)$ and $\mathbf{u}(\mathbf{x}(\lambda))$ as dragged back to $\mathbf{x}_0$ using the differential map $(g_{\mathbf{v}}^{-\lambda})_*$, which is obtained from the flow map $g_{\mathbf{v}}^{-\lambda}$, as discussed in Section 6.3. If $\mathbf{u}$ is invariant under this flow, then $(g_{\mathbf{v}}^{-\lambda})_*(\mathbf{u}(\mathbf{x}(\lambda)))$ is $\mathbf{u}(\mathbf{x}_0)$ and the Lie derivative vanishes.

Compute this Lie derivative in a fixed, but arbitrary, local coordinate system.

Denote by $\mathbf{x}(\lambda, \mu)$ the curve through $\mathbf{x}(\lambda)$ such that $\mathbf{x}(\lambda, 0) = \mathbf{x}(\lambda)$ and $[d\mathbf{x}(\lambda, \mu)/d\mu]|_{\mu=0} = \mathbf{u}(\mathbf{x}(\lambda))$.

$$u^i(\mathbf{x}(\lambda)) = u^i(\mathbf{x}_0) + \lambda \left.\frac{\partial u^i}{\partial x^j}\right|_{\lambda=0} \left.\frac{dx^j}{d\lambda}\right|_{\lambda=0} + \cdots$$

$$= u^i(\mathbf{x}_0) + \lambda \mathbf{v}(\mathbf{x}_0)[u^i] + \cdots \tag{10.17}$$

The components of the dragged vector in definition (10.16) for the Lie derivative are given by

$$\left[(g_{\mathbf{v}}^{-\lambda})_*(\mathbf{u}(\mathbf{x}(\lambda)))\right]^i = \left.\frac{d}{d\mu}\right|_{\mu=0} \left[g_{\mathbf{v}}^{-\lambda}(\mathbf{x}(\lambda, \mu))\right]^i$$

For the coordinates of the point $\mathbf{x}(\lambda, \mu)$ dragged back by the flow map $g_{\mathbf{v}}^{-\lambda}$, we find

$$\left[g_{\mathbf{v}}^{-\lambda}(\mathbf{x}(\lambda, \mu))\right]^i = x^i(\lambda, \mu) - \lambda \left.\frac{dx^i(\lambda', \mu)}{d\lambda'}\right|_{\lambda'=\lambda} + \cdots$$

The components of the tangent vector $\mathbf{u}(\mathbf{x}(\lambda, 0)$ are then

$$u^i(\mathbf{x}(\lambda, 0)) = \left.\frac{d}{d\mu}\right|_{\mu=0} \left[g_{\mathbf{v}}^{-\lambda}(\mathbf{x}(\lambda, \mu))\right]^i$$

$$= u^i(\mathbf{x}(\lambda)) - \lambda \left.\frac{dx^j}{d\mu}\right|_{\mu=0} \frac{\partial}{\partial x^j} \left(\left.\frac{dx^i(\lambda', \mu)}{d\lambda'}\right|_{\lambda'=\lambda} \right)\Bigg|_{\mu=0}$$

Substituting from (10.17) and using u^j for $(dx^j/d\mu)$

$$u^i(\mathbf{x}(\lambda, 0)) = u^i(\mathbf{x}_0) + \lambda \mathbf{v}(\mathbf{x}_0)[u^i] - \lambda \mathbf{u}(\mathbf{x}(\lambda))\left[v^i(\mathbf{x}(\lambda))\right] + \cdots$$

To the order needed, we can replace $\mathbf{u}(\mathbf{x}(\lambda))[v^i(\mathbf{x}(\lambda))]$ by $\mathbf{u}(\mathbf{x}_0)[v^i(\mathbf{x}_0)]$ in the preceding equation. We thus obtain

$$[L_{\mathbf{v}}\mathbf{u}]^i|_{\mathbf{x}_0} = \mathbf{v}(\mathbf{x}_0)[u^i] - \mathbf{u}(\mathbf{x}_0)[v^i]$$

Since $\mathbf{x}_0$ is an arbitrary point, by comparison with (10.5), we obtain

$$L_{\mathbf{v}}[\mathbf{u}] = [\mathbf{v}, \mathbf{u}] \tag{10.18}$$

The Lie derivative of the vector field $\mathbf{u}$ with respect to the vector field $\mathbf{v}$ is the Lie bracket or commutator of $\mathbf{v}$ with $\mathbf{u}$.

The Lie derivative gives yet another way of viewing invariance under flow mappings. That a constant of the motion F must satisfy $L_{\mathbf{v}_H}[F] = 0$, has already been mentioned. Similarly, the Lie derivative of the canonical 2-form $\boldsymbol{\gamma} = \mathbf{d}p_i \wedge \mathbf{d}q^i$ with respect to $\mathbf{v}_H$ must vanish under Hamiltonian flow.

For the Lie derivative of a form we have

$$L_{\mathbf{v}}[\omega] = \lim_{\lambda \to 0} \frac{1}{\lambda}\{g_{\mathbf{v}}^{\lambda *}\omega - \omega\} \tag{10.19}$$

Problem 10.2 is an exercise to obtain a formula similar to (10.18) for 1-forms. If $\mathbf{v}$ is a Hamiltonian flow vector $\mathbf{v}_H$, then g_H^{λ} is a canonical transformation and $g_H^{\lambda *}\boldsymbol{\gamma} = \boldsymbol{\gamma}$. Thus the Lie derivative with respect to $\mathbf{v}_H$ of the canonical 2-form $\boldsymbol{\gamma}$ vanishes; that is, $L_{\mathbf{v}_H}[\boldsymbol{\gamma}] = 0$.

10.3. LIE TRANSFORMATIONS

The primary motivation for considering transformations on phase space has always been to simplify the description of mechanical systems. Chapter 9 discusses in detail the simplifications that result from expressing completely integrable systems in terms of action-angle variables. This section is devoted to the development of an operator technique for solving a Hamiltonian system. The central quantity in this section is the Lie transformation operator to be defined below.

In Eq. (5.23) we considered the exponentiation of a matrix operator. In the simple example considered, the exponentiated operator, defined by a power series expansion, converged to a matrix operator with sines and cosines as its elements. In a similar fashion the exponentiation of differential operators (tangent vectors) can be defined.

Let $\mathbf{v}$ be a vector field defined on T^*M and α a scalar. Then $e^{\alpha\mathbf{v}}$ is defined by

$$e^{\alpha\mathbf{v}} = \sum_{n=0}^{\infty} \frac{1}{n!}(\alpha\mathbf{v})^n = \left(1 + \alpha\mathbf{v} + \frac{1}{2!}\alpha^2\mathbf{v}^2 + \cdots\right). \tag{10.20}$$

Whether or not this series converges to a well-defined operator on real-valued functions depends on α and $\mathbf{v}(\mathbf{x})$. If the series does not converge, then we simply view the series in (10.20) as a formal object, and $e^{\alpha\mathbf{v}}$ as a shorthand notation for writing the formal power series.

Differentiating (10.20) with respect to α, we see that $e^{\alpha\mathbf{v}}$ satisfies the differential equation for an operator $(dT/d\alpha) = \mathbf{v}T$. This suggests an even more general class of operators, which includes all those defined by (10.20).

The transformation E_v is defined by the differential equation:

$$E_{v,\alpha} = \mathbf{v}(\alpha)E_v; \qquad E_v(0) = 1. \tag{10.21}$$

In the case that $\mathbf{v}$ does not depend on α, then $e^{\alpha\mathbf{v}}$ is the solution to (10.21). In the definition (10.21) of E_v, we have used $(\partial E_v/\partial\alpha) = E_{v,\alpha}$ since $\mathbf{v}$, and hence E_v, depends on other parameters as well, such as time and phase space coordinates. The operator E_v is called the *Lie transformation* generated by the vector field $\mathbf{v}$. If $\mathbf{v}$ is the flow vector associated with a phase space function F, then we say that E_v is the Lie transformation associated with F. Since for $\alpha = 0$, E_v is just the identity, we know that for some neighborhood of $\alpha = 0$, $E_v(\alpha)$ will have an inverse that we denote in the usual way $E_v^{-1}(\alpha)$. The inverse transformation satisfies the differential equation $E_{v,\alpha}^{-1} = -E_v^{-1}\mathbf{v}$, as can be readily verified by differentiating $E_v^{-1}E_v f = f$ with respect to α. For brevity we have written $E_v f \equiv E_v[f]$ and we use this notation in the following discussion, unless the argument of the operator must be included in parentheses to preclude ambiguities. The utility of E_v depends largely on its behavior on products of functions and Poisson brackets. This behavior depends in turn on the properties of $\mathbf{v}$, which we now examine.

Any operation satisfying the usual product rule of differentiation is called a *derivation*. We show that an arbitrary flow-vector field $\mathbf{v}_F$ satisfies the product rule of differentiation, not only for products of functions but also for Poisson brackets. Let f and g be two arbitrary functions on phase space. Then we assert that

$$\mathbf{v}_F[fg] = f\mathbf{v}_F[g] + g\mathbf{v}_F[f] \tag{10.22}$$

and

$$\mathbf{v}_F[\{f, g\}] = \{\mathbf{v}_F[f], g\} + \{f, \mathbf{v}_F[g]\}. \tag{10.23}$$

Equation (10.22) follows immediately from the product rule of partial differentiation once we write $\mathbf{v}_F$ in terms of a coordinate basis, that is, $\mathbf{v}_F = v_F^i(\partial/\partial x^i)$. Equation (10.23) we prove in the following manner. Using

$\mathbf{v}_F[G] = \{G, F\}$ we write

$$\mathbf{v}_F[\{f, g\}] = \{\{f, g\}, F\}.$$

Apply the Jacobi identity for Poisson brackets to find

$$\mathbf{v}_F[\{f, g\}] = \{\{f, F\}, g\} + \{f, \{g, F\}\} = \{\mathbf{v}[f], g\} + \{f, \mathbf{v}[g]\},$$

which shows (10.23).

We use the properties (10.22), (10.23) of $\mathbf{v}_F = \mathbf{v}$ to prove the following relations for the Lie transformation E_v generated by $\mathbf{v}$.

$$E_v[fg] = E_v[f]E_v[g]. \tag{10.24}$$

$$E_v\{f, g\} = \{E_v f, E_v g\}. \tag{10.25}$$

Comparing (10.24) and (10.25) with (10.22) and (10.23) makes evident that the Lie transformation E_v is not a derivation.

To prove (10.24) and (10.25) we require an intermediate result. If $A(\alpha)$ satisfies the operator differential equation

$$A_{,\alpha} = \mathbf{v}A + B, \tag{10.26}$$

then a solution of (10.26), in terms of E_v satisfying (10.21), is given by

$$A(\alpha) = E_v \int_0^{\alpha} E_v^{-1} B \, d\alpha' + E_v(\alpha) A(0). \tag{10.27}$$

That (10.27) is a solution to (10.26) can be verified by straightforward differentiation.

Use (10.27) to prove (10.25) in the following way. Define the operator of (10.26) and (10.27) as $A(f, g) = E_v\{f, g\} - \{E_v f, E_v g\}$. Differentiate this result with respect to α.

$$A_{,\alpha}(f, g) = E_{v,\alpha}\{f, g\} - \{E_{v,\alpha} f, E_v g\} - \{E_v f, E_{v,\alpha} g\}$$

$$= \mathbf{v}[E_v\{f, g\} - \{E_v g, E_v g\}] = \mathbf{v}A(f, g).$$

The functions f and g are arbitrary and so $A_{,\alpha} = \mathbf{v}A$. Since $E_v = 1$ for $\alpha = 0$, $A(0) = 0$. From (10.27) we conclude that the solution is then $A \equiv 0$. Thus (10.25) has been shown. Equation (10.24) can be shown in a similar manner.

To obtain a further property of E_v, let c denote a constant function. Consider the function $g(\alpha) \equiv E_v c$. The function $g(\alpha) = E_v c$ satisfies the differential equation $g_{,\alpha} = \mathbf{v}g$ with the initial condition $g(0) = c$. The unique solution to this differential equation with this initial condition is $g = c$. Thus

$$E_v c = c \tag{10.28}$$

and the Lie transformation operator E_v acts like the identity operator when applied to constants.

If E_v is applied to the coordinate functions themselves, it then becomes a mapping on phase space. Let $\{x^i\}$ denote a set of coordinate functions for some region of phase space. Then

$$y^i = E_v x^i \tag{10.29}$$

denotes a new set of coordinate functions. In this we are viewing E_v as an active transformation on phase space.

If E_v is viewed as a mapping on phase space by (10.29), then how does the pull-back map on functions (0-forms) relate to E_v? The pull-back map is defined in (10.13) and for this specific application we write

$$f(\mathbf{x}) \equiv (E_v^* F)(\mathbf{x}) = F(E_v \mathbf{x}) = F(\mathbf{y}). \tag{10.30}$$

Equation (10.30) is (10.13) where we have denoted explicitly the pull-back map E_v^* and used $f(\mathbf{x})$ for $F^*(\mathbf{x}_0)$ of (10.13). How does E_v^* relate to E_v, which after all was defined as a mapping on phase space functions? To answer this question examine the differential equation that E_v^* must satisfy. Using (10.30) we compute

$$\begin{aligned}(E_v^* F)_{,\alpha}(\mathbf{x}) &= \frac{\partial F}{\partial y^i} E_{v,\alpha} x^i \\ &= \frac{\partial F}{\partial y^i} \mathbf{v}[E_v x^i] = \frac{\partial F}{\partial y^i} \mathbf{v}[y^i] = \mathbf{v}[F(\mathbf{y})] = \mathbf{v}[E_v^* F(\mathbf{x})].\end{aligned}$$

Since the phase point $\mathbf{x}$ and the function F are both arbitrary

$$E_{v,\alpha}^* = \mathbf{v} E_v^*,$$

with the initial condition that $E_v^*(0)$ is the identity mapping. Since the differential equation and initial conditions are the same for E_v^* and E_v, they are identical, and we will not further distinguish between them.

The utility of E_v in (10.29) is that it generates a canonical transformation! This is easily demonstrated using (10.25) to show the preservation of the Poisson brackets in $\{y^i, y^j\} = \gamma_{ij}$ [cf. (8.45)]. We see that

$$\{y^i, y^j\} = \{E_v x^i, E_v x^j\} = E_v\{x^i, x^j\} = E_v \gamma_{ij} = \gamma_{ij}. \qquad (10.31)$$

In addition to (10.25), we have also used (10.28) in the last equality of (10.31).

For the remainder of this section, and the next, we continue with the notation introduced in (10.30). We denote with lower-case letters phase-space functions defined on coordinates $\{x^i\}$, and with upper-case, functions defined on the coordinates $\{y^i\}$, which are related to the $\{x^i\}$ by a Lie transformation E_v. For example, a Hamiltonian $h(\mathbf{x})$ and its transformed counterpart are related by

$$h(\mathbf{x}) = (E_v H)(\mathbf{x}) = (E_v^* H)(\mathbf{x}) = H(E_v \mathbf{x}) = H(\mathbf{y}). \qquad (10.32)$$

One example of the use of a Lie transformation is to simplify a Hamiltonian so that it has a particularly simple form and thus facilitate obtaining integrals of the motion. Suppose that we can find an integral invariant $F(\mathbf{y})$ with respect to the transformed Hamiltonian $H(\mathbf{y})$. Let $f = E_v F$; then $\{f, h\} = \{E_v F, E_v H\} = E_v\{F, H\} = 0$. The transformed invariant f is a first integral with respect to the original Hamiltonian h.

10.4. LIE TRANSFORM PERTURBATION THEORY

Simplifying a system by a Lie transformation finds its most frequent application in perturbation theory. We assume the Hamiltonian $h(\mathbf{x})$ can be written in the form

$$h(\mathbf{x}, t) = h_0(\mathbf{x}, t) + \varepsilon h_1(\mathbf{x}, t) + \varepsilon^2 h_2(\mathbf{x}, t) + \cdots, \qquad (10.33)$$

and seek a canonical transformation which "removes" the perturbation pieces in (10.33). More precisely, we seek a new set of coordinates and a new Hamiltonian H' in which the departure from the solvable piece is pushed to a higher order.

The Lie transform, which accomplishes the task of pushing the perturbation to higher order, is constructed by finding a real-valued function $s(\mathbf{x}, t)$ on phase space, with its flow vector generating the desired Lie transformation. This function $s(\mathbf{x}, t)$ is frequently referred to as a *generating function*, but it is not to be confused with the generating functions discussed in

Section 8.5–8.7. The generating function $s(\mathbf{x}, t)$ of present interest is a function only of "old" coordinates.

Thus in outline: (i) Find $s(\mathbf{x}, t)$ satisfying differential equations determined by the Hamiltonian, which (ii) generates a flow $\mathbf{v}_s$, and (iii) use $\mathbf{v}_s$ in (10.21) to find E_s. The Lie transformation E_s, when applied to the Hamiltonian, increases the order of the perturbation and hence makes an increasingly smaller contribution to the solvable dynamics.

Before proceeding we remind the reader of the notational convention in (10.30). Lower-case quantities are functions of $\mathbf{x}$ and upper-case quantities are functions of the transformed coordinates $\mathbf{y}$. For example $h(\mathbf{x}, t)$ and $H(\mathbf{y}, t)$ are related by an equation like (10.30). The quantity H' is the new Hamiltonian, after the canonical transformation has taken place, and is to be distinguished from H. The function H is the old Hamiltonian written in new coordinates. The relationship is analogous to H and H' in (8.68), and each of these functions can be expressed in terms of "new" or "old" variables. Only in the case that the canonical transformation generated by $s(\mathbf{x})$ is time independent are H and H' the same.

Let $s(\mathbf{x}, t, \varepsilon)$ denote the phase-space function that is to serve as the generating function. We form the flow vector field $\mathbf{v}_s(\mathbf{x}, t, \varepsilon)$ and the corresponding Lie transformation $E_s(\mathbf{x}, t, \varepsilon)$, defined by (10.21) with α replaced by the expansion parameter ε.

$$E_{s,\varepsilon} = \mathbf{v}_s E_s; \qquad E_s(0) = 1. \tag{10.34}$$

Our primary aim is to obtain a relationship between H and H' [Eq. (10.39)] and to do so requires a formula for $E_{s,t}$, which we now obtain.

We proceed in this calculation by differentiating (10.34) with respect to time. The time derivative of the flow vector $\mathbf{v}_s$ is given by $(\partial \mathbf{v}_s/\partial t) = \mathbf{v}_{s,t} = \mathbf{v}_{s_{,t}}$. This follows directly by considering the time derivative of $\mathbf{v}_s[f]$, where f is an arbitrary, time-independent function on phase space.

$$\mathbf{v}_{s,t}[f] = \frac{\partial}{\partial t}(\mathbf{v}_s[f]) = \frac{\partial}{\partial t}(\{f, s\}) = \left\{f, \frac{\partial s}{\partial t}\right\} = \mathbf{v}_{s_{,t}}[f].$$

Since f is arbitrary we have the desired result.

Differentiate (10.34) with respect to time, use the fact that the partial derivatives commute, and obtain

$$E_{s,t,\varepsilon} = \mathbf{v}_s E_{s,t} + \mathbf{v}_{s,t} E_s. \tag{10.35}$$

Equation (10.35) is of the form (10.26) and we apply (10.27):

$$E_{s,t} = E_s \int_0^{\varepsilon} E_s^{-1} \mathbf{v}_{s,t} E_s \, d\varepsilon', \tag{10.36}$$

where we have made use of the fact that $E_{s,t}(0) = 0$, since $E_s(0) = 1$ for all t.

With formula (10.36) we can relate the Hamiltonians H and H' by considering an arbitrary phase-space density function $F(\mathbf{y}, t)$. Since density functions are invariant under a Hamiltonian flow, $F(\mathbf{y}, t)$ satisfies Liouville's equation.

$$F_{,t} + \{F, H'\} = 0. \tag{10.37}$$

Similarly

$$f_{,t} + \{f, h\} = 0, \tag{10.38}$$

since f and F are related by (10.30), which in this context says that the density at a point is invariant. Since $f = E_s F$, we substitute into (10.38) and find

$$E_{s,t} F + E_s F_{,t} + \{E_s F, E_s H\} = 0.$$

Substituting from (10.36), (10.37), and using (10.25) gives

$$\left(E_s \int_0^{\varepsilon} E_s^{-1} \mathbf{v}_{s,t} E_s \, d\varepsilon' \right) F - E_s\{F, H'\} + E_s\{F, H\} = 0.$$

Operate on the foregoing equation from the left with E_s^{-1} and differentiate with respect to ε to obtain

$$E_s^{-1} \mathbf{v}_{s,t} E_s F - \{F, H'_{,\varepsilon}\} + \{F, H_{,\varepsilon}\}$$

$$= E_s^{-1}\{E_s F, s_{,t}\} + \{F, H_{,\varepsilon} - H'_{,\varepsilon}\} = 0.$$

Using (10.25) and combining, we find

$$\{F, E_s^{-1} s_{,t} + H_{,\varepsilon} - H'_{,\varepsilon}\} = 0.$$

This result holds for arbitrary ε and arbitrary F. Consequently,

$$H'_{,\varepsilon} = H_{,\varepsilon} + E_s^{-1} s_{,t}. \tag{10.39}$$

This is the desired relationship for the Hamiltonians H and H'. Equation (10.39) is a general result for a Lie transformation generated by a function $s(\mathbf{x}, t, \varepsilon)$.

Let us now consider the Hamiltonian of interest to be given by (10.33) and let

$$s(\mathbf{x}, t, \varepsilon) = 0 + s_1(\mathbf{x}, t) + \varepsilon s_2(\mathbf{x}, t) + \cdots, \tag{10.40}$$

$$H' = h_0 + \varepsilon H_1' + \varepsilon^2 H_2' + \cdots. \tag{10.41}$$

Our first task is to calculate $\mathbf{v}_s$ and then E_s^{-1}. Let f be an arbitrary phase-space function. Then by substituting (10.40) into $\mathbf{v}_s[f] = \{f, s\}$ we find

$$\mathbf{v}_s = \mathbf{v}_{s_1} + \varepsilon \mathbf{v}_{s_2} + \cdots. \tag{10.42}$$

Assume E_s^{-1} can be written as $E_s^{-1} = 1 + \varepsilon E_1 + \varepsilon^2 E_2 + \cdots$, and substitute into $E_{s,\varepsilon}^{-1} = -E_s^{-1}\mathbf{v}_s$ along with (10.42). Matching coefficients of the arbitrary parameter ε, we find

$$E_s^{-1} = 1 - \varepsilon \mathbf{v}_{s_1} + \frac{\varepsilon^2}{2}\left(\mathbf{v}_{s_1}^2 - \mathbf{v}_{s_2}\right) + \cdots. \tag{10.43}$$

In a similar fashion find from (10.34)

$$E_s = 1 + \varepsilon \mathbf{v}_{s_1} + \frac{\varepsilon^2}{2}\left(\mathbf{v}_{s_1}^2 + \mathbf{v}_{s_2}\right) + \cdots. \tag{10.44}$$

For $H(\mathbf{y}) = E_s^{-1}h(\mathbf{y})$, we find

$$H = h_0 + \varepsilon\left[h_1 - \mathbf{v}_{s_1}h_0\right] + \varepsilon^2\left[h_2 - \mathbf{v}_{s_1}h_1 + \tfrac{1}{2}\left(\mathbf{v}_{s_1}^2 - \mathbf{v}_{s_2}\right)h_0\right] + \cdots. \tag{10.45}$$

Both sides of (10.45) are functions of the same coordinates. Differentiating (10.45) and (10.41) we can substitute into (10.39). Use (10.43) again and equate coefficients of like powers of ε to find

$$H_1' = h_1 + s_{1,t} + \{s_1, h_0\} = h_1 + \dot{s}_1.$$

In the last expression we have used the definition $\dot{s}_1 \equiv s_{1,t} + \{s_1, h_0\}$. To make $H_1' = 0$ we simply find s_1 such that $\dot{s}_1 = -h_1$. Proceeding to the next order, we find after a little algebra

$$H_2' = h_2 + \tfrac{1}{2}\dot{s}_2 + \tfrac{1}{2}\mathbf{v}_{s_1}\dot{s}_1.$$

If h_2 is zero, then we may choose $s_2 = 0$, implying $H_2' = -\frac{1}{2}\mathbf{v}_{s_1}h_1$. This procedure can in a straightforward fashion be carried to higher orders. With $H_1' = 0$, the new Hamiltonian has its perturbation second order in ε, and consequently the perturbation has been pushed to higher order. This result has been achieved for an arbitrary, perturbed Hamiltonian, not necessarily a completely integrable Hamiltonian, as in the canonical perturbation theory of Section 9.4. There is no attempt to perturb away from invariant toroids and so the issues of resonant surfaces and resonant denominators never arise. The complication arising here is that h_0 occurring in (10.41) may turn out to be a very complicated function in the new coordinates.

Applications and a discussion of many results presented here may be found in Kaufman (1978), Dewar (1976), and Cary (1977).

10.5. GROUP REPRESENTATIONS

In Section 10.1 we studied three different Lie algebras, which were related: (1) the algebra of Poisson brackets, (2) the algebra of vector fields, and (3) the algebra of adjoint operators on vector fields. With each Lie algebra there corresponds one or more Lie groups. Although we cannot here delve deeply into the theory of Lie groups as applied to dynamics, we indicate some of the relationships. A more complete account of the use of Lie groups in dynamics can be found in the treatise of Sudarshan and Mukunda (1974).

Briefly we remind the reader of the usual group axioms for a set of elements S to be a Lie group. The set S is a manifold and

1. There is a group composition law $S \times S \to S$ and this mapping is differentiable.
2. The group operation is associative.
3. There is an identity e on the group.
4. For every element there is an inverse.

Axiom 1, in terms of coordinates, means that if a, b, c are elements of a Lie group G, such that $ab = c$, then the coordinates on the manifold for the element c are differentiable functions of the coordinates for the elements a and b.

Specifically how do Lie groups relate to dynamics? We have seen where the algebras come from. They arise in a natural way because of the Lie algebra structure of Poisson brackets. This is all well and good, but where are the groups?

There are basically two answers to this question. The most direct case is when a group manifold is in fact the configuration space of the dynamical system under study. For example, a freely rotating rigid body has its configuration at any time completely specified by specifying a rotation matrix. The rotation matrix represents the rotation from some fiducial orientation necessary to produce the current configuration. Using Euler angles, such a rotation matrix is unique. These rotation matrices form a well-known Lie group denoted $SO(3)$. This group manifold describes the configuration space of the system. We return to this example in the next section.

A second answer to this query concerning the whereabouts of the group is that canonical transformations on the phase space of the dynamical system form a group.

Suppose that ρ, σ, τ are all canonical transformations on phase space T^*M for some dynamical system with configuration space M. We know that canonical transformations are invertible. The identity transformation is a canonical transformation. Mapping composition is associative, i.e. $(\rho \circ \sigma) \circ \tau = \rho \circ (\sigma \circ \tau)$. At the end of Section 8.4, we showed that the composition of canonical transformations gives a canonical transformation. Thus these canonical transformations fulfill all requirements and form a group.

In Section 10.1 we discussed different realizations of essentially the same Lie algebra. Similarly, there are different realizations of the same abstract Lie group. Group homomorphisms are defined similarly as well. A group homomorphism must be well defined and preserve the group composition law. If G_1 and G_2 are two groups and $\phi: G_1 \to G_2$ is a group homomorphism, then $\phi(gh) = \phi(g)\phi(h)$, for all $g, h \in G_1$.

We denote the identity element of the group as e and the Lie algebra associated with the group G as $\mathfrak{g}$. The Lie algebra $\mathfrak{g}$ is defined to be TG_e, the tangent space at the identity element. The Lie algebraic properties of $\mathfrak{g}$ are immediate from the Lie algebra of vector fields by restricting our attention to the single point $e \in G$. A given Lie algebra may be associated with several Lie groups, since $\mathfrak{g} = TG_e$ can only determine the properties of the group in a local neighborhood of e. The global properties of several groups may be quite different even though their properties in the neighborhood of the identity element are the same.

One-Parameter Subgroups in a Neighborhood of *e*

The basis of TG_e gives a coordinate mesh in the neighborhood of the identity in G. This coordinate mesh consists of the flow lines of the basis vectors in this neighborhood. These flow lines individually correspond to one-parameter groups of diffeomorphisms. These groups are all one-parameter subgroups of G and are curves representing one-parameter diffeomorphisms in the neighborhood of the identity. This one-parameter mapping is generated by the group composition law. For a sketch depicting some one-parameter subgroups, see Fig. 10.1. The one-parameter subgroups all pass through the identity element since all subgroups of G must contain e.

Let λ denote the path parameter of a one-parameter subgroup and let $a(\lambda)$ denote an arbitrary element in this subgroup. We can always choose the parameterization of such a curve so that $a(\lambda_1)a(\lambda_2) = a(\lambda_1 + \lambda_2)$. [For a proof see Saletan and Cromer (1972).] Since $\lambda_1 + \lambda_2 = \lambda_2 + \lambda_1$ such a parameterization shows that these one-parameter subgroups are commutative. Commutative groups are termed *Abelian*.

Since these one-parameter subgroups are Abelian, we denote the group composition law for these one-parameter subgroups in the form

$$a(\lambda_1)a(\lambda_2) = a(\lambda_2)a(\lambda_1) = a(\lambda_1 + \lambda_2). \tag{10.46}$$

As curves in G, the one-parameter subgroups all pass through e and at the identity such a curve has a tangent vector $\mathbf{v} \in \mathfrak{g} = TG_e$. Because of the Abelian nature of this one-parameter subgroup it is convenient to denote the elements in the subgroup as $a(\lambda) = \exp(\lambda\mathbf{v})$. At this point $\exp(\lambda\mathbf{v})$ is not to be viewed as the exponentiation of anything, but rather a convenient notation for the elements of a one-parameter, Abelian subgroup of G, with group composition law

$$\exp(\lambda_1\mathbf{v})\exp(\lambda_2\mathbf{v}) = \exp((\lambda_1 + \lambda_2)\mathbf{v}), \tag{10.47}$$

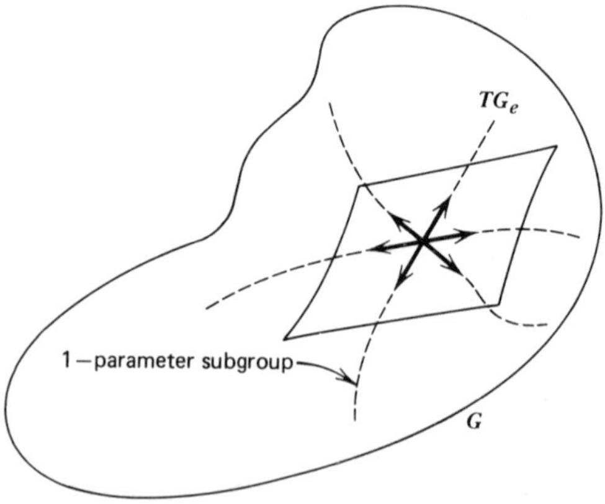

FIGURE 10.1. The tangent space TG_e at the identity e and some one-parameter subgroups corresponding to basis elements of $\mathfrak{g} = TG_e$.

and with tangent vector $\mathbf{v}$ at the identity given by

$$\left.\frac{d}{d\lambda}\right|_{\lambda=0} \exp(\lambda\mathbf{v}) = \mathbf{v}. \tag{10.48}$$

We can view "exp" as a mapping from the lines $\lambda\mathbf{v}$ in TG_e to one-parameter subgroups in G.

$$\exp\colon TG_e \to G, \tag{10.49}$$

where $\exp(\mathbf{v}) = a_{\mathbf{v}}(1)$. Since at $t = 0$, $a_{\lambda\mathbf{v}}(t) = a_{\mathbf{v}}(\lambda t)$ and

$$\frac{d}{dt}\left[a_{\lambda\mathbf{v}}(t)\right] = \lambda\mathbf{v} = \lambda\frac{d}{d\tau}a_{\mathbf{v}}(\tau) = \frac{d}{dt}\left(a_{\mathbf{v}}(\lambda t)\right)$$

we have that $\exp(\lambda\mathbf{v}) = a_{\lambda\mathbf{v}}(1) = a_{\mathbf{v}}(\lambda)$. Consequently,

$$\exp((\lambda_1 + \lambda_2)\mathbf{v}) = a_{\mathbf{v}}(\lambda_1 + \lambda_2) = a_{\mathbf{v}}(\lambda_1)a_{\mathbf{v}}(\lambda_2) = \exp(\lambda_1\mathbf{v})\exp(\lambda_2\mathbf{v}).$$

We emphasize that $\exp(\lambda\mathbf{v})$ denotes a group element and not an operator as earlier described in this chapter. We have chosen to use the notation $\exp(\lambda\mathbf{v})$, as opposed to $e^{\lambda\mathbf{v}}$, to remind the reader of this distinction.

Adjoint Representation

If the realization of a group is in terms of linear mappings on a vector space, then we term such a realization a *representation*.

Representations of dynamical symmetries find numerous applications in the study of dynamical systems, be they classical or quantum mechanical. It is always possible to construct such a linear representation. Doing so in the following paragraphs, we meet once again the adjoint algebra, but in a new guise. First we discuss some mappings induced by the group composition law.

Every element in a Lie group G induces mappings on G called left and right translation. Let $g \in G$ be an arbitrary element of G. Then $l_g\colon G \to G$ and $r_g\colon G \to G$ are defined by $l_g(h) = gh$ and $r_g(h) = hg$ respectively. These mappings are differentiable because the composition law for Lie groups is differentiable.

Using the right and left translation mappings, we define for each $g \in G$ the mapping $\mathscr{A}_g\colon G \to G$ given by $\mathscr{A}_g(h) = (r_{g^{-1}} \circ l_g)(h) = ghg^{-1}$. The mapping $\mathscr{A}_g$ leaves the identity element of the group fixed. Thus the derivative map $\mathscr{A}_{g*}$ maps $\mathfrak{g} = TG_e$ into itself and is known to be a linear map. We denote the space of linear operators on a vector space A by $L(A)$. We thus have a mapping from the group G into the space of linear maps on

$\mathscr{g}$, that is, $\mathscr{A}: G \to L(\mathscr{g})$ given by $\mathscr{A}(g) = \mathscr{A}_{g*e}$ (the derivative map of $\mathscr{A}_g$ at the identity).

To show that we have a representation for G, we need only show that $\mathscr{A}$ preserves group multiplication. Let $g, h \in G$ be arbitrary elements and let $\mathbf{v}$ be an arbitrary element in $\mathscr{g}$.

$$\begin{aligned}
\mathscr{A}(gh)[\mathbf{v}] = \mathscr{A}_{gh*e}[\mathbf{v}] &= \frac{d}{dt}\bigg|_{t=0} \mathscr{A}_{gh}(\exp(t\mathbf{v})) \\
&= \frac{d}{dt}\bigg|_{t=0} \{(r_{(gh)^{-1}} \circ l_{gh})(\exp(t\mathbf{v}))\} \\
&= \frac{d}{dt}\bigg|_{t=0} \{gh \exp(t\mathbf{v})(gh)^{-1}\} = \frac{d}{dt}\bigg|_{t=0} \{\mathscr{A}_g \circ \mathscr{A}_h(\exp(t\mathbf{v}))\} \\
&= \mathscr{A}_{ge*} \circ \mathscr{A}_{he*}[\mathbf{v}] = \mathscr{A}(g) \circ \mathscr{A}(h)[\mathbf{v}].
\end{aligned}$$

Since $\mathbf{v}$ is an arbitrary element

$$\mathscr{A}(gh) = \mathscr{A}(g) \circ \mathscr{A}(h).$$

The group composition law in $L(\mathscr{g})$ is mapping composition which is matrix multiplication. Thus, we have shown that the mapping $\mathscr{A}: G \to L(\mathscr{g})$ is a group homomorphism and provides a representation in terms of linear operators. This representation is termed the *adjoint representation* of the group G. From this choice of name one must suspect a connection with the former adjoint algebra. This connection we proceed to establish and elucidate the adjoint representation.

Recall from (10.10) that the adjoint algebra is an algebra of linear operators on vector fields. When an adjoint operator $\tilde{\mathbf{u}}$ is restricted to the identity element, it is a linear operator on TG_e. We show that the operators $\tilde{\mathbf{u}}$ are intimately related to those elements of $L(\mathscr{g})$ obtained by the group homomorphism $\mathscr{A}: G \to L(\mathscr{g})$.

Let $\mathbf{v}$ be an arbitrary element of $\mathscr{g}$ and let $\exp(t\mathbf{v})$ be the corresponding one-parameter subgroup of G. Let a be an arbitrary element of G such that $a = \exp(\mathbf{w})$. Then $\mathscr{A}(a) \in L(\mathscr{g})$ and can operate on $\mathbf{v}$.

$$\begin{aligned}
\mathscr{A}(a)[\mathbf{v}] = \mathscr{A}_{a*e}[\mathbf{v}] &= \frac{d}{dt}\bigg|_{t=0} \{\mathscr{A}_a(\exp(t\mathbf{v}))\} \\
&= \frac{d}{dt}\bigg|_{t=0} \{a(\exp(t\mathbf{v}))a^{-1}\} = \frac{d}{dt}\bigg|_{t=0} \{\exp(\mathbf{w})\exp(t\mathbf{v})\exp(-\mathbf{w})\}.
\end{aligned} \tag{10.50}$$

To combine the exponentials in (10.50), we use the Baker–Campbell–Hausdorff (BCH) formula. The BCH formula (10.52), and its appropriate application in the present context, is proved in many places, including Sudarshan and Mukunda (1974). The BCH formula is that for arbitrary vectors (operators) **u**, **v**, and arbitrary real parameters a, b

$$e^{a\mathbf{u}}e^{b\mathbf{v}} = e^{\mathbf{w}}, \tag{10.51}$$

where

$$\mathbf{w} = a\mathbf{u} + b\mathbf{v} + \frac{ab}{2}[\mathbf{u},\mathbf{v}] + \frac{a^2b}{12}[\mathbf{u},[\mathbf{u},\mathbf{v}]] + \frac{ab^2}{12}[\mathbf{v},[\mathbf{v},\mathbf{u}]] + \cdots. \tag{10.52}$$

Applying the BCH formulas (10.51) and (10.52) to (10.50), paying careful attention to the order of t, we find after successive application

$$\exp(\mathbf{w})\exp(t\mathbf{v})\exp(-\mathbf{w}) = \exp\big(t\mathbf{u} + 0(t^2)\big),$$

where

$$\mathbf{u} = \mathbf{v} + [\mathbf{w},\mathbf{v}] + \frac{1}{2!}[\mathbf{w},[\mathbf{w},\mathbf{v}]] + \cdots = \sum_{n=0}^{\infty}\frac{1}{n!}\tilde{\mathbf{w}}^n[\mathbf{v}] = e^{\tilde{\mathbf{w}}}[\mathbf{v}]. \tag{10.53}$$

In obtaining (10.53) we have used the series representation for the exponential of an operator as used in (5.24). The vector $t\mathbf{u} + O(t^2)$ is some element of $TG_e = \mathscr{g}$ and so the exp mapping applied to this vector makes sense. The tangent vector to the corresponding one-parameter subgroup will agree with the vector coefficient of the term linear in t. Thus we have from (10.50) and (10.53)

$$\mathscr{A}(a)[\mathbf{v}] = \mathscr{A}(\exp(\mathbf{w}))[\mathbf{v}] = e^{\tilde{\mathbf{w}}}[\mathbf{v}].$$

Since **v** is arbitrary we have

$$\mathscr{A}(\exp(\mathbf{w})) = e^{\tilde{\mathbf{w}}}. \tag{10.54}$$

This establishes the relationship between the adjoint algebra of operators $\tilde{\mathbf{u}}, \tilde{\mathbf{v}}, \tilde{\mathbf{w}}, \ldots,$ and the adjoint representation of the group elements $\exp(\mathbf{v}), \exp(\mathbf{u}), \exp(\mathbf{w}), \ldots.$ Since all elements in a neighborhood of the identity can be written in the form exp(**v**), all elements of the adjoint representation may be given as $e^{\tilde{\mathbf{v}}}$.

Algebra of Angular Momentum and Rotations

We return to a concrete example. Let a phase space have coordinates given by cartesian components of the position vector and momentum, that is, (p_i, x_i), and let f be an arbitrary function. We examine the algebra generated by the components of the angular momentum per unit mass.

$$L_i = [ijk]x_j p_k, \tag{10.55}$$

and

$$\{L_i, L_j\} = [ijk]L_k, \tag{10.56}$$

where $i, j, k = 1, 2, 3$ and repeated indices are summed as usual. The symbol $[ijk]$ is the permutation symbol of (1.35).

This is a subalgebra of the algebra formed by all analytic functions on phase space with the Poisson bracket as the Lie bracket. We form the adjoint algebra and exponentiate to construct the adjoint representation of the Lie group corresponding to the algebra of (10.56). From the algebra of (10.56) we obtain the algebra of the vector fields. We denote the flow vector of L_i as $\mathbf{L}_i$.

$$\begin{aligned}\mathbf{L}_i(f) = \{f, L_i\} &= -\{[ijk]x_j p_k, f\} = -[ijk]\big(x_j\{p_k, f\} + p_k\{x_j, f\}\big)\\ &= [ijk]\left(x_j\frac{\partial}{\partial x_k} + p_j\frac{\partial}{\partial p_k}\right)[f].\end{aligned} \tag{10.57}$$

Since f is an arbitrary function on phase space,

$$\mathbf{L}_i = [ikj]\left(x_j\frac{\partial}{\partial x_k} + p_j\frac{\partial}{\partial p_k}\right) \equiv \mathbf{X}_i + \mathbf{P}_i. \tag{10.58}$$

The operators $\mathbf{X}_i = [ikj]x_j(\partial/\partial x_k)$ and $\mathbf{P}_i = [ikj]p_j(\partial/\partial p_k)$ commute and thus

$$e^{t\mathbf{L}_i} = e^{t(\mathbf{X}_i + \mathbf{P}_i)} = e^{t\mathbf{X}_i}e^{t\mathbf{P}_i} = e^{t\mathbf{P}_i}e^{t\mathbf{X}_i}. \tag{10.59}$$

The exponentiated operator gives

$$e^{t\mathbf{X}_i}[f] = \sum_{n=0}^{\infty}\frac{t^n}{n!}\mathbf{X}_i^n[f]. \tag{10.60}$$

To see the meaning of this, we let $i = 3$. Then

$$\mathbf{X}_3[f] = \left(y\frac{\partial}{\partial x} - x\frac{\partial}{\partial y}\right)[f].$$

If we consider an infinitesimal rotation about the z-axis through an angle t, then from (3.42) we have that $x' = x - ty$, $y' = y + tx$, $z' = z$. From $f(x', y', z')$ we find

$$\left.\frac{df}{dt}\right|_{t=0} = \left(x\frac{\partial}{\partial y} - y\frac{\partial}{\partial x}\right)f(x, y, z) = -\mathbf{X}_3[f]. \tag{10.61}$$

Thus

$$e^{t\mathbf{X}_3}[f] = f(x + ty, y - tx, z); \tag{10.62}$$

is an infinitesimal rotation through the angle $-t$ about the z-axis. We see that the one-parameter subgroups correspond to the rotations; that is, $e^{t\mathbf{X}_i}$ corresponds to a one-parameter group of rotations about the ith axis in coordinate space. The one-parameter group $e^{t\mathbf{P}_i}$ does exactly the same thing for the momentum variables. The $\mathbf{L}_i$ operator is the generator of a rotation about the ith axis.

Let us now examine the adjoint algebra and its corresponding representation of the group obtained by exponentiation. Denote the adjoint operators as $\tilde{\mathbf{L}}_i$. Recall that the operators $\tilde{\mathbf{L}}_i$ are linear operators on the vectors in TG_e and thus are matrices. The matrices we can obtain by operating with $\tilde{\mathbf{L}}_i$ on the individual basis elements $\{\mathbf{e}_{x_i}, \mathbf{e}_{p_i}\}$.

$$\tilde{\mathbf{L}}_i[\mathbf{e}_{x_j}] = \left[\mathbf{L}_i, \mathbf{e}_{x_j}\right]. \tag{10.63}$$

Using (10.58) in (10.63) and allowing $\tilde{\mathbf{L}}_i[\mathbf{e}_{x_j}]$ to operate on an arbitrary function shows readily that

$$\tilde{\mathbf{L}}_i[\mathbf{e}_{x_j}] = -[ijk]\frac{\partial}{\partial x_k}. \tag{10.64}$$

In a similar fashion we obtain

$$\tilde{\mathbf{L}}_i[\mathbf{e}_{p_j}] = -[ijk]\frac{\partial}{\partial p_k}. \tag{10.65}$$

If we denote an arbitrary vector $\mathbf{u}$ as $\mathbf{u} = u_x^i\mathbf{e}_{x_i} + u_p^i\mathbf{e}_{p_i}$, $i = 1, 2, 3$ then (10.64) and (10.65) together show

$$\tilde{\mathbf{L}}_i[\mathbf{u}] = -[ijk]\left(u_x^j\frac{\partial}{\partial x_k} + u_p^j\frac{\partial}{\partial p_k}\right). \tag{10.66}$$

Let us be specific again by choosing $i = 3$. Then

$$\tilde{\mathbf{L}}_3 = \begin{bmatrix} 0 & a_3 \\ a_3 & 0 \end{bmatrix}, \qquad \text{where } a_3 = \begin{bmatrix} 0 & 1 & 0 \\ -1 & 0 & 0 \\ 0 & 0 & 0 \end{bmatrix}. \tag{10.67}$$

$$(\tilde{\mathbf{L}}_3)^2 = \begin{bmatrix} a_3^2 & 0 \\ 0 & a_3^2 \end{bmatrix} \quad \text{and} \quad a_3^2 = \begin{bmatrix} -1 & 0 & 0 \\ 0 & -1 & 0 \\ 0 & 0 & 0 \end{bmatrix}, \tag{10.68}$$

and

$$(\tilde{\mathbf{L}}_3)^3 = -\tilde{\mathbf{L}}_3. \tag{10.69}$$

To find the adjoint representation $\mathscr{A}$, we exponentiate $\tilde{\mathbf{L}}_i$. Again we choose $i = 3$, and use (10.67)–(10.69).

$$\begin{aligned} e^{t\tilde{\mathbf{L}}_3} &= \sum_{n=0}^{\infty} \frac{t^n(\tilde{\mathbf{L}}_3)^n}{n!} \\ &= \left[1 + t\tilde{\mathbf{L}}_3 + \frac{t^2}{2!}(\tilde{\mathbf{L}}_3)^2 - \frac{t^3}{3!}\tilde{\mathbf{L}}_3 - \frac{t^4}{4!}(\tilde{\mathbf{L}}_3)^2 + \cdots\right] \end{aligned}$$

Using $R_z(t)$ to denote the rotation matrix in (3.42) about the z-axis, we find that

$$e^{t\tilde{\mathbf{L}}_3} = \begin{bmatrix} R_z(-t) & 0 \\ 0 & R_z(-t) \end{bmatrix}. \tag{10.70}$$

Again we see the group to which the algebra of the $\tilde{\mathbf{L}}_i$ belongs is the rotation group.

As we step back and look at this example we see that the Lie group in question is a transformation group that transforms points in phase space into new points in phase space. A group is an abstract, algebraic quantity and to be more precise we should say that the group is being realized as a transformation group on phase space. This realization of the group leads us to a representation of the group as we have illustrated. In the preceding example of the rotation group, we began with a realization of the Lie algebra, which then led us to the adjoint representation.

Realization by Lie Transformations

To conclude this section and to further see the relationship with a previous discussion concerning invariants of Hamiltonian flow and Lie transformations, we consider the one-dimensional subalgebra on phase space generated by an arbitrary function h. According to (10.2) the one-dimensional vector algebra is generated by

$$\mathbf{v}_h = \frac{\partial h}{\partial p_i}\frac{\partial}{\partial q^i} - \frac{\partial h}{\partial q^i}\frac{\partial}{\partial q_i}. \tag{10.71}$$

The one-parameter subgroup associated with $\mathbf{v}_h$ is $\exp(\lambda \mathbf{v}_h)$. The Lie transformation associated with $\mathbf{v}_h$ is E_h, and since $\mathbf{v}_h$ is independent of λ, we can write the solution to (10.21) in the form $e^{\lambda \mathbf{v}_h}$. The product of two such operators by the BCH formula is

$$e^{\lambda_1 \mathbf{v}_h} e^{\lambda_2 \mathbf{v}_h} = e^{(\lambda_1 + \lambda_2)\mathbf{v}_h}$$

and there is an obvious isomorphism between the one-parameter group on phase space and the one-parameter group of operators. So we study the one-parameter group on phase space by studying the one-parameter group of Lie transformation operators.

Consider this exponential operator to operate on the functions q^j, p_j, and an arbitrary function $f(\mathbf{p}, \mathbf{q})$. Using (10.21)

$$q^j(\lambda) = e^{\lambda \mathbf{v}_h} q^j(0) \quad \text{and} \quad \frac{dq^j}{d\lambda} = \mathbf{v}_h[q^j] = \frac{\partial h}{\partial p_j}. \tag{10.72}$$

$$p_j(\lambda) = e^{\lambda \mathbf{v}_h} p_j(0) \quad \text{and} \quad \frac{dp_j}{d\lambda} = \mathbf{v}_h[p_j] = -\frac{\partial h}{\partial q^j}. \tag{10.73}$$

$$f(\lambda) = e^{\lambda \mathbf{v}_h} f(0) \quad \text{and} \quad \frac{df}{d\lambda} = \mathbf{v}_h[f]. \tag{10.74}$$

From (10.72)–(10.74) we have

$$\frac{df}{d\lambda} = \mathbf{v}_h[f] = \frac{dp_i}{d\lambda}\frac{\partial f}{\partial p_i} + \frac{dq^i}{d\lambda}\frac{\partial f}{\partial q^i}. \tag{10.75}$$

Thus we have that $f(\lambda) = e^{\lambda \mathbf{v}_h}[f] = f(\mathbf{p}(\lambda), \mathbf{q}(\lambda))$. Since f was arbitrary, for all functions on phase space we find the transformed f at $(\mathbf{p}, \mathbf{q})$ by looking at f (untransformed) at the transformed point $(\mathbf{p}(\lambda), \mathbf{q}(\lambda))$.

If we interpret h as the Hamiltonian and λ as t (time), then (10.75) becomes

$$\frac{df}{dt} = \mathbf{v}_h[f] = \{f, h\}, \tag{10.76}$$

which is the familiar equation of motion for f.

Therefore an alternative way of stating the fundamental problem of dynamics is that we seek the one-parameter subgroup of transformations on phase space that take the initial point $(\mathbf{p}(0), \mathbf{q}(0))$ to the point $(\mathbf{p}(t), \mathbf{q}(t))$.

If we consider a second one-parameter subgroup generated by f, and since,

$$\mathbf{v}_f[h] = -\mathbf{v}_h[f], \tag{10.77}$$

then f can be constant along one-parameter subgroups generated by h iff h is constant along one-parameter subgroups generated by f. Note further that the result for canonical transformations proved in (10.31) applies to $e^{t\mathbf{v}_h}$ and provides another view of the result that the Hamiltonian flow is a canonical transformation.

10.6. GROUPS AS CONFIGURATION SPACE

The last section was concerned with groups as represented by canonical transformations on phase space, and their representations. Groups can also enter dynamics by corresponding to the configuration manifold of the system. In this section we discuss the relationship between the geometrical quantities and the dynamical variables. We also add that not many systems have been studied in this way, that is, by identifying the configuration manifold with a Lie group. We exclusively consider the example of $SO(3)$ and rigid-body dynamics. More general considerations can be found in various monographs and research papers (Arnold, 1978; Miscenko and Fomenko, 1978; Ratiu, 1980; Dikii, 1972; Holm, 1981; Manakov, 1976).

Before considering $SO(3)$ specifically, we make some observations and establish notation. Since the Lie group manifold corresponds directly with the configuration space of the system, tangent spaces relate to velocity vectors and cotangent spaces to momentum vectors, as is true for a general configuration-space manifold.

The dual vector space to $\mathfrak{g} = TG_e$, we denote as $\mathfrak{g}^* = T^*G_e$. This is the space of linear, real-valued functionals on $\mathfrak{g}$. As before we refer to elements of $\mathfrak{g}^*$ as covectors. For $\xi \in \mathfrak{g}^*$, then $\xi\colon \mathfrak{g} \to \mathbb{R}$. For $\mathbf{v} \in \mathfrak{g}$ we denote the real number $\xi(\mathbf{v})$ in the form $\langle \xi, \mathbf{v} \rangle$, using angle brackets.

Left and right translation mappings on the group, as discussed in previous sections, lead to the adjoint representation on $\mathfrak{g}$. Similarly, we can obtain group representations on the dual space $\mathfrak{g}^*$. As the derivative map $\mathscr{A}_{g*}$ was used before, we now use the pull-back map on covectors. This map was defined for forms or covectors in Eq. (8.46). If l_g denotes left translation on G by $g \in G$, then the mapping $l^*_{gh}\colon T^*G_{gh} \to T^*G_h$ is defined by

$$\langle l^*_{gh}(\xi), \mathbf{v} \rangle = \langle \xi, l_{g*h}(\mathbf{v}) \rangle. \tag{10.78}$$

We remind the reader of the meaning in the symbols making up l^*_{gh}. The part l_g denotes left translation by g. The * as superscript denotes the pull-back map, the subscripted $_*$ denotes the derivative map. The subscript h denotes manifold (Lie group) point where the mappings on the tangent and cotangent spaces are relevant.

We have a similar definition for the pull-back map induced from right translation r_g. Once again we form the composition mapping $\mathscr{A}_g = r_{g^{-1}} \circ l_g$ and have the corresponding pull-back map $\mathscr{A}^*_g$, which maps $\mathfrak{g}^*$ into itself. The map $\mathscr{A}^*_g$ is linear and so is an element of $L(\mathfrak{g}^*)$, the linear maps on $\mathfrak{g}^*$. Hence, we have a mapping, denoted $\mathscr{A}^*$, from the group G into $L(\mathfrak{g}^*)$, $\mathscr{A}^*\colon G \to L(\mathfrak{g}^*)$. This map satisfies $\mathscr{A}^*(gh) = \mathscr{A}^*(h) \circ \mathscr{A}^*(g)$ and is called the *co-adjoint representation*.

Let us consider specifically $G = SO(3)$, the configuration manifold for a rigid body, which is free to move without forces acting on it. A point $g_0 \in G$ denotes a particular orientation the body has achieved in its motion. This motion is represented by a path or curve $g(t)$ in G. At g_0 this curve has some tangent vector $\boldsymbol{\omega}$. This vector $\boldsymbol{\omega}$ can be carried back from TG_{g_0} to $\mathfrak{g}$ by either right of left translation. We show below that right translation gives the angular velocity in the space system S, and left translation gives the angular velocity in the moving (body) system M.

As discussed in the paragraphs prior to (3.42), the order of the application of two successive rotations determines whether we are calculating components in the stationary or moving frame: R_2R_1 denotes stationary frame, and R_1R_2 denotes moving frame. Consider the one-parameter group $\exp(t\boldsymbol{\omega})$ and consider the curve in $SO(3)$ given by $g(t) = \exp(t\boldsymbol{\omega})g_0$. The order of this product implies that the rotations are being viewed in the space system. The tangent to this curve at g_0 is

$$\dot{g} = \left.\frac{d}{dt}\right|_{t=0} (\exp(t\boldsymbol{\omega})g_0).$$

Translated back to TG_e, $\dot{g}$ must be the angular velocity vector in the space system. We calculate

$$r_{g_0^{-1}*}(\dot{\mathbf{g}}) = \frac{d}{dt}\bigg|_{t=0} r_{g_0^{-1}}\big(\exp(t\boldsymbol{\omega})g_0\big) = \boldsymbol{\omega}. \tag{10.79}$$

We can demonstrate these results explicitly by considering $SO(3)$ as parameterized using the Euler angles. An arbitrary element of $SO(3)$ is given by the rotation matrix (3.43). Let f be an arbitrary function, then

$$\boldsymbol{\omega}[f] = \frac{d}{dt}\bigg|_{t=0} f\big(r_{g_0^{-1}}(g(t))\big). \tag{10.80}$$

The element g_0 denotes a fixed rotation R_0, $g(t)$ denotes an arbitrary rotation away from R_0, namely, $R(t)R_0$. Changing to this notation in terms of rotation operators

$$\boldsymbol{\omega}[f] = \frac{d}{dt}\bigg|_{t=0} f\big(R(t)R_0R_0^{-1}\big) = \frac{d}{dt}\bigg|_{t=0} f(R(t)). \tag{10.81}$$

Consider the following choices for f: $f = \pi^3{}_1$, $\pi^2{}_3$, and $\pi^1{}_3$, where $\pi^i{}_j$ is the projection map on matrices introduced in Chapter 6 prior to (6.10).

$$\frac{d}{dt}\bigg|_{t=0} \pi^3{}_1(R) = \dot{\theta}\sin\psi\cos\theta + \dot{\psi}\sin\theta\cos\psi. \tag{10.82}$$

$$\frac{d}{dt}\bigg|_{t=0} \pi^2{}_3(R) = -\dot{\theta}\cos\phi\cos\theta + \dot{\phi}\sin\phi\sin\theta. \tag{10.83}$$

$$\frac{d}{dt}\bigg|_{t=0} \pi^1{}_3(R) = \dot{\theta}\sin\phi\cos\theta + \dot{\phi}\cos\phi\sin\theta. \tag{10.84}$$

A set of basis vectors in the space (S) system, is the set of vectors $(\mathbf{e}_x, \mathbf{e}_y, \mathbf{e}_z)$ of (6.17)–(6.19). Operating on the functions $R^3{}_1$, $R^2{}_3$, and $R^1{}_3$, we find

$$\mathbf{e}_x\big(R^3{}_1\big) = \cos\phi\cos\theta\sin\psi + \sin\phi\cos\psi;$$

$$\mathbf{e}_y\big(R^3{}_1\big) = \sin\phi\cos\theta\sin\psi - \cos\phi\cos\psi; \qquad \mathbf{e}_z\big(R^3{}_1\big) = 0.$$

$$\mathbf{e}_x\big(R^2{}_3\big) = -\cos\theta; \qquad \mathbf{e}_y\big(R^2{}_3\big) = 0; \qquad \mathbf{e}_z\big(R^2{}_3\big) = \sin\phi\sin\theta.$$

$$\mathbf{e}_x\big(R^1{}_3\big) = 0; \qquad \mathbf{e}_y\big(R^1{}_3\big) = \cos\theta; \qquad \mathbf{e}_z\big(R^1{}_3\big) = \sin\theta\cos\phi.$$

Comparing $(\omega^1\mathbf{e}_x + \omega^2\mathbf{e}_y + \omega^3\mathbf{e}_z)[f]$ and (10.82)–(10.84) for $R^3{}_1$, $R^2{}_3$, and $R^1{}_3$, we obtain three simultaneous equations for the components $(\omega^1, \omega^2, \omega^3)$. The solution of these equations gives (4.20), as it should.

The curve $g_0\exp(t\boldsymbol{\Omega})$ represents the product of the rotation $\exp(t\boldsymbol{\Omega})$ with g_0, but now in the moving system because of the reverse order. We can show explicitly (see Problem 10.8) that the curve $g_0\exp(t\boldsymbol{\Omega})$ has angular velocity vector $\boldsymbol{\Omega}$ in the moving system, by translating the tangent vector back to $TG_e = \mathfrak{g}$ using $l_{g_0^{-1}*}$.

The moment-of-inertia tensor for the rigid body supplies a positive definite (kinetic energy $\geqslant 0$) metric for the manifold $SO(3)$. As in Chapter 4, we denote this operator as K. Then $K\dot{\mathbf{g}}$ is a covector since $2T = \langle K\dot{\mathbf{g}}, \dot{\mathbf{g}}\rangle = \dot{\mathbf{g}} \cdot K\dot{\mathbf{g}}$ [cf. (4.9) and (4.10)]. The covector $K\dot{\mathbf{g}}$ is the angular momentum for the rigid body and lies in the cotangent space $T^*G_{g_0}$ when $\dot{\mathbf{g}} \in TG_{g_0}$.

As with the earlier spatial metric, g_{ij}, the moment-of-inertia operator provides a mapping (isomorphism) between the tangent spaces and the cotangent spaces, K: $TG_g \to T^*G_g$. The angular momentum vectors (covectors) for a point g_0 can also be carried back to $\mathfrak{g}^*$ using the pull-back maps from left and right translation. These images are the angular momentum in the moving (body) and stationary (space) systems respectively.

$$\langle r^*_{g_0}(K\dot{\mathbf{g}}), \boldsymbol{\omega}\rangle = \langle K\dot{\mathbf{g}}, r_{g_0*}\boldsymbol{\omega}\rangle = \langle K\dot{\mathbf{g}}, \dot{\mathbf{g}}\rangle = 2T.$$

Thus $r^*_{g_0}(K\dot{\mathbf{g}}) = K\boldsymbol{\omega}$, which is the angular momentum in the S system.

As interesting as these ideas and their generalizations are, we do not pursue further here the Lie group formulation of rigid body dynamics, but refer the interested reader to the references given in the first paragraph of this section.

PROBLEMS

10.1. Show that the Lie bracket given by (10.5) satisfies the Jacobi identity:

$$[\mathbf{v}, [\mathbf{u}, \mathbf{w}]] + [\mathbf{u}, [\mathbf{w}, \mathbf{v}]] + [\mathbf{w}, [\mathbf{v}, \mathbf{u}]] = 0.$$

10.2. Using the definition for the Lie derivative of a form as given in (10.19), specialize this to a 1-form $\boldsymbol{\omega}$ and show that

$$L_{\mathbf{v}}\boldsymbol{\omega}[\mathbf{u}] = L_{\mathbf{v}}[\boldsymbol{\omega}[\mathbf{u}]] - \boldsymbol{\omega}[L_{\mathbf{v}}\mathbf{u}],$$

where $\mathbf{u}$ is another vector field.

10.3. Show that the maps $F \to -\mathbf{v}_F$ and $\mathbf{v} \to \tilde{\mathbf{v}}$ are Lie algebra homomorphisms. Are they isomorphisms?

10.4. Show that the polynomials $g_2 = x^1 x^2 \ln \lambda$ and $g_3 = (-1/3)(x^1/\lambda - \lambda x^2)^3$ generate the flow vectors

$$\mathbf{v}_2 = \ln \lambda \left(x^1 \frac{\partial}{\partial x^1} - x^2 \frac{\partial}{\partial x^2} \right) \qquad \text{and}$$

$$\mathbf{v}_3 = \left(\frac{x^1}{\lambda} - \lambda x^2 \right)^2 \left(\lambda \frac{\partial}{\partial x^1} + \frac{1}{\lambda} \frac{\partial}{\partial x^2} \right)$$

respectively. Show that through (10.29) these vectors generate the canonical transformation

$$y^1 = \lambda\left[x^1 + (x^1 - x^2)^2\right]; \qquad y^2 = \frac{1}{\lambda}\left[x^2 + (x^1 - x^2)^2\right].$$

Verify explicitly that this transformation is canonical.

10.5. Show that the group of transformations on phase space generated by the subalgebra with basis elements $\{q^i, p_i, 1\}$ is the group of translations.

10.6. Show that the co-adjoint map satisfies $\mathscr{A}^*_{gh} = \mathscr{A}^*_h \circ \mathscr{A}^*_g$.

10.7. Consider the product of two rotations in the form for components in the moving system, $g_0 \exp(t\boldsymbol{\Omega})$, such that this curve has the same tangent $\dot{\mathbf{g}}$ at g_0 as the given curve $g(t)$. Show that left translation of $\dot{\mathbf{g}}$ to $\mathscr{g}$ gives $\boldsymbol{\Omega}$.

10.8. Consider the matrix representation of $SO(3)$ given by the three-dimensional rotation matrices, where an arbitrary element is given by the rotation matrix (3.43). Show that a basis for the tangent space at the identity is given by the matrices

$$m_x = \begin{bmatrix} 0 & 0 & 0 \\ 0 & 0 & -1 \\ 0 & 1 & 0 \end{bmatrix}; \qquad m_y = \begin{bmatrix} 0 & 0 & 1 \\ 0 & 0 & 0 \\ -1 & 0 & 0 \end{bmatrix};$$

$$m_z = \begin{bmatrix} 0 & -1 & 0 \\ 1 & 0 & 0 \\ 0 & 0 & 0 \end{bmatrix}.$$

Show that these are the same matrices as $\tilde{\mathbf{e}}_x, \tilde{\mathbf{e}}_y, \tilde{\mathbf{e}}_z$ of the adjoint representation, where $\mathbf{e}_x, \mathbf{e}_y, \mathbf{e}_z$ are given in (6.17)–(6.19). By using this matrix representation show that $\Omega^i m_i = l_{g_0^{-1}*}(\dot{\mathbf{g}})$ gives the components Ω^i given by (4.21), where $\dot{\mathbf{g}}$ is the tangent vector to the curve $g(t)$ at g_0.

10.9. Consider the Hamiltonian for a perturbed simple harmonic oscillator written in the canonical (action-angle) variables (I, θ):

$$H = I - \tfrac{1}{6} I^2 \cos^4 \theta.$$

In this case the energy is small and approximately equal to I. Thus I itself may be taken as the small expansion parameter. Find to lowest order, by Lie perturbation techniques, the canonical transformation that eliminates all but the lowest-order term in the preceding Hamiltonian.

REFERENCES

Arnold, V. I., *Mathematical Methods of Classical Mechanics*, Springer-Verlag, New York, 1978.

Berry, M. V., "Nearly Integrable and Integrable Systems," in S. Jorna, *Topics in Nonlinear Dynamics*, American Institute of Physics, New York, 1978.

Cary, J. R., *J. Math. Phys.* **18**, 2432, 1977.

Chirikov, B. V., *Phys. Rep.* **52**, 263, 1979.

Choquet-Bruhat, Y., and C. DeWitt-Morette, *Analysis, Manifolds, and Physics*, North-Holland, New York, 1982.

Deprit, A., J. Henrad, J. Price, and A. Rom, *Celest. Mech.* **1**, 222, 1962.

Dewar, R. L., *J. Phys.* **A9**, 2043, 1976.

Dikii, L. A., *Funk. Anal. Ego Priloz.* **6**, 83, 1972.

Flanders, H., *Differential Forms*, Academic, New York, 1963.

Ford, J., "The Statistical Mechanics of Classical Analytic Dynamics," in E. G. D. Cohen, ed., *Fundamental Problems in Statistical Mechanics*, Vol. III, North-Holland, Amsterdam, 1975.

Goldstein, H., *Classical Mechanics*, Addison-Wesley, Reading Mass., 1980.

Henon, M., *Quart. Appl. Math.* **27**, 291, 1969.

Henon, M., and C. Heiles, *Astron. J.* **69**, 73, 1964.

Hermann, R., *Differential Geometry and the Calculus of Variations*, Academic, New York, 1968.

Kaufman, A. N., "The Lie Transform: A New Approach to Classical Perturbation Theory," in S. Jorne, ed., *Topics in Nonlinear Dynamics*, American Institute of Physics, New York, 1978.

Manakov, S. V., *Funk. Anal. Ego Priloz.* **10**, 93, 1976.

Miscenko, A. S., *Izvestija* **12**, 371, 1978.

Misner, C. W., K. S. Thorne, and J. A. Wheeler, *Gravitation*, W. H. Freeman, San Francisco, 1973.

Poincaré, H., *Les Methodes Nouvelles de la Mecanique Celeste*, Gauthier-Villars, Paris, 1892; Dover Press, 1957; NASA Translation TTF-450, Washington, 1976.

Ratiu, T., *Indiana U. Math. J.* **29**, 609, 1980.

Saletan, E. J., and A. H. Cromer, *Am. J. Phys.* **38**, 892, 1970.

Saletan, E. J., and A. H. Cromer, *Theoretical Mechanics*, Wiley, New York, 1971.

Schutz, B. F., *Geometrical Methods of Mathematical Physics*, Cambridge University Press, Cambridge, 1980.

Sudarshan, E. C. G. and N. Mukunda, *Classical Dynamics: A Modern Perspective*, Wiley, New York, 1974.

Thirring, W., *Classical Dynamical Systems*, Springer-Verlag, New York, 1978.

Whitman, K. J., *Rep. Prog. Phys.* **40**, 1033, 1977.

INDEX